Praise for *Wigmore on Alcohol*

It will certainly be the new Alcohol Bible for decades to come, and be of great value to active practitioners and researchers.

—Dr. Kurt M. Dubowski, George Lynn Cross Distinguished Professor Emeritus of Medicine, University of Oklahoma

This book is a must for anyone who is working on medicolegal issues involving alcohol. It will be appreciated by lay persons interested in this topic as well as scientific and legal experts involved in alcohol-related medicolegal issues.

—Thomas L. Pazdernik, PhD, University of Kansas Medical Centre

In short, Wigmore on Alcohol is a wonderful research tool that will prove a valuable resource in both legal and scientific libraries.

—Justice Rick Libman, Judge of the Ontario Court of Justice

I regret that Wigmore on Alcohol was not available to me during my years of active practice in forensic science.

—Doug Lucas, MSc, PhD (Hons), Retired Director of the Centre of Forensic Sciences, Toronto

In summary, this work provides scientist and non-scientist, student and expert alike with a source of relevant articles to assist them in learning about, understand or deepening their knowledge of alcohol toxicology.

—*Canadian Society of Forensic Science Journal*

I worked in the insurance industry for many, many years and I have attended mediations, discoveries, and trials. I have also published articles and presented at seminars. Your book would have been of such a help as a reference had it been available. ... It certainly one that band should be used by the Insurance Industry, the Law Society and Police Departments across the country.

—Carolan Shannon, RN, Claims Consultant

Praise for *Wigmore on Cannabis*

Medical and legal practitioners will find this an indispensable tome with up-to-date scientific evidence related to cannabis and its use. Academic law libraries will also benefit from having this practical book in their collections, as it emphasizes the depth of analysis that has been undertaken and will continue to be required as new cannabis-related issues arise in our courts.

—Margo Jeske, Canadian Law Library Review

I found *Wigmore on Cannabis* to be an extremely useful, detailed, and accurate reference book that assisted me to gain comprehensive knowledge of this complex substance. With over 5,000 peer reviewed scientific studies being assessed meant every possible angle of the drug was covered and especially useful was the information on the effect of cannabis on driving skills. The indexing and short abstract of each article was extremely user-friendly and easy to understand.

—Caro Smit, Founder and Director of South Africans against Drunk Driving-SADD, Alcohol and Drug Educator and Counselor, Psychiatric Social Worker, Road Safety Expert

In summary, I believe the reader will find this a very useful book for the breadth and depth of the studies and areas covered, and also for the many tables that elegantly summarize and provide an easier way to understand the take-away messages for a range of studies Jim has categorized in his second *Wigmore On* book.

—Robert M. Langille, PhD, Forensic Toxicologist

The science of cannabis is a rapidly evolving area of research. This book is a valuable tool for finding published literature and keeping up to date on a variety of topics related to cannabis.

—Karen Woodall, PhD, Assistant Professor, Forensic Toxicology, University of Toronto

Wigmore on Alcohol and Wigmore on Cannabis, written by James Wigmore, are two fundamental and essential texts for any coroner and forensic toxicologist who wants to carry out his profession with seriousness and credibility.

—Giampiero Cortis, Tossicologo Forense, Sardinia, Italy

This book was available at our recent Criminal Lawyers Conference in Toronto! It's a must have if you practice Criminal Law and it is awesome. Albeit I am not sure James Wigmore sleeps.

—Ann Marie Stewart, CLA

Praise for *Wigmore on Nicotine*

This is great, I recommend to all tobacco control advocates and health professionals. Kudos James Wigmore.

—Ampomah Isaac Benefo, Youth Tobacco Prevention Advocate

Long overdue important book.

—Dr. Ernst Snyman, Family Practictioner

Mr. Wigmore has many excellent books he sells. He is an intellectual authority in forensics. I highly recommend law enforcement officers and departments to have these references in their station libraries.

—Yvonne G. MSCM

WIGMORE ON CARBON MONOXIDE

Also by James G. Wigmore

Wigmore on Alcohol: Courtroom Alcohol Toxicology for the Medicolegal Professional, 2nd edition

Wigmore on Cannabis: The Forensic Toxicology of Marijuana for Lawyers and Other Medicolegal Professionals

Wigmore on Nicotine and Its Drug Delivery Systems. The Medicolegal Aspects of Our Most Addictive and Dangerous Legal Drug

WIGMORE ON CARBON MONOXIDE

Medicolegal and Forensic Aspects of an Ancient, Silent Killer and Other Toxic Gases in Fires, Exhausts, and Explosions

JAMES G. WIGMORE
Forensic Toxicologist

UNIVERSITY OF TORONTO PRESS
Toronto Buffalo London

Wigmore on Carbon Monoxide: Medicolegal and Forensic Aspects of an Ancient, Silent Killer and Other Toxic Gases in Fires, Exhausts, and Explosions

Irwin Law
An imprint of University of Toronto Press
Toronto Buffalo London
utppublishing.com

Printed and bound by CPI Group (UK) Ltd, Croydon, CR0 4YY

ISBN 978-1-4875-7060-6 (paper) | ISBN 978-1-4875-7061-3 (PDF)
ISBN 978-1-4875-7062-0 (EPUB)

Library and Archives Canada Cataloguing in Publication

Title: Wigmore on carbon monoxide : medicolegal and forensic aspects of an ancient, silent killer and other toxic gases in fires, exhausts and explosions / James G. Wigmore, forensic toxicologist.
Other titles: Carbon monoxide
Names: Wigmore, James G., author
Description: Includes bibliographical references and index.
Identifiers: Canadiana (print) 20250274841 | Canadiana (ebook) 20250274876 | ISBN 9781487570606 (paper) | ISBN 9781487570620 (EPUB) | ISBN 9781487570613 (PDF)
Subjects: LCSH: Carbon monoxide—History. | LCSH: Carbon monoxide—Toxicology.
Classification: LCC RA1247.C2 W54 2026 | DDC 615.9/1—dc23

Cover design: Rafael Chimicatti
Cover image: Unsplash.com/Marek Piwnicki

The manufacturer's authorised representative in the EU for product safety is Mare Nostrum Group B.V., Mauritskade 21D, 1091 GC Amsterdam, The Netherlands. Email: gpsr@mare-nostrum.co.uk

We wish to acknowledge the land on which the University of Toronto Press operates. This land is the traditional territory of the Wendat, the Anishnaabeg, the Haudenosaunee, the Métis, and the Mississaugas of the Credit First Nation.

University of Toronto Press acknowledges the financial support of the Government of Canada, the Canada Council for the Arts, and the Ontario Arts Council, an agency of the Government of Ontario, for its publishing activities.

Canada Council for the Arts | Conseil des Arts du Canada

ONTARIO ARTS COUNCIL
CONSEIL DES ARTS DE L'ONTARIO
an Ontario government agency
un organisme du gouvernement de l'Ontario

Funded by the Government of Canada | Financé par le gouvernement du Canada

Canada

WIGMORE ON CARBON MONOXIDE

Medicolegal and Forensic Aspects of an Ancient, Silent Killer and Other Toxic Gases in Fires, Exhausts, and Explosions

JAMES G. WIGMORE

> Carbon monoxide poisoning has been a problem for humans ever since our ancestors learned to build fires in unventilated shelters.
>
> —N. Cobb and R.A. Etzel, "Unintentional Carbon Monoxide" (1991)[1]

> Carbon monoxide is relatively harmless to invertebrates, micro-organisms and plants. ... J.S. Haldane (1895) showed that cockroaches would live for long periods in a mixture of 80% CO and 20% O_2.
>
> —J.B. Haldane, "Carbon Monoxide as a Tissue Poison" (1927)[2]

> Carbon Monoxid [*sic*] is the most widespread and important toxic agent of modern civilization, preeminently greater than classic plumbism, on account of its multitude of applications in industry.
>
> —R.T. Legge, "Carbon Monoxid Poisoning" (1928)[3]

---

1 N. Cobb and R.A. Etzel, "Unintentional Carbon Monoxide: Related Deaths in the United States, 1979 through 1988," *Journal of the American Medical Association,* 266 (5): 659–663, 1991.

2 J.B. Haldane, "Carbon Monoxide as a Tissue Poison," *Biochemistry Journal*, 21 (5): 1068–1075, 1927.

3 R.T. Legge, "Carbon Monoxid Poisoning," *California and Western Medicine*, 29 (6): 401, 1928.

This book is dedicated to my lovely and wonderful wife, partner, and soulmate of many years, Penelope Connor.

Summary Table of Contents

Detailed Table of Contents

CHAPTER 4
LOCATION AND/OR SOURCES OF CARBON MONOXIDE POISONING *185*

List of Tables

CHAPTER 5

CHAPTER 6

Preface

> Every society has ancient myths of demons who come on cold nights to take the lives of the young and elderly. Poorly ventilated charcoal fires lead to heart disease, mental health problems and death.
>
> —M. Guarnieri, "Cave Painting Hazard?" (1999)[1]

On a bright day on 10 May 1553, three newly built ships commanded by Sir Hugh Willoughby sailed down the Thames from Radcliffe, saluting King Edward VI with gun and cannon, shouts and flags and drums, as they headed past Greenwich. This was the first English Arctic expedition that was seeking the great riches of the Orient by sailing around the northwest coast of Norway and around Russia. During a great storm, two ships were separated from the other, and, after a fruitless search, Sir Willoughby decided to winter with his two remaining ships in a protected safe harbour that was isolated and uninhabited. In the winter of 1554, the two ships' entire companies were gathered below decks in one ship to conserve heat. All the openings, gunports, and hatches were sealed to keep out the cold, but, unfortunately, it did not allow the toxic carbon monoxide (CO) from fires and lamps to escape either.

Several months later in the spring of 1554, all the bodies of the crew, including the ship dogs, were found in the ship, with no sign of trauma, all appeared to have died quietly in natural positions. It is now known that the deaths were not due to witchcraft or black magic but to that great,

---

Preface

1 M. Guarnieri, "Cave Painting Hazard?" *Science*, 283 (5410): 2017, 1999.

ancient, silent enemy, CO.[2] CO has plagued humanity for thousands of years ever since the invention of fire as it is produced by the burning or smoldering of any hydrocarbon such as wood or coal. It is invisible and non-irritating and slowly causes death by hypoxia. Its toxic effects are described in more detail in this book and include the following facts:

- CO has a more than two hundred times greater binding to hemoglobin (Hb) than oxygen and forms carboxyhemoglobin, depriving the tissues of oxygen.
- CO also impairs the dissociation of the remaining oxygen from Hb causing greater hypoxia, especially to the brain and heart.
- CO binds to myoglobin as well, with an affinity that is 60 times greater than oxygen, which causes cardiac dysfunction and arrhythmias.
- CO also binds to other hemoproteins such as cytochrome a3, which results in cellular respiratory dysfunction.
- CO promotes the formation of free radicals and lipid peroxidation, especially in the brain.[3]

Humanity's fight against CO suffered a serious setback during Victorian times when coal/water gas, containing a high CO concentration was deliberately piped into homes, offices, and theaters for inexpensive and easy illumination and heat. The slow leaks and deliberate and accidental exposures caused untold numerous deaths and chronic illnesses. The famous author Charles Dickens was said to have suffered from acute chronic CO poisoning due to his exposure to coal gas lighting in theatres where he performed during his American tour. The chronic exposure to CO may have indirectly caused the Gothic horror genre and superstitions about the supernatural and ghosts, as exemplified by Edgar Alan Pole.[4]

It was not until coal gas was replaced by natural gas and electricity, and until cars had catalytic converters, that deaths, suicides, and injuries from CO decreased dramatically. But after many years of apparent peace,

2 E.C. Gordon, "The Fate of Sir Hugh Willoughby and His Companions: A New Conjecture," *Geographical Journal*, 152: 243, 1986.

3 A.A. Gozubuyuk, H. Dag, A. Kacar, Y. Karakurt, and V. Arica, "Epidemiology, Pathophysiology, Clinical Evaluation, and Treatment of Carbon Monoxide Poisoning in Child, Infant and Fetus," *Northern Clinics of Istanbul*, 4 (1): 100, 2017.

4 M. Turner, "Edgar Allan Poe and CO: The Dangers of Carbon Monoxide Poisoning," *Catalyst*, 11 December 2019.

we seem to have forgotten our ancient, invisible, enemy and the deadly tricks that it has played on humanity since the discovery of fire. The titles of the many scientific studies on CO poisoning reflect this memory lapse even in the twenty-first century:

- "Carbon Monoxide Poisoning: A New Incidence for an Old Disease" (N.B. Hampson and L.K. Weaver, 2007)
- Carbon Monoxide Poisoning in Tents: A Review" (S. Leigh-Smith, 2004)
- "An Outbreak of Carbon Monoxide Poisoning after a Major Ice Storm in Maine" (W.R. Daley et al., 2000)
- "Occult Carbon Monoxide Poisoning: A Cause of Winter Headaches" (P.S. Heckerling, 1986)
- "Chronic and Occult Carbon Monoxide Poisoning: We Don't Know What We're Missing" (J. Wright, 2002)
- "Carbon Monoxide Poisoning: Undetected by Both Patients and Their Doctors" (A. Harper and J. Croft-Baker, 2004)
- "Carboxyhemoglobin Levels in Patients with Flu-Like Symptoms" (M.C. Dolan et al., 1987)
- "Carbon Monoxide Poisoning: Easy to Treat but Difficult to Recognize (M.Z. Balzan, G. Agius, and A.G. Debono, 1996)
- "Carbon Monoxide Poisoning: A Disease of a Thousand Faces" (J. Fisher, 1999)

This book covers our fascinating relationship with CO and how it has evolved over the years.

Introduction

> Because CO is a small, light, molecule with almost the same density as air, its distribution is subject to the movement of ambient air such as drafts, winds and air currents. This property can cause great variations in its concentration over short distances. This in turn can create considerable confusion, in cases where-for example-two people are found in relatively close proximity with one dead from CO poisoning and the other only slightly affected.
>
> —Doug Lucas, "A Life of Crime: My Career in Forensic Science" (2018)[1]

Alcohol and carbon monoxide (CO) are the classic drugs/poisons of forensic toxicology. Many famous alcohol experts such as Wilfred Derome, Dr. Kurt Dubowski, A.R. Alha, and Doug Lucas were also involved with CO. The famous eminent English forensic pathologist, Sir Bernard Spilsbury, committed suicide with coal gas. I originally started at the Centre of Forensic Sciences in Toronto in 1976 as a Breathalyzer repair technician and repaired over two hundred Breathalyzer models 900 and 900A in Ontario. I eventually obtained my four-year science degree part-time from the University of Toronto and became a forensic alcohol toxicologist. But for several years, I also was involved in CO investigations.

WILL O' THE WISP

It has been more than 40 years since I was a young scientist and analyzed my first CO fatal poisoning case with a senior toxicologist Frank McAuley.

Introduction

1 D.M. Lucas, *A Life of Crime: My Career in Forensic Science* (Boca Raton, FL: CRC Press, 2018), 53.

The case involved a young woman who accidentally locked herself in the trunk in a car that was running in the family garage. We conducted a reconstruction to determine how this happened with members of the Toronto police. The original car was found to have no catalytic converter and so emitted a high CO concentration, which was responsible for her death as the garage door was also closed. We were so intent on the reconstruction and obtaining the readings that we overlooked that the family dog was still inside the locked house. CO is such a "will o' the wisp" that it could leak anywhere. Fortunately, the garage was almost airtight, and no CO leaked inside the home. The dog, who we could hear barking inside, was safe.

CANARY IN THE COAL MINE?

My director of the lab at that time was the brilliant and wise Doug Lucas, and he wrote about one of his early cases of CO poisoning of a man that illustrates the vagaries of CO poisoning. The faulty gas furnace had caused the death of a middle-aged man in his bedroom. When Lucas arrived at the scene, he found a lethal CO concentration in the bedroom, but, beside the window, there was a healthy canary in a cage near the window. This is the complete opposite to what would have been expected. What saved the canary was that the cage was placed beside a tiny hole in the window that provided the canary with fresh air.[2]

WIGMORE ON ... BOOKS

This is the fourth book in my Wigmore On ... series of books on the medicolegal aspects of alcohol, cannabis, nicotine, and now carbon monoxide. But all four volumes have the same basic and unique structure since the first edition of *Wigmore on Alcohol*, published in 2011. To date, I have reviewed nearly fourteen thousand published studies on the medicolegal aspects of these drugs. All the studies are based on humans and not animals as animal studies have little merit and relevance in criminal court. As my long-time valued colleague Rita Charlebois replied when she was asked on the stand why she did not conduct any animal studies: "Because, sir, rats don't drive." I have personally reviewed and read all the studies and have abstracted them in my own words. It is 100 percent artificial intelligence (AI) free and has no possibility of AI hallucinations, which is important in writing reports and courtroom testimony.

2 Lucas, *A Life of Crime*.

For *Wigmore on Carbon Monoxide*, I have divided the over four hundred studies into six chapters:

1. Historical Aspects and the Pharmacokinetics of Carbon Monoxide
2. Carbon Monoxide Analysis in Living Persons
3. Types/Effects of Carbon Monoxide Exposure
4. Locations and/or Sources of Carbon Monoxide Poisoning
5. Postmortem Carbon Monoxide
6. Other Toxic Gases

The studies are then divided into 53 other subsections such as 3.05 CO Exposures to Fires, 3.09 Delayed Prolonged Sequelae, 4.12 Indoor Skating Rinks and Arenas, and 5.03 Fatal COHb. Each study or reference has a unique five-digit number assigned to it, of which the first digit is the chapter, followed by two digits for the section in that chapter, and, finally, the last two digits denote the position that the reference appears in that section. For example, a reference number of **41408** means that the reference is located in Chapter **4,** in section **14** and is the **08** reference number there. If you look up that reference, you will find:

Reference Number: 41408

DEITCHMAN, S., J. DECKER, AND L. SANTIS. "A Novel Source of Carbon Monoxide Poisoning: Explosives Used in Construction." *Annals of Emergency Medicine*, 32: 381–84, 1998 (15 references)

Abstract: During a municipal sewer project, a new manhole was built and finished at 2 p.m. At 3:40 p.m., 265 pounds of explosives were detonated about 50 feet south of the manhole. At 4:30 pm, a 45-year-old worker entered the manhole and collapsed. Two workers, aged 32 years old and 25 years old, retrieved their co-worker. The 32-year-old worker made it to the surface, but the 25-year-old worker collapsed into the bottom of the manhole. The 25-year-old worker died and had a COHb concentration of 33%. The other two workers had a COHb of 33% and 23% and were administered HBO_2 and released several days later. A gas probe six feet below the surface showed a CO of 600 ppm and hydrogen sulfide (H_2S) of 12 ppm, which was probably false as the detector also recorded high levels of CO as H_2S. Later analysis showed a CO gradient with the greatest CO concentration (1.901 ppm and no H_2S) near the bottom of the sewer. A concentration of 19.5% oxygen and 3% carbon dioxide (CO_2) were also detected. A test detonation of the same explosives used found the production of 27–29 litres of CO was produced from each kilogram detonated.

> The danger of CO exposure from use of explosives in confined areas such as underground mines is well known. We believe this to be the first report of a fatal CO poisoning in the construction industry from occupational exposure to CO migrating through soil and accumulating in a confined space after surface detonations. Although the physiology of CO intoxication is understood, poisonings and deaths still occur. These incidents transpire in part because new technologies and applications present new or unrecognized sources of CO exposure.

This reference is about the little-known fact that construction explosions such as those in roadwork or sewers produce CO, which can migrate through the soil (that is, will o' the wisp) and poison construction workers or homeowners living nearby. The authors' last names appear in capitals, followed the title of the paper and the location and year in which the paper was published. The number of tables, figures, and references in the paper appears in brackets. Next appears the abstract, which is a short summary of the paper that I wrote in not-jargon-filled or overly technical language. Finally, in italics, is a direct quote from the author(s), which I feel is important for you to obtain the significance of this paper in their own words.

As far as possible you should also obtain the original paper from an online search at Google Scholar, the US National Library of Medicine at PubMed, or the publisher's websites.

HOW TO READ *WIGMORE ON CARBON MONOXIDE*

I would recommend that readers initially peruse the entire book by reading and reflecting on the carefully selected quotes at the beginning of each chapter and section as they will provide a historical and cultural overview of CO and provide a general knowledge of its effects and how pervasive it is globally. And it is a lot of fun. I would then read the chapter and section introductions as they are almost small scientific treatises on the various topics. I would then read the abstracts that I have written on each reference and the author's quotes as well. Finally, the original scientific study can be read as well online.

ENJOY!

Acknowledgements

> If I have seen further than others, it is by standing upon the shoulders of giants.
>
> —Sir Isaac Newton[1]

It has been 14 years since Jeff Miller and his small crew at Irwin Law took in a novice author and decided to publish my first book, *Wigmore on Alcohol*. Jeff must have saw something in my rough draft that caught his eye and saw merit in my work. Some publishers only wanted to publish my book if I changed the title as they thought it may be confused with *Wigmore on Evidence*, which was written by Professor John Henry Wigmore at Northwestern University in Chicago in 1904 and his Wigmore chart. I was fortunate to have the same last name as John Henry in my profession as a forensic toxicologist for, even in Canada, his book was referred to in criminal cases and by the Supreme Court of Canada. Numerous times after I testified in criminal court, the judge would make a witty remark: "I agree with *Wigmore on Evidence* and Wigmore's evidence." It was on the basis of this pun that I wanted to use my surname in the title. It was my name after all. After its publication, one New Jersey prosecutor told me that, when you go to criminal court, you need only two books, *Wigmore on Evidence* in one hand and *Wigmore on Alcohol* in the other. As alcohol is by far the major drug involved in criminal cases.

Acknowledgements

1 Quoted in C. Chen, "On the Shoulders of Giants," in C. Chen, *Mapping Scientific Frontiers: The Quest for Knowledge Visualization*, 135–166 (London: Springer London, 2003).

Jeff Miller encouraged me to keep publishing, and so a book on cannabis was published in 2018, in time for the legalization of recreational cannabis in Canada. One defence lawyer told me ruefully that a lot of my book seems to have been incorporated into the new law. I also appeared on the Health Canada Cannabis expert panel on "Your Cannabis Questions Answered, Get the Honest Facts." And, for several years, I appeared on media about cannabis. During the pandemic, I wrote a book on an epidemic nicotine, which was published in 2023. Nicotine is by far the most addicting and harmful drug to society, more than alcohol or cannabis.

My magnum opus, the massive second edition *of Wigmore on Alcohol* (expanded to 911 pages), was published in 2024 initially by Irwin Law but now an imprint of the University of Toronto Press, which I am pleased to join and continue with a Canadian publisher.

The fourth drug/poison I dealt with was the great ancient and silent killer, carbon monoxide (CO). I feel that nicotine and CO have been somewhat ignored and neglected by forensic toxicology, and I am sure you will agree.

I would like to acknowledge, from my Centre of Forensic Sciences days, the director, Doug Lucas, who was also an expert on CO and alcohol, and a senior scientist, Frank McAuley, who was a "god" of CO and took me under his wing.

I would also like to acknowledge the amazing staff at the University of Toronto Press, especially Josephine Mo, who has been of great support and encouragement and even attended my free public lectures at the Toronto Public Libraries, and my wonderful copy editor Stacy Belden.

I have already dedicated this book to Penelope Connor, my wonderful wife, partner, and soulmate of many years, and so I would like to acknowledge our great children, Alexander (and spouse Ginny) and Kathleen (and spouse Kris), and our five grandchildren, Chloe, Simon, and Benji and Calvin and Alice (respectively) who continue to inspire me, love me, and keep me properly grounded. It is in part for their future that I write so that they will not forget the lessons of the past and the ancient silent killer who lies in wait.

CHAPTER 1

Historical Aspects and the Pharmacokinetics of Carbon Monoxide

> CO is a funny little molecule. The triple bond that ties its carbon to its oxygen is the strongest covalent bond known on Earth.
>
> —A. Katsnelson, "The Good Side of Carbon Monoxide" (2019)[1]

Carbon monoxide (CO) has had a major impact, yet a hidden role, on the history of humans for thousands of years. Pharmacokinetics is the science of the effects of the body on the drug/poison. As such, it includes the absorption, distribution, and elimination of CO.

1.01 HISTORICAL ASPECTS

> [A]ll of us would have been dead in a few minutes by carbonic oxide gas poisoning generated by the presence of burning coal in a closed room and the laymen would have ascribed our deaths to ghosts and legends of haunted houses.
>
> —K.P. Banerjee, "My Story of a Demon-
> The Carbonic Oxide Gas" (1902)[2]

Our ancestors were exposed to the effects of CO produced by their fires and torches in caves or other dwellings. Some of the effects may have

1 **Chapter 1**
A. Katsnelson, "The Good Side of Carbon Monoxide," *ACS Publication,* 5 (10): 1632–1635, 2019.

2 K.P. Banerjee, "My Story of a Demon—The Carbonic Oxide Gas: My Punishment for Want of Foresight." *Indian Medical Gazette*, 37 (9): 373, 1902.

caused visions and hallucinations, resulting in some of the amazing, artistic cave paintings (10101–2, 10114). Other CO effects could have been the death of entire families or tribes during the night, which was thought to be due to evil spirits (10113, 10114). Even the rich were not immune to the effects of CO, as several Byzantine emperors died of it (10103, 10119). Arctic explorers were especially prone to CO poisoning, and many died not of starvation or polar bears but, rather, due to CO in poorly ventilated ships or dwellings (10104–5, 10124–5). Other famous people who were thought to have died of CO poisoning are listed in Appendix 5, which included the famous French writer Emile Zola (10117).

The so-called "Golden Age" of CO poisoning occurred during the Industrial Revolution, when coal/water gas containing high concentrations of toxic CO was pumped into thousands of homes, offices, and theatres. Small leaks of such gas caused chronic illnesses to many individuals, including Michael Faraday and Edgar Allan Poe (10115–16). The numerous case reports of death due to coal gas in this section shows the enormous impact that CO had during this time (10106–12, 10114, 10122–3). Oxygen therapy was developed later in the nineteenth century (10118, 10120–1), and measures to abate the risks of CO were adapted during the construction of the Hudson River tunnel (10126).

Reference Number: 10101

GUARNIERI, M. "Cave Painting Hazard?" *Science*, 283: 2017, 1999 (1 figure)

Abstract: Light was required for prehistoric cave painters in order for them to paint deep in caves. But light was produced by fires and torches, which would produce CO in the ill-ventilated caves. It is known that low levels of CO exposure in mines caused visions and hallucinations in miners.

> Every society has ancient myths of demons who come on cold nights to take the lives of the young and elderly. Poorly ventilated charcoal fires lead to heart disease, mental health problems and death. Scientists who investigate the mystery at Pont D'Arc may want to consider that some artists may have died for their art.

Reference Number: 10102

FUENTE-FERNANDEZ, O. "Environmental Conditions around Fires inside Paleolithic Caves: The Hearths of Tito Bustillo (Ribadesella, Asturias,

Spain)." *Journal of Archaeological Science: Reports*, 45: 103590, 2022 (3 tables, 18 figures, 80 references)

Abstract: Temperature, relative humidity, carbon dioxide, atmospheric pressure, and air flow were measured throughout the year in the Tito Bustillo Cave. In addition, a fire was started in two hearths in which of seven kilograms of firewood from the surrounding area were burned. Ambient gases and CO concentration were measured during the fire. The CO concentration was 50 ppm between two to five metres from the hearth and decreased to traces at 15 metres from the hearth. The ambient air conditions in the area of the hearth would be extremely polluted in the summer as the natural air movement would not have been sufficient to ventilate the smoke. General air circulation in a cave with a small entrance is hardly perceptible.

As other kinds of studies are showing, not all hearths inside caves should be considered equally, as they may be related to domestic or productive activities. However, there is no doubt that the Paleolithic groups in Tito Bustillo Cave would be able to use the Living Area during winter conditions when ventilation is more active and can remove the smoke from the hearth through a large entrance at greater velocity.

Reference Number: 10103

LASCARATOS, J.G., AND S.G. MARKETOS. "The Carbon Monoxide Poisoning of Two Byzantine Emperors." *Clinical Toxicology*, 36 (1–2): 103–107, 1998 (2 figures, 21 references)

Abstract: The Byzantine Emperor Julius the Apostate suffered from CO poisoning while stopping at a small town of Celts (Lukretia) in 363 CE, where he ordered his servants to bring some hot coals to heat his room. He fell asleep due to the fumes but was rescued and survived. His successor Emperor Jovian suffered fatal CO poisoning when a brazier containing coals was lit in his tent due to the extreme cold. The emperor did not realize anything was amiss as he had drank copious amounts of wine before he succumbed to the CO fumes.

Poisoning with CO, a chemical asphyxiant, is caused in part by its binding to hemoglobin and the conversion of oxyhemoglobin to carboxyhemoglobin. In the past, CO exposures had been one of the main causes of acute poisoning because of the common practice of brazier use for indoor heating, especially in Southeast Europe. While the use of braziers for home heating has significantly declined during the latter half of the 20th century, CO remains a very frequent cause of poisoning from fires, explosions of pyrites in closed areas, and leakage of gas in badly ventilated garages and industrial furnaces. In the US, CO poisoning constitutes the most common cause of fatal poisoning.

Reference Number: 10104

GORDON, E.C. "The Fate of Sir Hugh Willoughby and His Companions: A New Conjecture." *The Geographical Journal*, 152: 243–247, 1986 (13 references)

Abstract: In 1553, Sir Hugh Willoughby attempted to reach the Orient by means of the Northeast passage off the coast of Norway in three ships, which were scattered in a storm. One ship under the command of Richard Chancellor reached the Bay of St. Nicholas in Russia and eventually Moscow. The other two ships vanished. Russian fishermen found the two ships and the bodies of the entire crews (including several dogs) at anchor in the mouth of a river in Lapland in the spring of 1554. All the bodies were found in one ship; none were buried on shore. The deaths were unexpected and sudden and seem to have affected everyone at the same time, so it was not a result of exposure, scurvy, or starvation. Sea coal, which is coal that has been cast up on the shore, was a common source of fuel on sixteenth-century ships, but it is more smelly and inefficient than wood and generates a lot of CO. The ships had run out of wood or driftwood for burning and resorted to sea coal. CO poisoning became a recognized hazard in the Arctic regions, and members of several expeditions had been poisoned by it. Another English Arctic expedition in 1596 nearly met the same fate in their hut onshore, which was heated by sea coal with the chimney blocked to conserve heat. Fortunately, before all the men lost consciousness, one person managed to open the door before collapsing, and none died.

> The accident might have happened in the following way: On one particularly cold day, the entire company was assembled below decks on one ship in order to conserve fuel. The crew had carefully sealed all openings and gunports to keep out the cold. Down in the hold, the cook and his mater were preparing the meal, using sea coal for the first time instead of wood. The smoke and heat from the cooling fire rose rapidly through the cook room hood and chimney and out of an open hatch to the main deck where the group had assembled. The hatch to the weather deck, through which

the smoke usually escaped to the outside, remained tightly closed. The sea coal gave off lethal carbon monoxide, and the concentration of the gas rose rapidly in the unventilated space. Unconsciousness occurred without warning and the men collapsed and died where they were.

Reference Number: 10105

JOHNSON, F.R. "Unsolved Mysteries of the Arctic by Vilhjalmur Stefansson." *Isis*, 32: 212–214, 1940.

Abstract: The fate of Arctic explorers and other mysteries are discussed. The death of S.A. Andrée and his companions, who disappeared in their polar balloon flight in 1897 and whose bodies were found 33 years later on White Island off Spitzbergen, was due to CO poisoning caused by a defective primus stove.

This book then, abundantly illustrates the maxim that man learns but slowly through hard experience; yet it also demonstrates with equal clarity how much, during the last century, has been learned about the Arctic regions. The Arctic is much less mysterious than it used to be; in fact, even the specific mysteries that form the subject of this book have been cleared up through Dr. Stefansson's skilful application of logical reasoning and expert knowledge to their solution.

Reference Number: 10106

DEXTER, R. "Art. VII-A Case of Poisoning by Coal Gas." *Journal of Nervous and Mental Disease*, 3: 412–413, 1876.

Abstract: A servant girl went to bed in her room and thought she had turned off the coal gas light but, instead, turned it too far, and, although the flame was out, the gas remained on. In the morning, her employer smelled gas and knocked on the door with no response. He burst into the room and found the girl unconscious and frothing at the mouth. The doctor arrived and administered 15 drops of spirits of ammonia in water every 10 minutes.

In this case there was not any other medicinal agent given; ammonia and fresh air were the only agents brought into use. The restoration to the normal physiological conditions of the circulation, respiration, nervous oppression, cerebral congestion, condition of the pupil and consciousness were apparently simultaneous.

Reference Number: 10107

ANONYMOUS. "Gas Poisoning and Cold Weather." *British Medical Journal*, 1781: 381–382, 1895.

Abstract: A 20-year-old domestic servant and two girls who were six and seven years of age were exposed to coal gas fumes in their small bedroom. The six-year-old girl was still alive but unconscious and died shortly after being resuscitated. The other two victims were already dead. There was no smell of gas. The source of the CO was due to coal gas leaking from the main pipeline through the ground and into the bedroom. Coal gas loses its smell once it passes through the ground.

> In this case both of the postmortem appearances ascertained by Drs Medlinand and Stoboe, and the other circumstances of the case seem to leave little doubt that death was due, as found by the jury, to poisoning by carbon monoxide. The most essential piece of evidence concerns the cherry red colour of the blood, which was found of a cherry red colour.

Reference Number: 10108

WHITELEGGE, A. "The Use of Water Gas and Other Gases in Factories-Carbonic Oxide Poisoning." *Public Health*, 17: 82–85, 1904 (1 reference)

Abstract: Coal gas contains 4–12% CO. Dawson gas, Mond gas, power gas, producer gas, and blast furnace gas contains 10–25% CO. Carbureted water gas contains 30% CO, and uncarbureted water gas contains 50% CO. Between 1899 to 1903, at least 51 cases of CO poisoning that occurred in factories were reported, including 17 deaths.

> Carbonic oxide poisoning may occur in other ways without inhalation of the particular gases names. Thus danger of this kind may arise in laundries from the use of gas irons, and in workrooms from defective gas fittings and from gas stoves especially when no provision is made for the products of combustion to be carried away by a flue or a chimney. The absence of a chimney in a room greatly increases the risk. Carbonic oxide is found also in lime kilns, cement works and where braziers and coke fires are used in confined spaces.

Reference Number: 10109

LOCKIE, S. "A Case of Poisoning by the Inhalation of Coal-Gas." *British Medical Journal*, 25 September 1875, 392–393.

Abstract: A 18-year-old hotel employee was found unconscious in his cupboard bed in the smoking room of a hotel. One vent of a coal gas gaselier was on full, and the smell of gas in the room was strong. He was administered a limited amount of oxygen from a tube in his mouth. He did not recover consciousness and died two days later. It was concluded that the administration of oxygen had no beneficial effect whatsoever.

> This case supports that view already proved by the cases recorded by M. Tourdes, that a proportion of coal-gas to atmospheric air less than that at which the mixtures becomes explosive, may be sufficient to cause death, if respired for a sufficiently long time. As recorded by this gentleman, a family of six persons were poisoned by a atmosphere in which a stove had been a long time the seat of active combustion, and in which a candle had burned to its end.

Reference Number: 10110

CHALKE, H.D., AND J.R. DEWHURST. "Accidental Coal-Gas Poisoning: Loss of Sense of Smell as a Possible Contributory Factor with Old People." *British Medical Journal*, 19 October 1957, 915–917 (4 tables, 1 reference)

Abstract: Between 1946 and 1955, over 70% of the coal gas fatalities in the United Kingdom involved people 60 years of age or older. The ability to detect the odour of town gas at a level that was 50 parts per 10,000 was tested in 61 persons over the age of 65 years. Thirty-three per cent could not reliably detect the odour of gas.

> The use of odour as a danger signal for accidental escapes of unburnt gas is clearly unreliable for old people living alone. In this country, there are 1 1/2 million one-person households; nearly two-thirds of them are occupied by persons over 60, but the proportion varies widely throughout the country. The carbon monoxide content of domestic gas also varies, and it is possible that the level of odour of the gas is different in different areas. It is suggested therefore, that further information should be sought in other parts of the country on this ability of older persons to detect and recognize the smell of gas.

Reference Number: 10111

ABBOTT, S.W. "Poisoning by Carbon Monoxide." *Public Health*, 147: 562–563, 1902.

Abstract: In the thirteen years prior to the introduction of water gas to Boston (1876–8), the number deaths due to illumination gas was eight. In the 13 years after the introduction of water gas (1889–1901), the number of deaths was 459.

Table: Number of Deaths Due to Water Gas in Boston between 1892 and 1901

Year	Number of deaths
1892	14
1893	23
1894	25
1895	33
1896	42
1897	45
1898	58
1899	51
1900	25
1901	14

Source: Adapted from Abbott 1902.

The increase and decrease in the deaths from this cause in Boston is accounted for by the fact that the percentage of water-gas in the supply furnished to that city gradually increased to about 95% in 1889 and 1899, then decreased to less than 50 per cent in the remaining years.

Reference Number: 10112

DOW, D.C. "Carbon Monoxide Poisoning." *Massachusetts Medico-Legal Society*, 201: 583–585, 1929.

Abstract: A medical examiner recounts the growing number of CO poisoning cases due to coal gas (6–8% CO) and the cheaper water gas (30% CO) for lights and heat. In one apartment with a small gas heater (radiant type), five adults were found dead due to CO poisoning as well as two children who survived but died later.

It is unusual for even Medical Examiners to come into cases of carbon monoxide poisoning in such wholesale lots, but this instance shows the danger of whole families being wiped out by a poison which is odorless, which is non-irritant, and which -according to Haldane-can saturate the blood up to thirty per cent before symptoms become manifest. Carbon monoxide poisoning is on the increase because of the growing use of illuminating gas for cooking and heating purposes, and the substitution of water gas for coal gas and the added use of the gasoline engine. In New York City alone there are over four hundred such deaths a year, either accidental or with intent, and it is claimed that as an industrial poison, carbon monoxide has superseded lead. The so-called water gas, containing as it does about thirty per cent carbon monoxide, as against six to eight per cent in coal gas, renders it a much more potent poison, and is-somewhat at least-responsible for the great increase of the number of victims. If the people want cheap gas, their demands can be satisfied.

Reference Number: 10113

HELLIE, R. "The Russian Smoky Hut and Its Probable Health Consequences." *Russian History*, 28: 171–184, 2001 (64 references).

Abstract: The smoky hut (*kurnaia izba*) was the home of most Russians until quite recently. To conserve fuel and preserve heat through the long winter, stoves and fireplaces did not vent outside (that is, they had no chimneys) but directly into the hut. In huts with ceilings, the smoke, gas, and fumes filled the hut right down to the floor. When the smoke got too bad, a hole in the roof or walls was opened and then closed when the smoke had cleared. Softwoods are more toxic than hardwoods and burn less clearly and produce twice as much particulate matter. Over time, the Russian stove emissions would have destroyed the respiratory system of the peasants and shortened their lifespan by increasing the number of deaths from cardiopulmonary disease. It is also unknown how many families died of CO poisoning while sleeping with the fires smoldering.

Beside the CO emission, equally or more harmful must have been the particulate emissions of the Russian stove. One can deny CO because one can't see it, but the particulate emissions were the substances that blackened the peasant hut and cannot be denied under any circumstances. Wood smoke particles have a potentiating effect in that they worsen the toxicity of the chemical present. Particles in wood smoke appear to suppress the immune system, inter alia, by suppressing the activity of macrophages, a type of white blood cell that destroys foreign disease particles. Thus, it is not accidental that smoke inhalation injuries are followed by pathogenically significant bacterial infections.

Reference Number: 10114

SCHNEIDER JR., F. "An Investigation of a Haunted House." *Journal of the American Medical Association*, 60: 1883–1884, 1913.

Abstract: An investigation was conducted by Dr. Franz Schneider Jr., of the Massachusetts Institute of Technology into a reported haunted house in the Black Bay district of Boston. The slumber of the family living there was disturbed by a strange sensation of oppression or powerlessness, and the children appeared pale and sluggish in the morning. These mental disturbances were due to a defective hot air furnace that leaked CO into the house.

The symptoms in Schneider's cases pointed to the carbon monoxid as the probable offender. Sensations of oppression and other mental disturbances are said to be typical of acute carbon monoxid poisoning. Any textbook description of the toxicology of carbon monoxid will bring to mind the possibility of loss of psychic powers, the confused sensations and other features which suffice to explain the sense of oppression that persistently entered into the delusions of the inmates of the haunted house. The belief in walking spirits is easily nourished by person in whose minds the real noises would be likely to become exaggerated during the intoxication.

Reference Number: 10115

HARE, E.H. "Michael Faraday's Loss of Memory." *Proceedings from the Royal Society of Medicine*, 67: 617–618, 1974 (10 references)

Abstract: When the famous scientist Michael Faraday was 48 years old in 1839, he suffered an illness that prevented him from doing any work or research for three years. Until his death at 75 years of age, he complained of bad memory and recurrent attacks of giddiness and confusion.

Between the ages of 30 and 40, Faraday spent much of his working time in the underground laboratory at the Royal Institution engaged in researches on steel and glass which involved the continual use of a furnace. There is no evidence that he ever suffered from acute carbon monoxide poisoning but exposure to low concentrations of carbon monoxide certainly seems a possibility and this, taken with the recent evidence that such exposure can cause arteriosclerosis, affords an explanation for what may have been the initial pathological lesion in Faraday's case.

Reference Number: 10116

TURNER, M. "Edgar Allan Poe and CO: The Dangers of Carbon Monoxide Poisoning." *Catalyst*, 11 December 2019, 4 (2 figures, 7 references)

Abstract: There are many theories about what caused Edgar Allan Poe's psychological problems and early death, including alcoholism, heavy metal poisoning, and rabies. CO, the silent killer, is the probable cause. Poe was exposed to CO on a regular basis as lights were powered by coal gas at the time and produced a lot of CO. Pictures of Poe's face show the tell-tale slanting of the eyes and mouth—a hallmark of low-grade CO exposure. His paranoia, delirium, and hallucinations are also signs of long-term CO exposure.

> Ultimately, there is no way to prove that either Poe's inspiration for his works or his ultimate demise were results of chronic carbon monoxide poisoning. His cause of death was officially listed simply as phrenitis or swelling of the brain. However, understanding the subtle symptoms of CO exposure could be the key to saving a life in the present day. And even it isn't true, it is interesting to consider that the morbid and fantastical worlds imagined in Poe's works could be rooted in science after all.

Reference Number: 10117

ANONYMOUS. "The Cause of M. Zola's Death." *British Medical Journal*, 4 October 1902, 1082.

Abstract: The famous French author Emile Zola and his wife returned to their house in Paris in the evening, after a fire of patent fuel (that is, charcoal briquettes) had been lit in their bedroom all day. The servant noticed that the chimney was not drawing as it appeared to have been obstructed by falling mortar. The servant left the window open and let the fire to smoulder and go out. When Zola entered the bedroom, he closed the window and went to sleep with his wife and two dogs. At 9:30 the next morning, a servant broke into the bedroom when there was no response and found Zola dead on the floor and his wife and two dogs unconscious but alive. Another example of CO as a will o' the wisp.

> From the accounts of the tragic accident by which M. Zola lost his life it seems almost certain that he was poisoned by carbon monoxide, and an examination by M. Girard of samples of blood taken from the body, as well as from Madame Zola and a dog which was present in the room furnished conclusive proof.

Reference Number: 10118

STERNBACH, G.L., AND J. VARON. "Claude Bernard: On the Origin of Carbon Monoxide Poisoning." *Resuscitation*, 58: 127–130, 2003 (1 figure, 23 references)

Abstract: French physiologist Claude Bernard (1813–78) was the first to publish an accurate description of the physiology of CO poisoning.

> Through his experimentation with carbon monoxide, Bernard arrived at the fundamental facts that oxygen is not carried loosely in solution by the blood but is fixed and that it was the red blood cells that were responsible for the blood's respiratory function (hemoglobin had yet to be described as the blood's oxygen carrying element). In carbon monoxide poisoning the ability to provide oxygen to the tissue became impaired as a poison that troubles the blood by displacing oxygen, the delivery of oxygen by the blood being indispensable for life functions.

Reference Number: 10119

KASHFI, K., AND K.K. PATEL. "Carbon Monoxide and Its Role in Human Physiology: A Brief Historical Perspective." *Biochemical Pharmacology*, 204: 115230, 2022 (1 figure, 59 references)

Abstract: The formal discovery of CO occurred in 1772, by Joseph Priestly who produced CO by heating chalk to produce carbon dioxide and reduce it to CO over a hot iron. Coal fumes are described by Aristotle in 350 BCE as causing pulmonary toxicity, neurotoxicity, and death. The cherry-red skin of CO poisoning was described by Marcellas Donato in 1570. CO has been implicated in the death of Cleopatra, the suicide of the Roman emperor Lucius Seneca, and the execution of the Roman empress Fausta in 326 CE. Until the eighteenth century, CO deaths were thought to be due to supernatural causes, including the devil. The first therapeutic action of CO occurred in 1795 when Adair Crawford reported that CO inhalation had a tranquil effect. It was also prescribed for local and general anaesthesia, asthma, tuberculosis, and tumour growth inhibition.

> CO is a primordial and flammable gas that has been present in the earth's atmosphere from the beginning of time and historically has been regarded as an industrial toxin. Since it is odorless, it is often referred to as a silent killer because it is hard to detect and its intoxication can be fatal. Despite

its toxicity, CO has many physiological roles and therapeutic applications. In the realm of therapeutics, new delivery systems are being developed so that CO can be released where it's needed. Many discoveries in science have had an element of serendipity, curiosity, and a prepared mind; the CO arena is no exception. Acknowledgement of the pioneers in any field leads to a better understanding of the issues and a reminder that the past is never obsolete.

Reference Number: 10120

SEKHAR, K.C., AND S.S.C. CHAKRA RAO. "John Scott Haldane: The Father of Oxygen Therapy." *Indian Journal of Anaesthesia*, 58: 350–352, 2014 (1 figure, 12 references)

Abstract: John Scott Haldane (1860–1936) became professor of physiology at Oxford University in 1887. In 1906, Haldane and Joseph Priestley discovered that the respiratory reflex was triggered by an excess of carbon dioxide and not a lack of oxygen. He studied asphyxia in miners and determined that CO was the main cause of death in mining disasters. He suggested that miners carry mice or canaries with them into the mines as their high metabolism would detect CO earlier before it became too toxic for humans. This practice was used in mines until 1986.

John Scott Haldane was a versatile genius who solved several problems of great practical significance. His ability to look beyond the laboratory and investigate theory added crucial findings in the field of respiratory physiology. His work on high altitude physiology, diving physiology, oxygen therapy and carbon monoxide poisoning led to a sea change in clinical medicine and improved safety and reduced mortality and morbidity in many high-risk situations.

Reference Number: 10121

MEGAS, I.-F., J.P. BEIER, AND G. GRIEB. "The History of Carbon Monoxide Intoxication." *Medicina*, 57: 400, 2021 (1 figure, 46 references)

Abstract: A literature review was conducted on PubMed and Google Scholar of CO intoxication. Between 1900 and 1945, 38 articles were found, which increased to 1,850 entries between 1946 and 1975 and 2,035 studies from 1976 to 2000. Currently, between 2001 and 2020, 2,564 studies were found. The limitation of the search before 1900 was that CO was known as carbonic oxide, and so many earlier papers were not included.

CO intoxications can occur in all living beings and has been a problem since fire and its vapors have existed on Earth. The etiology, diagnosis and specific treatment for CO intoxications have been studied very carefully over the years. Recovery, healing and therapeutic strategies of affected patients have been developed only in the 20th century and have been refined since then. As the controversial debate of HBO therapy shows, the destination of this journey has not been reached even today.

Reference Number: 10122

CHAUMONT, F. "Case of Poisoning by Coal Gas." *Lancet*, 25 October 1873, 592.

Abstract: A coal gas company laid down pipes and a syphon six feet from a house where an 82-year-old woman and her 30-year-old granddaughter and 38-year-old grandson lived. The pipes and syphon were not tested for leaks. Coal gas leaked into the poorly ventilated house and caused the death of the grandmother and grandson. The granddaughter was violently sick and had a severe headache and lost consciousness but survived. An examination the next day showed that the syphon was cracked, and the loose gravel soil in front of the house was highly charged with gas. The gas company attempted to shift the blame to other causes.

At the inquest, held at Itchen Ferry, the jury brought in a verdict in accordance with the evidence: that the deceased met their deaths from the effects of poisoning by coal gas, caused by a defect in the syphon of the company's main. The survivor received compensation from the company for the personal injury she had suffered.

Reference Number: 10123

BANERJEE, K.P. "My Story of a Demon: The Carbonic Oxide Gas." *Indian Medical Gazette*, 37: 373, 1902.

Abstract: The doors and windows of a sleeping room of a family, which included a husband, wife, and two children, were tightly closed, and a coal fire was burning due to the cold. An hour later, the wife complained of an intense headache. One child vomited, and one tumbled out of bed. The husband felt giddy and developed a severe headache and eventually passed out. The maidservant rushed outside to vomit and left the door ajar, which saved their lives. Another servant entered the room and opened all the doors and windows and poured cold water on their heads to revive them. The husband's headache lasted for four days.

> Had not the maidservant providentially left the door ajar in going out, all of us would have been dead in a few minutes by carbonic oxide gas poisoning generated by the presence of burning coal in a closed room and the laymen would have ascribed our deaths to ghosts and legends of haunted houses. Every medical man must have read in books and some might have seen cases, but my practical experience teaches me never to keep a fire in a closed room, and I shall be happy if my professional brethren reap any advantage at my cost, as this was my punishment for want of foresight. It may be noted that there was no smoke from the fire.

Reference Number: 10124

ROSE, L.A. "Exploring a Secret Land: The Literary and Technological Legacies of Richard E. Byrd." *Virginia Magazine of History and Biography*, 110: 175–208, 2002 (14 figures, 77 footnotes)

Abstract: A short biography of Richard E. Byrd (1888–1957), an Arctic and Antarctic explorer and American naval officer, is given. He first gained international recognition when he claimed to have flown over the north pole from Spitsbergen, Norway, in 1926. In 1934, he spent five months alone in the Antarctic in a hut with a gasoline generator and stove. Both leaked CO fumes. He eventually fixed the stovepipe of his stove with surgical tape and so saved himself from more CO poisoning and survived.

> For days Byrd, badly poisoned by fumes whose origins he could not precisely determine at first, lay with a pounding head and aching eyes in his sleeping bag, drifting frequently into unconsciousness. He knew that the gasoline generator was the major source of his distress, but he needed to use it to power the radio for his regular talks with Little American. For long weeks, he did not realize that his stove was also leaking fumes. He was, in effect, living in a gas chamber. Dizzy, nauseous, filled with increasing fear, he had just enough energy to rum the generator and prepare whatever food he could keep down just to stay alive. As the engine slowing wound down because of erratic maintenance, the main radio transmitter quit, and Byrd had to switch to an emergency transmitter that required painful cranking by hand.

Reference Number: 10125

SCHLEDERMANN, P. "The Muskox Patrol: High Arctic Sovereignty Revisited." *Arctic* 56: 101–106, 2003 (5 figures, 17 references)

Abstract: The Muskox Patrol describes the sending of Royal Canadian Mounted Police (RCMP) officers to the High Arctic between 1922 and 1933 to safeguard Canada's claim there. Three RCMP officers nearly died of CO poisoning when fumes from the coal stove leaked into the shed where they had been sleeping.

> The second time danger struck; the three officers were nearly killed from carbon monoxide poisoning. In the blubber shed, they had jury-rigged a cooking stove to get some heat. During the day, all seemed well. Lee was the last to turn in, but when he lay down on the bunk he felt poorly. About to put out the candle, he noticed that the flame was very dim. Weakly he jumped down from the bunk and headed for the door—fell heavily against it and tumbled outside. The sound of the crash woke Anstead, and both he and Michelson managed to get outside. Gas, formed when the coal stove was banked up for the night, had been seeping out into the little shed. As Lee points out, it had been a close call.

Reference Number: 10126

HAMPSON, N.B. "Preventing Carbon Monoxide Poisoning in the Hudson River Tunnel in 1921: Recounting History." *Undersea and Hyperbaric Medicine*, 48 (1): 89–96, 2021 (3 tables, 7 figures, 16 references)

Abstract: The world's longest tunnel for passenger cars (8,300 feet) was built beneath the lower Hudson River to connect Manhattan and New Jersey in 1920. Numerous human experiments were conducted by Yandell Henderson and Howard W. Haggard and their colleagues as to the effect of CO from car exhaust in the tunnel. Earlier, John Scott Haldane had recommended the maximum ambient CO concentration should be 100 parts per million (ppm). It was determined, however, than no adverse effects of CO were observed at 400 ppm for one hour. Construction of the Holland Tunnel, named after the chief engineer who died during construction, was finished in 1927. There is a ventilation building at either end of the tunnel, with a total of 84 massive fans, including one-half that blow clean air into the tunnel and one-half that blow the exhaust out of the tunnel. However, only 56 fans are operational at a time as 28 fans are reserved for emergency use. The CO level in the tunnel is monitored 24 hours a day, and, in its first day of operations, the average ambient CO in the tunnel was 69 ppm (the peak was 180 ppm). Between 1976 and 2019, the total CO emissions of automobiles in the United States decreased 90% due to the catalytic converter.

Table: Blood Carboxyhemoglobin (COHb) Levels Measured after Subjects Breathed Various CO Concentrations for One Hour by Henderson and Colleagues in 1921[3]

CO concentration (ppm)	Number of exposures	COHb (%)
200	2	11, 12
300	3	10, 13, 14
400	11	12, 14, 14, 17, 18, 20, 21, 22
500	1	–
600	9	16, 16, 16, 17, 18, 18, 21, 25, 26
800	4	22, 24, 24, 27
900	1	34
1,000	1	38

Source: Adapted from Hampson 2021.

As noted, ventilation was planned for 1,900 vehicles per hour and a transit time of 45–60 minutes. Today, more than 100,000 vehicles pass through the tunnel daily, and a trip can take up to an hour. Despite a traffic volume more than twice anticipated, CO levels remain well controlled.

1.02 ABSORPTION

Ambivalence is like carbon monoxide-undetectable yet deadly.

—Cherie Carter-Scott, "Quotefancy.com"

Since the air-blood gradient of CO is much higher than the blood-air gradient, the uptake of CO from the ambient air is much faster than the elimination of CO from the blood. The absorption of CO is increased with an increase in altitude and in ventilation rate (10201).

3 Y. Henderson et al., "Physiological Effects of Automobile Gas and Standards of Ventilation for Brief Exposures," *Journal of Industrial Hygiene*, 3: 79, 1921.

Reference Number: 10201

PAN, K.-T., G.S. LEONARDI, AND B. CROXFORD. "Factors Contributing to CO Uptake and Elimination in the Body: A Critical Review." *International Journal of Environmental Research and Public Health*, 17 (2): 528, 2020 (2 tables, 57 references)

Abstract: CO enters the human body through the lungs and is transported by the blood into the tissues/muscle system where it can be stored until the ambient CO concentration is less than the muscle/tissue concentration, and the process is reversed. The CO is removed from the tissue/muscle by the blood and released into the lungs. Various factors can affect the absorption (uptake) and elimination of CO.

Table: Factors That Affect the Absorption of CO

Factor	CO absorption increased	CO absorption decreased
Increase ambient CO	x	
Longer exposure duration	x	
Ambient oxygen increase		x
Increase in altitude	x	
Exercise/Work	x	
Ventilation rate increase	x	
Diffusion capacity of CO (DLCO) increase	x	
Increase in blood volume	x	
Anaemia	x	

Source: Adapted from Pan, Leonardi, and Croxford 2020.

Among the environmental factors, the rate of CO uptake increases by raising the CO concentration or reducing the oxygen concentration of the inhaled gas. Moreover, the altitude can alter the rate of CO uptake due to the different partial pressure of oxygen at different altitudes. The duration of CO exposure is an important factor. If people were exposed to CO for a long time, even if the concentration of CO were low, it would also have an adverse impact and reduce the rate of CO elimination. That is why attention is needed not only for acute CO poisoning but also chronic CO poisoning, which is often ignored.

1.03 DISTRIBUTION

One reason why in the past the public has not been more cognizant of the dangers of carbon monoxid as compared to some other industrial poisons is that the gas is colorless, tasteless, and practically odorless and does not, like chlorine, or a number of other gases produce immediately such irritating effects, when breathed in low concentrations. If the amount of carbon monoxid present in the air being breathed is small, the victim commonly has no warning that he is being exposed to a dangerous poison until the gas has so far affected him that he is powerless to aid himself.

—J.B. Clemens and W. Gilman, Special Subcommittee on Public Health Relations (1926)

Once CO is inhaled, it appears in the arterial blood in a matter of 30 seconds and in venous blood within minutes (10301, 10306), and it rapidly equilibrates throughout the body (10301, 10303, 10304). The affinity of hemoglobin for CO is 250 times that of oxygen (10305), but small amounts of CO dissolve in the plasma (10307–8). The half-life of CO in the body is about five hours for exposure to normal air, one and a half hours for oxygen, and a half hour for hyperbaric oxygen therapy (HBOT) (10305).

Reference Number: 10301

GARVICAN, L., C.M. BURGE, A.J. COX, S.A. CLARK, D.T. MARTIN, AND C.J. GORE. "Carbon Monoxide Uptake Kinetics of Arterial, Venous, and Capillary Blood during CO Rebreathing." *Experimental Physiology*, 95: 1156–1165, 2010 (1 table, 5 figures, 26 references)

Abstract: CO rebreathing is a reliable method to measure blood volume or the total hemoglobin mass. The arterial, muscle venous and capillary blood COHb concentrations were measured every 1 to 10 minutes for 40 minutes after rebreathing CO for 2 and 40 minutes by 7 male and 2 female subjects (mean ages 38.4 and 25.1 years respectively). The amount of CO exhaled was also determined. The arterial and capillary blood COHb concentrations were nearly identical, but the venous blood COHb was significantly lower at 1, 2, and 3.5 minutes. CO mixing time was complete by most of the subjects within 10 minutes. Approximately 1.7% of the CO dose was bound to myoglobin.

> Carbon monoxide enters the circulatory system via diffusion at the alveoli and binds to Hb in competition with oxygen, forming carboxyhemoglobin (HbCO). In addition, CO may diffuse to extravascular compartments binding with other molecules, e.g. myoglobin (Mb), although all to a lesser extent than Hb. Over time, the distribution of HbCO throughout all circulatory compartments equilibrates, and sampling of HbCO from this point on can be used to estimate Hbmass from the circulation via exhalation.

Reference Number: 10302

ENIGNUS, V.A., M.J. HAZUCHA, M.V. SMITH, AND P.A. BROMBERG. "Prediction of Carboxyhemoglobin Formation due to Transient Exposure to Carbon Monoxide." *Journal of Applied Physiology*, 76: 1739–1745, 1994 (2 tables, 3 figures, 19 references)

Abstract: Fifteen male subjects (mean age 26.5 years) were exposed to 6,683 ppm carbon-18 monoxide for 3.69 to 6.65 minutes. Arterial and venous blood samples were collected at one-minute intervals until 10 minutes after the start of exposure. The blood COHb concentrations were determined by an IL-282-CO-oximeter within one hour after collection. The appearance of COHb in arterial blood was less than 30 seconds after exposure, where, in venous blood, it was one minute. The mean maximum arterial-venous difference was 6.2% (range 2.3–12.1%). Equilibrium between venous and arterial blood occurred within 10 minutes after exposure.

> Increases in venous and arterial HbCO were inaccurately predicted during and immediately after HbCO formation, however. Venous blood was overestimated during CO uptake because of a delayed appearance of HbCO. Individual subjects differed markedly in the degree of delay of HbCO appearance in venous blood. Arterial delta HbCO was consistently underestimated either by the CFKE or by predictions based on venous blood samples. Thus, exposure of such organs as brain or heart to HbCO may be substantially higher than expected during transient high-level CO exposure.

Reference Number: 10303

TOUGHER, M., E.J. GALLAGHER, AND J. TYRELL. "Relationship between Venous and Arterial Carboxyhemoglobin Levels in Patients with Suspected

Carbon Monoxide Poisoning." *Annals of Emergency Medicine*, 25: 481–483, 1995 (1 figure, 8 references)

Abstract: An arterial blood sample was collected, followed by a venous blood sample collected within five minutes and were analyzed for COHb with an Oximeter Model IL-482 in 48 male and 13 female patients (mean age 36 years) suspected of CO poisoning between 1991 and 1993. The correlation between arterial and venous COHb was excellent (r = 0.99). The mean arteriovenous difference was 0.15% COHb with 95% of the difference ranging from +2.4% COHb to –2.1% COHb.

Venous COHb levels predict arterial levels with a high degree of accuracy. Patients with suspected CO poisoning can be screened with the use of venous blood without the need for arterial puncture.

Reference Number: 10304

BRUCE, E.N., AND M.C. BRUCE. "A Multicompartment Model of Carboxyhemoglobin and Carboxymyoglobin Responses to Inhalation of Carbon Monoxide." *Journal of Applied Physiology*, 95: 1235–1247, 2003 (2 tables, 9 figures, 39 references)

Abstract: A model was developed to predict the distribution of CO in the body due to acute inhalation. The model includes a lung compartment, arterial and venous blood, muscle, and non-muscle soft tissues and was validated in human subjects. Arterial blood initially has a higher COHb concentration than venous blood (an overshoot) but can rapidly equalize within 10 minutes. Measurable amounts of CO can diffuse to the circulatory system and into the tissue compartments in as little as 10 minutes. The model allows for a small flux of CO into the muscle compartments containing myoglobin. It is estimated that a 70-kilogram man can have 135 grams of myoglobin, which suggests that muscle could be a significant storage site for CO, which could increase the half-time of washout of CO by 30%.

The distribution of inhaled CO into muscle tissue as well as into multiple circulatory compartments complicates the prediction of washout profiles after cessation of the exposure. If alveolar ventilation is constant, the washout profile will not be monoexponential. Furthermore, because of the slow transfer of CO between Mb and Hb, washout of the muscle tissue compartment will be much slower than washout of the circulation. Consequently, the washout profile is expected to exhibit a long tail, and the time required for full washout may be very much longer than twice the half-life of the washout.

Reference Number: 10305

ANIOL, M.J. "Carbon Monoxide Toxicity." *Canadian Family Physician*, 38: 2123–2134, 2147, 1992 (4 tables, 2 figures, 28 references)

Abstract: J.B.S. Haldane a respiratory physiologist who investigated numerous mining accidents at the end of the nineteenth century determined that up to 95% of the miners died not by mechanical injuries or burns but, rather, by CO poisoning. He suggested that rescue workers carry a mouse or canary with them to detect CO. Free heme in a solution has an affinity to CO of 25,000 times compared to oxygen, but, when the heme is part of hemoglobin, the affinity is reduced to 220–50 times. The ratio of cardiac carboxymyoglobin compared to COHb is three to one. Thus, a COHb of 10% could indicate a carboxymyoglobin level of 30%. The dissociation of carboxyhemoglobin is also slower than COHb, and this could account for the rebound of COHb after oxygen therapy. The CO half-life is five and one-third hours for air, one and one-half hours for normobaric oxygen therapy (NBOT), and 26 minutes for HBOT.

Carbon monoxide is a very dangerous gas; exposure can be fatal. the toxicity resulting from exposure can be explained by the reduction of hemoglobin's oxygen-carrying capacity, a leftward shift of the oxyhemoglobin dissociation curve, the possibility of effects from myoglobin binding and the effects of lipid peroxidation following globular cellular ischemia. Those surviving carbon monoxide poisoning may go on to develop late neurological sequelae.

Reference Number: 10306

HAMPSON, N.B. "Myth Busting in Carbon Monoxide Poisoning." *American Journal of Emergency Medicine*, 34: 295–297, 2015 (1 table, 1 figure, 17 references)

Abstract: A medical doctor with over 25 years of research in the field of CO poisoning disproved four popular myths. The myths are (1) the COHb levels correlate with symptoms of acute CO poisoning; (2) without fuel-burning appliances in the home, there is no risk of CO poisoning (however CO easily passes from one attached townhouse and apartment to another); (3) fresh arterial blood samples are needed for the accurate determination of CO; and (4) CO poisoning predisposes a person to long-term risk of cardiac death.

These are only a few of many myths about CO poisoning, and they serve to illustrate that myths likely exist in every discipline of medicine. The key to discovering and disproving myths is to go to the literature and find the original citation. It often does not exist or does not say what everybody believes. If you take the time to look for the original evidence, it is not difficult to become a myth buster.

Reference Number: 10307

CHALMERS, A.H. "Simple, Sensitive Measurement of Carbon Monoxide in Plasma." *Clinical Chemistry*, 37: 1442–1445, 1991 (3 figures, 13 references)

Abstract: CO in the blood is mainly bound to hemoglobin, but small amounts dissolve in the plasma. It is suggested that the dissolved CO is free to bind with cytochrome a3, which prevents the tissue from using oxygen. A spectrophotometer method is described to determine the CO concentration in plasma by the addition of CO-free hemoglobin (Hb) and to determine the resultant COHb concentration. The half-life of CO in water was 52 minutes but was longer in plasma. No significant CO loss was found in plasma stored in a sealed tube at four degrees Celsius for five hours. The plasma CO concentration for 17 men and 18 women ranged from 0.14 to 0.60 milligrams per litre (mg/L) (mean 0.36 mg/L). There was no marked difference in dissolved CO in plasma for the eight smokers compared to the non-smokers in the group, although the mean COHb was 4.6% for the smokers and 1.0% for the non-smokers.

A simple, sensitive method for estimating carbon monoxide in plasma is described. In this method, the carbon monoxide in plasma is trapped with hemoglobin and subsequently estimated by dithionite reduction. The method has an intra-and intraassay precision (CV) of 10.7% and 12.8% respectively, at a concentration of 1.12 mg of carbon monoxide per litre and has a detection limit of 0.1 mg/L.

Reference Number: 10308

ERNST, A., AND J.D. ZIBRAK. "Carbon Monoxide Poisoning." *New England Journal of Medicine*, 22: 1603–1608, 1998 (2 tables, 1 figure, 66 references)

Abstract: The rate of absorption of CO through the lungs depends on the minute ventilation, duration of exposure, and relative concentration of CO and oxygen in the environment. Less than 1% of the CO is oxidized to carbon dioxide (CO_2) in the body, the rest is eliminated through the lungs as an unchanged gas. Ten to 15% of CO is bound to proteins, including myoglobin and cytochrome-c-oxidase, and less than 1% of CO exists as an absorbed gas in solution (plasma). CO toxicity results from cumulative tissue hypoxia and direct CO damage at the cellular level. Thus, the small fraction of CO dissolved in plasma has an important role as experiments involving the transfusion of blood with highly bound COHb, but no dissolved CO resulted in clinical symptoms.

Approximately 600 accidental deaths due to carbon monoxide are reported annually in the United States, and the number of intentional carbon monoxide-related deaths is 5 to 10 times higher. The rate of accidental death caused by carbon monoxide from motor vehicles is higher in the northern United States and peaks during the winter months. The intentional deaths occur year-round without significant peaks.

1.04 ELIMINATION

> [U]nder physiological conditions the molecule is inert. CO goes in as CO and comes out as CO. In the body it undergoes no metabolism.
>
> —A. Katsnelson, "The Good Side of Carbon Monoxide" (2019) **(10501)**

Unlike virtually all the other drugs and toxins, CO is not metabolized or significantly eliminated by the liver. It is not converted into any metabolite in the body, and, unlike alcohol, nicotine, and delta-9-tetrahydrocannabinol (THC), it remains CO. The elimination of CO from the blood was found to be biphasic, which was faster shortly after exposure rather than later (10401) and is prolonged in persons with CO produced by methylene chloride intoxication (10404). The half-life of CO is prolonged in patients with lung dysfunction (10408). Increased ventilation rates using CO_2 can shorten the half-life to about 30 minutes (10405). The half-life using NBOT without CO_2 is about 74 minutes (10406).

Reference Number: 10401

BRUCE, M.C., AND E.N. BRUCE. "Analysis of Factors That Influence Rates of Carbon Monoxide Uptake, Distribution, and Washout from Blood and Extravascular Tissues Using a Multicompartment Model." *Journal of Applied Physiology*, 100: 1171–1180, 2006 (1 tables, 6 figures, 25 references)

Abstract: A multicomponent mathematical model was developed based on 15 human subjects who inhaled CO over four to six minutes and had multiple blood COHb measured over 4.5 hours while breathing room air.

Table: Mean COHb Half-Times at 5 to 259 Minutes Post-Exposure

Time from end of CO exposure (minutes)	Mean COHb half-life (minutes)
5–67	208
35–67	247
35–259	302
131–259	358
"True"	358

Source: Adapted from Bruce and Bruce 2006.

We found that CO washout from venous blood was biphasic and the postexposure times at which COHb samples were obtained significantly influenced the calculated CO half times. The first, more rapid phase of CO washout from the blood reflected the loss of CO to the expired air and to a slow uptake by the muscle compartment, whereas the second, slower washout phase was attributable to CO flow from the muscle compartment back to the blood and removal for blood via the expired air. When the model was used to predict the effects of varying exposure conditions for these subjects, the CO exposure duration, concentration, peak COHb levels, and subject-specific parameters each influenced washout times. Blood volume divided by ventilation correlated better with half-time predictions than did cardiac output, muscle mass or ventilation, but it explained only ~ 50% of half-time variability. Thus, exposure conditions, COHb sampling times, and individual parameters should be considered when estimating CO washout rates for poisoning victims.

Reference Number: 10402

PETERSON, J.E., AND R.D. STEWART. "Absorption and Elimination of Carbon Monoxide by Inactive Young Men." *Archive Environmental Health*, 21: 165–171, 1970 (5 tables, 1 figure, 11 references)

Abstract: Eighteen healthy male students (ages 24–42 years) were exposed to concentrations of CO ranging from 1 to 1,000 ppm for 0.25 to 24 hours. Blood COHb concentrations were measured by the CO oximeter and gas chromatograph. During the experiment, activity was sedentary, mainly reading or playing chess. The half-life of CO for ambient conditions ranged from 128 to 409 minutes (mean 320 minutes). When two subjects were administered 100% oxygen, the mean half-life was 80.3 minutes, and when the oxygen was administered at three atmospheres, the half-life was 23.3 minutes.

Excretion or elimination of CO from the blood was found to be a half-life phenomenon. In 39 separate experiments, the half-life of COHb in the blood was found to range from 128 to 409 minutes with an average of 320 minutes. Variations in the half-life of COHb in blood did not appear to be related to the amount of COHb in the blood, to the duration of exposure, to the number of exposures, or to the concentration inhaled.

Reference Number: 10403

OZTURAN, I.U., E. YAKA, S. SUNER, A.E. OZBEK, C. ALYESIL, N.O. DOGAN, S. YILMAZ, AND M. PEKEDEMIR. "Determination of Carboxyhemoglobin Half-Life in Patients with Carbon Monoxide Toxicity Treated with High Flow Nasal Cannula Oxygen Therapy." *Clinical Toxicology*, 57: 617–623, 2019 (1 table, 3 figures, 34 references)

Abstract: The COHb half-life was determined in 33 patients admitted to hospitals in Turkey between 2017 and 2018 and treated with high-flow nasal cannula (HFNC) (not a face mask) oxygen therapy. Venous blood samples were collected every 10 minutes, and COHb was determined with a radiometer ABL 735 gas analyzer. The mean initial COHb was 22.5%. A target flow rate of 60 litres per minute and temperature of 37 degrees Celsius was employed. The mean half-life was 36.8 minutes, which is shorter than other NBOT. HFNC was more comfortable for the patient and allowed for easier talking and eating than the mask.

Oxygen delivered via high flow nasal cannula is a safe and comfortable method to treat acute CO toxicity and is effective in reducing COHb half-life. HFNC may be a promising alternative method of delivering oxygen in the setting of CO toxicity.

Reference Number: 10404

HOWE, K., D. GRIFFEN, AND C. JAEGER. "Human Carbon Monoxide Factory." *American Journal of Emergency Medicine*, 34: 2251-e3, 2016 (1 figure, 22 references)

Abstract: Methylene chloride (dichloromethane) is an industrial solvent that is rapidly absorbed by the lungs or through the skin. Methylene chloride is lipophilic and distributed throughout the body. It is oxidized to CO in the liver by the cytochrome P450 pathway. In methylene chloride poisoning, the half-life of CO is prolonged to 13 hours without oxygen therapy rather than four to six hours. A 49-year-old man was found unconscious by his wife, slumped over a tank of paint-stripping fluids. Approximately four hours after admission, his blood COHb was 9.2%, and after four hours of NBOT, it increased slightly to 10.9%, and, four hours later, it had only decreased to 10.0%. The patient left after 16 hours and had no complaints at follow-up.

Table: Blood COHb Concentrations of a Patient with Methylene Chloride Exposure after NBOT

Time of NBOT	Blood COHb
Initial	9.2%
4 hours	10.9%
8 hours	10.0%

Source: Adapted from Howe, Griffen, and Jaeger 2016.

The half-life of COHb is prolonged in methylene chloride toxicity as the conversion to CO can last several hours beyond the exposure. In addition, methylene chloride can be stored in tissues, particularly adipose tissue, and converted to CO over time. The following case demonstrates that patients exposed to methylene chloride can essentially become CO factories with the production of CO within the body continuing long after exposure to the toxin.

Reference Number: 10405

ZAVORSKY, G.S., J. TESLER, J. RUCKER, L. FEDORKO, J. DUFFIN., AND J.A. FISHER. "Rates of Carbon Monoxide Elimination in Males and Females." *Physiological Reports*, 2: e12237 (10pp), 2014 (4 tables, 3 figures, 43 references)

Abstract: Eight female (mean age 29 years) and nine male (mean age 27 years) subjects inhaled 1,100 ppm CO for a mean duration of 31 minutes to obtain a peak blood COHb concentration of 10% on three occasions. The subjects were then administered 100% oxygen, oxygen plus CO_2, or CO_2 to increase the ventilation rate to four times resting. The CO half-life was determined from multiple blood samples.

Table: Mean COHb Half-Life in Men and Women with 100% Oxygen, Oxygen and CO_2, and CO_2 Alone

Washout condition	Men half-time (minutes)	Women half-time (minutes)
100% Oxygen	68	52
Oxygen and CO_2	50	40
CO_2 (four times resting ventilation rate)	32	25

Source: Adapted from Zavorsky et al. 2014.

This study indicates that the rate of clearance of CO influenced by gender only to the extent that the relationship of total body CO stores to VE differ from that of males. These factors, not merely gender, must be taken into account when assessing clearance of CO. In particular, conditions that alter HbTOT and VE must be considered regardless of gender. For example, anemia will tend to shorten $t_1/_2$ whereas polycythemia (which is common in female smokers) will tend to lengthen it. Similarly, any condition which alters VA will also affect $t_1/_2$. For example, anxiety induced hyperventilation will lower $t_1/_2$, whereas hypoventilation due to emphysema, or a reduced level of consciousness will raise $t_1/_2$. Furthermore, limitations in cardiac output will affect $t_1/_2$ with any type of treatment. Thus, a prediction of $t_1/_2$ from standard tables, not taking gender into account (i.e., most text books) will be erroneous. Less so if gender is taken into account. Optimal predictions will require in addition, a thorough clinical assessment.

Reference Number: 10406

WEAVER, L.K., S. HOWE, R. HOPKINS, AND K.J. CHAN. "Carboxyhemoglobin Half-Life in Carbon Monoxide Poisoned Patients Treated with 100% Oxygen at Atmospheric Pressure." *Chest*, 117: 801–808, 2000 (4 tables, 1 figure, 23 references)

Abstract: A retrospective chart review was conducted on 93 CO-poisoned patients administered 100% oxygen in a face mask or endotracheal tube between 1985 and 1995 in which at least two blood COHb measurements were conducted. The mean half-time was 74 ± 25 minutes, which was much shorter than the 130 minutes reported with earlier clinical trials.

Table: Sources of CO in the 59 Accidental and 34 Suicide Cases

Source	Number
Furnace	59
Auto	18
Propane buffer (industrial)	3
Propane heater (industrial)	3
Gas saw (industrial)	2
Propane forklift (industrial)	1
Charcoal grills	6
Fires	4
Water heater	1
Auto (suicide)	33
Charcoal grill (suicide)	1

Source: Adapted from Weaver et al. 2000.

As expected, we found an inverse relationship between the COHb t1/2 and the PaO_2. The COHb t1/2 was not influenced in our study by the patients' ages, genders, smoke inhalation, history of LOC, concurrent tobacco smoking, whether intubated or breathing O_2 by face mask or the initial COHb measurement.

Reference Number: 10407

THORNTON, D., R. GALLAGHER, D. GALLAGHER, J. HUNTER, R. CASTELLI, AND M. SCOTT. "Trends and Characteristics of Cases When Serial Carboxyhemoglobins Are Obtained." *Undersea and Hyperbaric Medecine*, 46: 655–658, 2019 (1 table, 1 figure, 13 references)

Abstract: Serial blood COHb measurements were determined in 106 patients with CO exposures between 2010 and 2015. The mean number of samples collected was 2.6 (range 2–9). The mean blood COHb concentrations were 8.9%, 2.8%, 1.8%, 1.1%, and 0.75%; initially, 353, 663, 1095, and 2023 minutes later.

Serial COHb levels were not infrequent in this study. No clinically significant increase in COHb was identified by serial testing. Serial COHb tests were more likely to be ordered on patients with burns and those admitted to hospital and ICU. Further studies should examine the clinical utility of such practices.

Reference Number: 10408

WEAVER, L.K., AND K. DERU. "Carboxyhemoglobin Half-Life during Hyperbaric Oxygen in a Patient with Lung Dysfunction: A Case Report." *Undersea and Hyperbaric Medecine*, 44: 173–177, 2017 (3 tables, 1 figure, 14 references)

Abstract: An 81-year-old man was found unconscious in his garage due to an improperly ventilated natural gas heater. He was administered 100% oxygen and later intubated at the emergency department (ED). His initial blood COHb was 34.4% and was 5.9% eighteen minutes before HBOT. After a compression interval of 17 minutes and 27 minutes at three atmospheres absolute, his COHb was 3.3%. He had concomitant respiratory failure and lung dysfunction and was mechanically ventilated. It was estimated that the COHb half-life was 242 minutes for 100% oxygen and 49–58 minutes for HBOT.

We present a case report in which the COHb t1/2 we calculated is greater than the mean reported by the single prior study of 12 CO-poisoned patients, likely explained by our patient's concomitant respiratory failure, lung dysfunction mechanical ventilation and long exposure time. The often-cited COHb t1/2 of 23 minutes at 3ATA likely underestimates the actual COHb t1/2 in CO poisoned patients, especially those with long exposure durations or cardiopulmonary dysfunction.

1.05 ENDOGENOUS PRODUCTION

Low levels of endogenous CO are produced in the liver by the oxidative degradation of heme proteins such as hemoglobin and myoglobin (10502, 10505). The median endogenous blood COHb concentration in healthy persons is approximately 1.3% (10506)

Hemoglobin → Heme + Globin → Biliverdin + Fe^{2+} + CO → Bilirubin+ COHb

Certain medical conditions such as pregnancy and hemolytic anemia (10503), sepsis (10504), chronic obstructive pulmonary disease (COPD) (10506), and obstructive sleep disorder (10507). At low concentrations, CO can act as a neurotransmitter and reduce inflammation (10501).

Reference Number: 10501

KATSNELSON, A. "The Good Side of Carbon Monoxide." *ACS Central Science*, 5: 1632–1635, 2019 (3 figures)

Abstract: CO poisoning, which is due to the incomplete combustion of fuels, sends about 50,000 people to the hospital in the United States each year and kills about 400. However, every day about 10 millilitres of CO is produced by the human body by the metabolism of heme by the heme oxygenase enzyme. At low concentrations, CO interacts with multiple heme-containing proteins and is a neurotransmitter. CO can reduce inflammation, decrease oxidative stress, and prevent cell death. CO may help treat sepsis, sickle cell disease, lung fibrosis, and ulcerative colitis. CO can be bound to a carrier that can release it in the body like a drug.

CO is a funny little molecule. The triple bond that ties its carbon to its oxygen is the strongest covalent bond known on Earth, under physiological conditions, the molecule is inert. CO goes in as CO and comes out as CO. In the body, it undergoes no metabolism. So, getting it to bind to a carrier would be challenging.

Reference Number: 10502

MANNAIONI, P.F., A. VANNACCI, AND E. MASINI. "Carbon Monoxide: The Bad and Good Side of the Coin, from Neuronal Death to Anti-Inflammatory Activity." *Inflammatory Research*, 55: 261–273, 2006 (1 table, 6 figures, 125 references)

Abstract: The physiological presence of CO was first detected in the blood of a dog by M.N. Grehant in 1894. The average blood hemoglobin concentration of the Weddell seal in the Antarctic is 40–50% higher than the average Hb concentration in adult humans. Hence, due to the greater Hb breakdown, the seal's blood contains about six times the COHb concentration as non-smoking humans. The amount of exhaled CO was increased in patients with hemolytic anemia and other diseases, which increased the rate of hemoglobin breakdown.

On the good side of the coin, we briefly outline the endogenous generation of CO and the leading role of heme-oxygenases (HO)—in relation to the place of CO in biology and medicine. The main focus of this section is on the growing literature on CO and inflammation. Here we report on in-vitro and in-vivo studies on the modulation afforded by exogenously administered/endogenously produced CO in a variety of experimental and clinical settings of inflammation.

Reference Number: 10503

HYSON, B.E., R.S. WEHBIE, N.A. HAIKAL, AND S.C. BISHOP-FREEMAN. "Endogenous Carbon Monoxide Due to Hemolytic Anemia: A Forensic Red Herring." *Journal of Forensic Science*, 67: 1294–1299, 2022 (1 table, 1 figure, 26 references)

Abstract: A 60-year-old woman, living alone, phoned emergency medical services after experiencing nausea, dizziness, and fatigue for two days. She was removed from her home, and her fingertip pulse CO-oximeter showed a carboxyhemoglobin concentration of 25%. Nearly two hours later at the ED, her blood COHb was 16%. She was administered oxygen, Ringers lactate, and blood products, and her blood COHb 11 hours later was 6.8%. But her condition deteriorated, and she was transferred to the Cardiac Intensive Care Unit where she suffered a fatal cardiac arrest and was declared dead 19 hours after presenting to the ED. Her post-mortem subclavian blood COHb was 5%. No source of CO could be detected in her home, and successive ambient CO readings showed 0 ppm. It was determined that she suffered from warm antibody autoimmune hemolytic anemia, which was the cause of her death.

Endogenous CO elevation as the result of a disease state is caused by an increase in red blood cell destruction. The elevated endogenous CO and COHb may not always clinically impact a person's health. Higher than normal endogenous CO production rates have been observed in pregnancy (30–40%) increase, sepsis, cardiac disease, critically ill patients undergoing dialysis, and hemolytic anemia. Of these conditions, hemolytic anemia was found to be associated with the highest rate of endogenous CO production at two to eight times higher than normal. Interestingly, exhaled CO is not necessarily affected in the setting of hemolytic anemia but is increased in smokers and may be affected by other disease states. There are also rare instances of extracorporeal membrane oxygenation (ECMO) patients having elevated COHb saturation levels after developing severe hemolytic anemia due to the ECMO treatment.

Reference Number: 10504

ZEGDI, R., D. PERRIN, M. BURDIN, R. BOITEAU, AND A. TENAILLON. "Increased Endogenous Carbon Monoxide Production in Severe Sepsis." *Intensive Care Medicine*, 28: 793–796, 2002 (1 table, 2 figures, 10 references)

Abstract: The breath CO concentration was measured in 24 mechanically ventilated patients with severe sepsis and five controls by an infrared CO analyzer. The mean exhaled breath CO concentration for the sepsis group was 1.53, 1.2, and 1.09 ppm for day 1, 2, and 3 for the sepsis group and 0.54, 0.54 and 0.54 ppm for the control group. The exhaled breath CO concentration decreased over three days but remained higher in the sepsis group than the control group. On the first day, the endogenous CO production was significantly lower in the sepsis patients who did not survive (8.5 microlitres per hour) compared to patients who survived (14.7 microlitres per kilogram per hour).

Carbon monoxide had long been regarded as a waste product of heme metabolism. Anti-inflammatory properties, however, such as the inhibition of proinflammatory cytokines production have recently been attributed to CO. In the present study, survivors of sepsis initially had a higher endogenous CO production, however, was too small to determine whether the level of endogenous CO production was an independent predictor of survival in severe sepsis.

Reference Number: 10505

OWENS, E.O. "Endogenous Carbon Monoxide Production in Disease." *Clinical Biochemistry*, 43: 1183–1188, 2010 (3 figures, 81 references)

Abstract: Endogenous CO is produced by the oxidative degradation of the heme proteins, such as hemoglobin or myoglobin by the stress response enzyme, heme oxygenase (HO). The major site of heme catabolism is the liver, which is the major source of endogenous CO, followed by the spleen and brain. The body typically produces 0.42 millilitres (mL)/hour or 0.007 mL/minute of CO. Endogenous CO is eliminated mainly by exhalation, and only about 10% is oxidized into CO_2 by cytochrome C oxidase in the mitochondria. Any condition or disease that increases red blood cell destruction, or hemolysis, will increase the endogenous CO production. Endogenous CO production also increases during inflammatory diseases or infection due to the induction of the HO enzyme. Women have fluctuating endogenous CO concentrations, which vary from about 0.62 mL/hour in the progesterone phase to 0.32 mL/hour in the estrogen phase.

It is also possibly that endogenous CO could be used as a marker of airway inflammatory diseases. Increased arterial COHb measured by CO-oximeter was reported in patients with a number of inflammatory lung diseases including bronchial asthma, pneumonia, idiopathic pulmonary fibrosis, pyelonephritis, and active rheumatoid arthritis. In contrary, there was no increase in patients with asbestos related inflammatory disease including asbestosis. Arterial COHb levels and eCO are increased in patients with chronic obstructive pulmonary disease (COPD). Endogenous CO correlate well with COPD severity, where it was proposed as an indicator of oxidative stress process in COPD.

Reference Number: 10506

DOGAN, N.O., S.K. CORBACIOGLU, F. BILDIK, I. KILLCASLAN, G.P. GUNAYDIN, Y. CEVIK, V. ULKER, O. HAKOGLU, AND E. GOKCEN. "Determining the Diagnostic Value of Endogenous Carbon Monoxide in Chronic Obstructive Lung Disease Exacerbations." *Journal of the Pakistan Medical Association*, 64: 1037–1041, 2014 (2 tables, 2 figures, 15 references)

Abstract: The arterial blood COHb concentrations were determined in 81 healthy controls and 28, 50, and 50 patients with slight, intermediate, and severe COPD respectively. The median COHb was 1.30% and 1.70% in the controls and COPD patients respectively.

Table: The Partial Pressure of Oxygen (pO_2), Partial Pressure of CO_2 (pCO_2), and % COHb in Patients with Slight, Intermediate, and Severe COPD

	Slight COPD median (interquartile range)	Intermediate COPD median (interquartile range)	Severe COPD median (interquartile range)
pO_2 (mmHg)	54.9 (52.8–68.6)	49.5 (43.0–64.8)	44.3 (33.3–56.6)
pCO_2 (mmHg)	38.5 (35.3–44.0)	40.8 (32.8–48.5)	56.9 (55.0–70.0)
COHb (%)	1.6 (0.95–2.0)	1.18 (1.38–2.20)	1.9 (1.5–3.0)

Source: Adapted from Dogan et al. 2014.

COHb values were higher in patients with severe COPD exacerbations, and there was a weak negative correlation with pO_2 and positive correlation with pCO_2 in terms of COHb values. Although it seems impossible at this stage to conclude that these high COHb values can help us in predicting clinical severity of COPD exacerbations, the study indicated that endogenous COHb levels in former smokers presenting with severe COPD exacerbation were higher than non-smoking controls and mild exacerbations.

Reference Number: 10507

KIS, A., M. MESZAROS, D.L. TARNOKI, A.D. TARNOKE, Z. LAZAR, P. HORVATH, L. KUNOS, AND A. BIKOV. "Exhaled Carbon Monoxide Levels in Obstructive Sleep Apnoea." *Journal of Breath Research*, 13: 1752–1763, 2019 (1 table, 4 figures, 40 references)

Abstract: Obstructive sleep apnea (OSA) is a common disorder in which there are repetitive episodes of completer or partial collapse of the upper airways during sleep. This intermittent repetitive hypoxemia leads to airway inflammation and oxidative stress. CO is a marker of oxidative stress. Exhaled CO concentrations were measured in the evening and morning using an electrochemical device (Pico Smokerlyser) on 60 never-smoking OSA patients, 14 ex-smoking patients, 39 current smoking OSA patients, 10 never-smoking asthmatic patients, 16 COPD patients with OSA, and 20 never-smoking control individuals. Exhaled CO is used clinically to assess smoking practices and monitor smoking cessation.

Table: Evening Breath Exhaled Carbon Monoxide (e-CO) Concentrations in Various Groups of Patients with OSA

Group	e-CO in evening (ppm)
Controls	1.95
Severe OSA	2.43
Never-smoking OSA	1.98
Former-smoking OSA	3.07
Current-smoking OSA	13.13
COPD and OSA	18.25

Source: Adapted from Kis et al. 2019.

In summary, OSA is not a major bias to interpret exhaled carbon monoxide levels in clinical practice. The mild but significant rise in severe OSA possibly reflects airway oxidative stress, however more sensitive tests, such as exhaled volatile organic compound analysis should be carried out if someone aims to monitor this in OSA. Nevertheless, the results in smoker populations points out the need for standardizing the timing of measurements when evaluating eCO levels in smokers.

CHAPTER 2

Carbon Monoxide Analysis in Living Persons

This chapter summarizes the main methods for determining carbon monoxide (CO) poisonings and concentrations in blood, breath, and pulse CO-oximetry methods.

2.01 BLOOD CARBOXYHEMOGLOBIN METHODS OF ANALYSIS

[C]arbon monoxide gives the blood a nice gleaming arterial colour.

— Claude Bernard (1813–78), cited by G.L. Sternbach and J. Varon, "Claude Bernard: On the Origin of Carbon Monoxide Poisoning" (1991) **(40404)**

Co-oximeters are acceptable to measure carboxyhemoglobin (COHb) concentrations in hospital patients (20101–2). Additional accuracy can be obtained from gas chromatograph (GC) and gas chromatograph/mass spectrometer (GC/MS) methods (20103).

Reference Number: 20101

KALES, S.N., F. PENTIUC, AND D.C. CHRISTIANI. "Pseudoelevation of Carboxyhemoglobin Levels in Firefighters." *Journal of Occupational Medicine*, 36: 752–756, 1994 (3 tables, 18 references)

Abstract: Endogenous CO production via heme catabolism results in a blood COHb of less than 1%. Differences in the measured blood COHb concentration between the manual spectrophotometric method and the CO-oximeter method were found in a group of firefighters in Boston. CO-oximeter results were lower and more uniform than the manual spectrophotometric method. Using the CO-oximeter, all 24 non-smoking firefighters had a blood COHb of less than 3% compared to 35% of the same firefighters who showed a blood COHb greater than 4% and = 26% of the firefighters who had a blood COHb of 10% or greater by the manual method.

Table: A Comparison of Blood COHb Percentage between the CO-Oximeter and Manual Methods

Subject	Percentage of COHb (CO-oximeter) (%)	Percentage of COHb (manual) (%)
Firefighter 1	0.7	4
Firefighter 2	0.7	4
Staff 1	0.2	< 4
Staff 2	0.6	< 4
Staff 3	0.8	4
Staff 4 (smoker)	3.2	7
Staff 5 (smoker)	8.4	13

Source: Adapted from Kales et al. 1994.

The fact that such clinically important spurious results could occur despite the use of a larger reference laboratory argues for standardization of laboratory techniques for COHb among laboratories. Occupational and emergency physicians should be aware of the potential for errors in the measurement especially when lower levels of COHb are suspected. Finally, the results of this study revealing errors in the performance of a common test should make us more vigilant regarding the possibilities of error in less frequently ordered biomarkers of more infrequent occupational and environmental exposures, especially when important medical, legal and work implications may follow from the results obtained.

Reference Number: 20102

WIDDORP, B. "Analysis of Carbon Monoxide." *Annals of Clinical Biochemistry*, 39: 378–391, 2002 (7 tables, 2 figures, 78 references)

Abstract: A review of the measurement of blood COHb for clinical purposes is presented. Simple photometric methods are prone to interferences by other hemoglobin pigments and are not recommended. Automated spectrophotometric instruments such as CO-oximeters have acceptable accuracy for COHb that is greater than 5% but are expensive to maintain. GC methods are preferable for postmortem blood.

> For fresh blood samples, spectrophotometric techniques that can estimate total hemoglobin, percentage oxyhemoglobin, and percentage COHb simultaneously are ideal. There can be no doubt that CO-oximeters, with their ease of operation, freedom from spectral interferences, and built-in software that calculates the parameters listed above, will remain the preferred instrument in hospital laboratories for many years to come. They are sensitive and precise enough to detect chronic exposure to carbon monoxide from, for example, faulty heaters.

Reference Number: 20103

WEAVER, L.K., K. DERU, B. FLETCHER, AND S. DANIEL. "Three Cases of Clinically Significant, Inaccurate Carboxyhemoglobin Measurement." *Undersea and Hyperbaric Medicine*, 49: 171–177, 2022 (7 tables, 2 figures, 78 references)

Abstract: Two clinical cases are reported in which the blood COHb measured at the hospital were false negatives and one case that was a false positive. The first case involved a 20-year-old man who developed headache, confusion, and near syncope while operating a gasoline-powered pressure washer in an enclosed space. His initial blood COHb was reported as 1.8%, which was ignored, and he was still treated with hyperbaric oxygen therapy. A later analysis showed his blood COHb was 20.1%. A 45-year-old man presented with several others showing symptoms of CO poisoning at a conference. His blood COHb was reported as 2%, whereas all the others had elevated COHb. He was released, but he returned later with the same symptoms and requested another COHb test. Another blood sample collected 84 minutes later was 17% and after three hours of normobaric oxygen therapy was 7.1%. The final case involved an 83-year-old non-smoking man who lived in an electrically powered house who was diagnosed with COVID pneumonia but whose blood COHb was reported as 7.1%. His companion was symptomatic and had a blood COHb of 3.0%. An additional analysis of the blood sample by GC/MS showed his blood COHb to be 2%.

COHb levels are not always accurate. Clinicians should use clinical judgment to manage their patients including rejecting laboratory values that do not fit the clinical picture.

2.02 BREATH CO METHODS OF ANALYSIS

The Chief Medical Officer has repeatedly urged doctors to remain aware of the risk of carbon monoxide poisoning during the winter months. We suggest the use of the breath analyzer at the triage station in the emergency department.

—A.J. Cunningham and P. Hormbrey, "Breath Analysis to Detect Recent Exposure to Carbon Monoxide" (2002) **(20203)**

One method of detecting breath CO concentrations is by using an electrochemical cell that converts the CO into an electrical current that can be measured (20202–3). Breath is always a rapid, specific, and non-invasive method of determining blood CO concentrations.

$$CO + H_20 \rightarrow CO_2 + 2H^+ + 2e^-$$

Reference Number: 20201

HALDANE, J. "The Detection and Estimation of Carbonic Oxide in Air." *Journal of Physiology*, 20: 521–522, 1896 (1 table, 2 references)

Abstract: A calorimetric/spectroscopic method was described to determine the CO concentration in air. A CO-free blood sample is shaken in the air, and the COHb measurement can be converted into a percentage of carbonic acid in the air.

A poisonous proportion of carbonic oxide in air may be detected most rapidly of all by watching the symptoms of a mouse carried into the suspected atmosphere. This method has recently been successfully employed in coal-mines by Mr. W.N. Atkinson, H.M. Inspector of Mines for North Staffordshire.

Reference Number: 20202

EBERHARDT, M., A. POWELL, G. BONFANTE, V. RUPP, J.R. GUARNACCIA, M. HELLER, AND J. REED. "Noninvasive Measurement of Carbon Monoxide Levels in ED Patients with Headache." *Journal of Medical Toxicology*, 2: 89–92, 2006 (2 tables, 9 references)

Abstract: Breath CO concentrations were measured in 170 emergency department (ED) patients with headache (no fever, no trauma) and 98 matched controls with no headache using a micro CO breath analyzer between 2003 and 2005. Overall, 12 of the 170 people in the headache group had elevated breath CO (7.1%) compared to one (1%) of the controls. Blood was collected from nine of the 12 people in the headache group and confirmed an elevated blood COHb of between 5.4 and 9.3%. Two of the sources of CO were a faulty car exhaust and a poorly ventilated fireplace.

> Noninvasive measurement of CO levels in ED patients with Headache is rapid and specific. During winter months, elevated CO levels are present in over 7% of ED patients with headaches.

Reference Number: 20203

CUNNINGHAM, A.J., AND P. HORMBREY. "Breath Analysis to Detect Recent Exposure to Carbon Monoxide." *Postgraduate Medical Journal*, 78: 233–238, 2002 (2 tables, 3 figures)

Abstract: A handheld CO breath analyzer was used to measure the end-expiratory breath CO concentration of 382 persons attending an ED who also filled out a questionnaire. The end-expiratory breath CO concentration decreased by 50% for each successive blow taken 30 seconds apart, and, thus, in this study, a single breath measurement was employed. The mean breath CO concentration in the non-smoking group (n = 276) was 1.26 parts per million (ppm) (range 0–6 ppm). The mean breath CO concentration in the smoking group (n = 90) was 16.4 ppm (range 1–68 ppm). The mean time since last smoking for this group was 3.14 hours (range 0–24 hours). Motor vehicle use was associated with a mean 0.38 ppm increase in breath CO compared to non-users.

Table: Mean Breath CO Concentration in Smokers According to Number of Cigarettes Smoked

Number of cigarettes smoked	Mean breath CO (ppm)
1–10	11.9
11–20	17.7
21–30	32.8
31–40	44.0

Source: Adapted from Cunningham and Hormbrey 2002.

The Chief Medical Officer has repeatedly urged doctors to remain aware of the risk of carbon monoxide poisoning during the winter months. We suggest the use of the breath analyzer at the triage station in the emergency department. Where there are features compatible with carbon monoxide exposure and no other obvious cause, a breath sample should be taken before any oxygen is given. It is essential that the sample is properly obtained to avoid the need for repeating samples at short intervals and obtaining a falsely low reading. A breath carbon monoxide greater than 6 ppm requires a specific consideration of the possibility of carbon monoxide poisoning. In this assessment potential exogenous and endogenous sources of carbon monoxide should be considered in addition to factors that increase vulnerability to carbon monoxide poisoning and may lower the threshold at which it becomes harmful. The time since potential exposure must also be considered.

2.03 PULSE CO-OXIMETERS

In conclusion, occult CO poisoning is an important frequently overlooked cause of winter headache in urban emergency department patients.

—P.S. Heckerling, "Occult Carbon Monoxide Poisoning: A Cause of Winter Headache" (1987) **(30105)**

Fingertip pulse oximetry is widely used in emergency rooms (ERs), but the earlier devices that only had two wavelengths of light to determine the COHb concentration gave false positive results compared to the eight wavelength devices (Masimo Rad-57) used today (20301–9). Pulse oximetry has allowed for the more rapid detection of occult CO poisoning and faster oxygen treatment (20303–4).

Reference Number: 20301

VEGFORS, M., AND C. LENNMARKEN. "Carboxyhaemogloinaemia and Pulse Oximetry." *British Journal of Anaesthesia*, 66: 625–626, 1991 (1 figure, 5 references)

Abstract: Fingertip pulse oximetry is widely used in ERs, but the devices using only two wavelengths of light: 940 nanometre (nm) (oxyhemoglobin [HbO]) and 660 nm (hemoglobin [Hb]) measures of carboxyhemoglobin (COHb or HbCO) (strongly absorbs at 660 nm) as HbO_2. A 26-year-old man attempted suicide by venting the exhaust of his car into the driver's compartment and then crashed when he drove his car into town. He was administered oxygen and taken to hospital where he complained of a headache and was disorientated. The two-wavelength pulse oximetry device showed a HbO_2 of 100% and no HbCO, but the blood results showed an HbO_2 of 70% and COHb of 30%.

This report confirms that pulse oximetry may be misleading during carbon monoxide poisoning as the pulse oximeter does not differentiate between HbO and HbCO. Diagnosis of carbon monoxide poisoning therefore relies on clinical evidence and must be confirmed by measuring the HbCO concentration with a multiple band CO-oximeter.

Reference Number: 20302

ZOLLER, T., E.H. MFINANGA, T.B. ZUMBA, P.J. ASILIA, E.M. MUTABAZI, D. WIMMERSBERGER, F. MHIMBIRIA, F. HARAKA, AND K. REITHER. "Symptoms and Functional Limitations Related to Respiratory Health and Carbon Monoxide Poisoning in Tanzania: A Cross Sectional Study." *Environmental Health*, 21 (1): 38, 2022 (6 tables, 2 figures, 36 references)

Abstract: The carboxyhemoglobin concentration (SpCO) was determined in 9,997 participants of the Tanzania Lung Health study using the fingertip Masimo Rad-57 pulse oximeter. The effects of CO exposure were determined in 808 participants, and SpCOs were measured. Seventy-five percent of the participants were never smokers, 11.5% were former smokers, and 13.1% were active smokers and removed from the data of the study. About 38% of the participants had shortness of breath. Non-smokers exposed to second-hand smoke (SHS) had a higher mean SpCO (8%) compared to those not exposed to SHS (6%). Participants who cooked using gas or electricity had the lowest median SpCO (5%), which increased to 7% for those who used charcoal and 11.5% for those using wood. The mean SpCO increased from 6% for those with no shortness of breath to 8% and 9% for those with increasing shortness of breath. Those who had severe physical limitations at work or home due to shortness of breath had a median SpCO of 11%.

In conclusion, the data show unexpected high levels of SpCO indicative of chronic CO poisoning in study participants. A substantial burden of disease and a relevant socioeconomic impact caused by respiratory symptoms and CO exposure in people living in urban and rural areas in Tanzania must be assumed.

Reference Number: 20303

HAMPSON, N.B. "Noninvasive Pulse CO-Oximetry Expedites Evaluation and Management of Patients with Carbon Monoxide Poisoning." *American Journal of Emergency Medicine*, 30: 2021–2024, 2012 (1 table, 1 figure, 19 references)

Abstract: The Masimo Rad-57 is a handheld pulse CO-oximeter that measures COHb by transilluminating the fingertip with multiple wavelengths of near-infrared light. It has been available on the market since 2005. It has been used by many paramedics and firefighters. A study was conducted between 2008 and 2011 of 1,606 patents who had COHb concentration measured in the blood as usual and 105 patients screened with pulse CO-oximetry who were treated for CO poisoning with hyperbaric oxygen (HBO_2). On average, patients with pulse-CO-oximetry were treated with HBO one hour earlier than those not screened. The initial COHb concentration was also higher in the pulse CO-oximetry screened group (25.9% versus 21.9%).

> Pulse CO-oximetry is associated with more rapid diagnosis and initiation of hyperbaric oxygen therapy in CO-poisoned patients compared with laboratory CO-oximetry.

Reference Number: 20304

COULANGE, M., A. BARTHELEMY, F. HUG, A.L. THIERRY, AND L. DE HARO. "Reliability of New Pulse CO-Oximeter in Victims of Carbon Monoxide Poisoning." *Undersea and Hyperbaric Medicine*, 35: 107–111, 2008 (1 table, 1 figure, 13 references)

Abstract: The Masimo Rad-57 pulse CO-oximeter measures spCO by using eight wavelengths of light from a sensor placed on the middle or ring finger. During a seven-month study period at an ED, COHb concentrations were determined in the blood, using Instrumentation Laboratory 682-CO-oximeter, and compared with the Masimo Rad-57 results in 12 patients. The mean COHb was 13.9% in the blood, and the mean spCO was 15% (not statistically significant).

Table: A Comparison of Blood COHb (> 10.2%) and Masimo Rad-57 Pulse CO-Oximeter Results (SpCO) and Clinical Signs

Number	Clinical signs	Blood COHb (%)	SpCO (%)
1	Minor	10.2	13
2	Severe	10.7	9
3	Minor	11.0	15
4	Severe	11.3	10
5	Minor	11.9	12
6	Severe	17.6	18
7	Severe	18.0	20
8	Absent	20.1	25
9	Minor	20.8	18
10	Minor	31.6	35

Source: Adapted from Coulange et al. 2008.

Our data indicate that pulse CO-oximetry could be useful for the management of COP in two possible situations. First, in hospital emergency departments, pulse CO-oximetry would accelerate detection time of COHb levels. More reliable and cost effective than conventional laboratory techniques, CO-oximetry could be used for routine measurement of COHb in all patients admitted to the ED. Routine screening would be an excellent tool to reduce the misdiagnosis rate especially for patients who present non-specific symptoms. Pulse CO-oximetry could also be used to perform serial or continuous readings to ascertain therapeutic compliance.

Reference Number: 20305

ROTH, D., D.R.N. HUBMANN, H. HERKNER, W. SCHREIBER, AND A. LAGGNER. "Victim of Carbon Monoxide Poisoning Identified by Carbon Monoxide Oximetry." *Journal of Emergency Medicine*, 40: 640–642, 2011 (1 figure, 6 references)

Abstract: A 28-year-old woman attended hospital as a result of a headache, vomiting, diarrhea, and palpitations. The normal pulse oximetry showed normal results, but the CO-oximeter Masimo Radical-7 showed a COHb of 28%. The patient was treated with 100% oxygen continuous positive airway pressure ventilation. The local fire department detected an ambient CO concentration of 10,000 ppm in her apartment caused by a faulty gas continuous flow water heater. A damaged gas stove was also detected in the neighbour's apartment.

> If our patient had been sent home without the diagnosis of CO poisoning, she would have been in danger of further poisoning. As CO poisoning was diagnosed in a timely fashion, the fire department could locate the defective device and this may have helped to prevent other potential medical emergencies in the neighboring flat. It is impossible to know if the cause of our patient's symptoms would have been identified without the use of a CO-oximeter. As blood gas analysis is performed in only 12% of patients admitted to our ED, there would be a good chance that she would have been treated for some form of gastric flu and discharged.

Reference Number: 20306

CHEE, K.J., D. NILSON, R. PARTRIDGE, A. HUGHES, S. SUNER, A. SUCOY, AND G. JAY. "Finding Needles in a Haystack: A Case Series of Carbon Monoxide Poisoning Detected Using New Technology in the Emergency Department." *Clinical Toxicology*, 46: 461–469, 2008 (2 tables, 11 references)

Abstract: The Masimo Rad-57 pulse CO-oximeter uses a fingertip probe and relies on multi-wavelengths to measure both carboxyhemoglobin concentration (SpCO) and oxygen concentration (SpO_2). The SpCO concentration was measured in 74,880 patients at triage between 2005 and 2006. Seven patients who presented with vague complaints were diagnosed with occult CO poisoning. The COHb ranged from 9% to 36%. In addition, 15 patients who were smokers had SpCO levels between 11% and 21% and blood COHb between 10% and 17%.

Table: A Comparison of SpCO and Blood COHb Concentrations in Patients with Occult CO Poisoning

Case #	Chief complaint	SpCO (%)	Blood COHb (%)	Interval between SpCO and blood (minutes)
1	Headache and palpitations	33	25	120
2	Weakness and dizziness	14	28	18
2a	Spouse of 2, headache and body aches	21	24	12
3	Syncope	33	26	38
4	Headache	33	36	13
5	Nausea, vomiting	19	17	43
5a	Spouse of %, no complaints	10	9	93

Source: Adapted from Chee et al. 2008.

We have presented cases of occult CO toxicity in which the diagnosis was significantly facilitated by non-invasive pulse CO-oximetry. While no conclusions can be drawn directly regarding the true accuracy of the device, given the various time delays between the non-invasive SpCO measurements and the COHb measurement from a venous blood draw, these cases suggest that non-invasive pulse CO-oximetry could be a major screening tool at ED triage for identifying unsuspected CO poisoning among patients with nonspecific symptoms.

Reference Number: 20307

ONODERA, M., Y. FUJINO, S. KIKUCHI, M. SATO, K. MORI, T. BEPPU, AND Y. INOUE. "Utility of the Measurement of Carboxyhemoglobin Level at the Site of Acute Carbon Monoxide Poisoning in Rural Areas." *Scientifica*, 2016 (1): 6192369, 2016 (2 figures, 19 references)

Abstract: The SpCO levels were measured in six male and four female CO poisoning victims (mean age 49.4 years) using a Masimo Rad-57 pulse CO-oximeter. The blood COHb was measured later at hospital using a blood gas analyzer. The mean CO exposure time of the victims was 115.1 minutes, and the mean SpCO level was 18.9% at the scene. The mean blood COHb was 13.4%, and the mean transport time was 76.2 minutes.

The SpCO level was found to correlate with the exposure time at the site of CO poisoning in rural areas. In-field SpCO measurement is rapid and repeatable and may be useful in gaining an understanding of the extent of exposure in the critical early moments of medical response. However, at this time further direct correlation with COHb levels as measured in the hospital is limited and further study is warranted.

Reference Number: 20308

HARLAN, N., L.K. WEAVER, AND K. DERU. "Inaccurate Pulse CO-Oximetry of Carboxyhemoglobin due to Digital Clubbing: Case Report." *Undersea and Hyperbaric Medecine*, 43: 59–61, 2016 (2 figures, 12 references)

Abstract: An 18-year-old male smoker with mild cystic fibrosis and hand (digital) clubbing was found unconscious inside a motor vehicle in an apparent suicide attempt. He was intubated and transported to hospital where his venous blood COHb was 33%. He was sedated with propranolol and fentanyl and received 100% oxygen. Four hours later, his blood COHb had dropped to 0.9%. But the SpCO measured by three Masimo Rad-57 monitors were 11%, 9%, and 9% respectively. The patient received three hyperbaric oxygen treatments over 24 hours. The SpCO measurement was still 10% even after the first hyperbaric treatment.

To our knowledge, this is the first reported evidence that non-invasive carbon monoxide monitoring (SpCO) may be distorted by digital clubbing, just as regular pulse oximetry readings may be. The acceptable perfusion index suggest that this inaccuracy was not caused by poor fit between the probe and the patient's finger. We considered the hospital measurement of COHb by CO-oximeter to be the gold standard because the measurement is from blood and adheres to all federal and hospital-required guidelines for precision.

Reference Number: 20309

WEAVER, L.K., S.K. CHURCHILL, K. DERU, AND D. COONEY. "False Positive Rate of Carbon Monoxide Saturation by Pulse Oximetry of Emergency Department Patients." *Respiratory Care*, 58: 232–240, 2013 (4 tables, 2 figures, 21 references)

Abstract: The SpCO concentration measured by a pulse oximeter (Masimo Rad-57) was compared to the clinical blood COHb measured in 1,363 patients presenting at the ER of the Intermountain Medical Centre in Utah during 2008. The manufacturers claim the results to be accurate with three percentage points of the COHb percentage saturation. Of the 1,363 patients, 614 (45%) were men, 84% were light skinned, 201 had nail polish, 14 were in shock, and 4 had CO poisoning. The mean SpCO was 2% (range 0–31%) compared to 3.2% (range 0.2–35%) for the blood COHb. If a three-percentage-point difference or greater is considered to indicate that the SpCO is in error, then 9% were false positive (+3% to +19%), and 18% met the criteria for a false negative (-3% to -13%). For the four patients with CO poisoning, the blood COHbs were 35%, 27%, 8.7%, and 8.4% compared to 31%, 17%, 4%, and 2% for the SpCO results respectively. The risks for a false positive SpCO result were related to being female and having a lower perfusion ratio. Methemoglobin, body temperature, and blood pressure appear also to affect the SpCO.

> While the Rad-57 pulse oximeter functions within the manufacturer's specifications, clinicians using the Rad-57 should expect some SpCO readings to be significantly higher or lower than COHb measurements and should not use SpCO to direct triage or patient management. An elevated SpCO could broaden the diagnosis of CO poisoning in patients with non-specific symptoms. However, a negative SpCO level in patients suspected of having CO poisoning should never rule out CO poisoning and should always be confirmed by COHb.

2.04 AMBIENT CO DETECTORS/ALARMS

> The London fire brigade advises members of the public to buy a CO detector for all rooms that contain a fuel burning appliance. CO detectors sound a loud alarm if a harmful level of CO gas is reached within an area. It is the only method to detect CO gas within the household enabling safe evacuation and preventing harmful CO effects.
>
> —J. Ashcroft et al., "Carbon Monoxide Poisoning" (2019) (30213)

Battery-powered home CO detectors have been commercially available since 1993. Currently, there are three main types of alarms (20403). Over 95% of severe CO poisonings occurred in homes without a functioning CO detector (20407).

Reference Number: 20401

JASLOW, D., J. UFBERG, J. UKASIK, AND P. SANANMAN. "Routine Carbon Monoxide Screening by Emergency Medical Technicians." *Academic Emergency Medicine*, 8: 288–291, 2001 (12 references)

Abstract: CO readings were taken in 264 homes in Wilmington, Delaware, over the three winter months of 1999–2000 by emergency responders using a battery-powered CO sensor that triggers an alarm at 35 ppm. There were nine (3.4%) positive results. Two hundred and twelve surveys were also conducted of whom only 54 (37%) of the participants could name a CO poisoning symptom and only 27 (13%) had a CO sensor.

Emergency medical services personnel can perform routine CO screening and detect occult elevations in CO levels during 911 responses. Public knowledge of CO poisoning is limited and the use of home CO detectors is rare in this study population.

Reference Number: 20402

RAUB, J.A., M. MATHIEU-NOLF, N.B. HAMPSON, AND S.R. THOM. "Carbon Monoxide Poisoning: A Public Health Perspective." *Toxicology*, 145: 1–14, 2000 (1 table, 115 references)

Abstract: Carbon monoxide is responsible for a large percentage of accidental poisonings and deaths globally. Outdoor CO concentrations are highest near street intersections, in congested traffic, and in ill-ventilated parking lots and tunnels. Electronic CO alarms are designed to be low cost like residential smoke detectors but to provide protection from CO poisoning. The most popular alternating current (AC)-powered CO alarms have a heated metallic sensor that reacts with CO. The battery-powered CO alarms have a chemically treated gel disk that darkens with exposure. There are also small, inexpensive CO detection cards, but they require frequent visual checks of the colour change and do not sound an alarm. CO detectors should not be installed where the air movement is high or where there are extremes of temperature or humidity. According to UL Standard 2034, an 85-decibel alarm must be activated within 189 minutes of exposure to 70 ppm CO, 50 minutes at exposure to 150 ppm, and 15 minutes at 400 ppm CO, which would result in a blood COHb of approximately 10%.

To be in compliance with the US Environmental Protection Agency and World Health Organization ambient air standards, it would be necessary to set alarms to register 9 ppm CO for 8 hours or 25 ppm for 1 hour. Current alarms provide warning only against CO levels that protect most healthy persons from experiencing the adverse effects of CO exposure (e.g., loss of ability to react). Despite these limitations, CO alarms can be reliable and effective, continue to improve and should be recommended for use in homes in addition to smoke detectors and fire alarms.

Reference Number: 20403

WRIGHT, J. "Chronic and Occult Carbon Monoxide Poisoning: We Don't Know What We're Missing." *Emergency Medicine Journal*, 19: 386–390, 2002 (3 tables, 52 references)

Abstract: A review was conducted on subacute or occult CO poisoning and how to prevent it. There are three major types of domestic CO alarms. First, a biometric-a gel-coated disk will change colour and darken in the presence of CO (mimicking the binding of CO and Hb), which is detected by a sensor. Low levels of CO can be detected, but the detector requires 48 hours or more to reset. These devices can be battery powered and use little electricity. Second, the oldest of the domestic sensors are metal oxide semiconductors. Heated tin dioxide reacts with CO. Since the tin needs to be heated, it requires a lot of power and typically needs to be plugged into the wall socket. It cannot detect CO levels below 100 ppm and commonly gives false alarms. Third, electrochemical sensors are typically three platinum electrodes immersed in electrolyte solution and produces a small electrical current in response to CO. These devices can detect low levels of CO and, initially, are very accurate, but the life span may only be two years.

Many devices on the market have created significant false alarm problems as well as life-threatening failures to go off when dangerous levels of CO are present. Manufacturers and government agencies throughout the world are still trying to develop the ideal device which will have increased resistance to false alarms and false negatives, improved accuracy and increased alarm reliability and repeatability. There is no doubt, however, that lives will continue to be lost and there will be large numbers of people chronically exposed to low levels of CO unless there is a dramatic increase in the number of CO detectors installed in UK homes.

Reference Number: 20404

CHRISTENSEN, G.M., P.D. CRESWELL, J. THEOBALD, AND J.G. MEIMAN. "Carbon Monoxide Detector Effectiveness in Reducing Poisoning, Wisconsin, 2014–2016." *Clinical Toxicology*, 58: 1335–1341, 2020 (4 tables, 26 references)

Abstract: In Wisconsin, there are approximately 5,000 ER visits each year due to CO exposure, and nearly 80% of the patients were exposed at home. In 2011, a law in Wisconsin mandated that all one- and two-family dwellings with fuel-burning appliances, fireplaces, or an attached garage must have a CO detector near sleeping areas and on each floor. Between 2014 and 2016, an analysis was conducted on 408 cases of CO poisoning reported to the Wisconsin Poison Centre, of which 52 (12.8%) had no CO alarm, 76 (18.6%) had a CO alarm that did not sound, and 280 (68.6%) had a CO alarm that sounded. The odds of CO poisoning were 3.2 times higher in those households that had no CO detector compared to those with operating CO detectors, and the severity of CO poisoning also increased by 0.34 points in households without a CO detector.

> Our analyses support previous studies that find CO detectors can prevent CO poisoning while also suggesting that the presence of these devices may account for changes in the severity of poisoning (i.e., lower average severity scores). Mandating CO detectors in residences has been found to be cost effective in reducing medical costs in the United States. Policymakers and stakeholders should regard CO detectors as paramount in the prevention of CO poisoning and consider programs to ensure compliance in areas where laws requiring CO detectors exist.

Reference Number: 20405

RUPERT, D.J., J.A. POEHLMAN, S.A. DAMONB, AND P.N. WILLIAMS. "Risk and Protective Behaviors for Residential Carbon Monoxide Poisoning." *Injury Prevention*, 19: 119–123, 2013 (3 tables, 20 references)

Abstract: Residential CO poisoning accounts for most of the non-fatal injuries and almost half of the deaths. The knowledge, attitudes, and beliefs regarding CO in the home was determined in four focus groups in Chicago, Illinois. Participants correctly described CO as an odourless and colourless gas, but many were confused between natural gas and CO. Participants also correctly identified many of the symptoms of CO poisoning such as headache and drowsiness. Most reported having at least one CO alarm in the house as most municipalities in Chicago have housing codes that require CO alarms. Most participants stated that, in responding to a CO alarm, they would leave the house and call the fire department or gas company, but some stated that they would open the windows and investigate the CO alarm for a malfunction before belatedly leaving the house.

Many participants were unsure where CO alarms should be placed in the home and how many should be installed. Most participants placed CO alarms near furnaces or gas appliances, many of which were in basements or utility rooms. Others placed the alarms in more appropriate locations such as in bedrooms, hallways, or dining rooms. Several participants stated that CO alarms should be located near smoke alarms. None of the participants acknowledged a connection between the location of a CO alarm and one's ability to hear it.

Reference Number: 20406

KRENZLOCK, E.P., R. ROTH, AND R. FULL. "Carbon Monoxide: The Silent Killer with an Audible Solution." *American Journal of Emergency Medicine*, 14: 484–486, 1996 (2 tables, 11 references)

Abstract: A study was conducted of sixty 911 calls regarding CO detector alarms and 41 calls of suspected CO poisoning but with no CO detectors in Pittsburgh from January to March 1995, which were investigated by an advance cardiac life support paramedic unit. Detectable CO concentrations were found in 69.3% of all calls and 80% of the calls involving a CO alarm.

Table: Comparison of Outcomes in Homes with and without CO Detectors

Outcome	CO alarm	No CO alarm
Percentage of homes with CO > 100 ppm (%)	1.6	12.2
Percentage of victims symptomatic (%)	11.7	63.4
Number of CO victims hospitalized	2	24

Source: Adapted from Krenzlock et al. 1996.

CO detectors with audible alarms were effective in alerting patients to the presence of abnormal amounts of CO in their dwelling. Subsequently they were exposed to less CO and they experienced a lower incidence of CO-related symptoms. Patients with CO alarms rarely needed transport to a health care facility, which essentially eliminated the medical expenses incurred by patients exposed to higher CO concentrations. CO detectors are an effective way to reduce morbidity by limiting or preventing exposure to CO.

Reference Number: 20407

LAVONAS, E.J., W.P. KERNS II, C.A. TOMASZEWSKI, T.H. BLACKWELL, P.N. GALASKA, T.L. HAY, G.E. MCCORMICK, AND A.S. BROWN. "Use of Carbon Monoxide Alarms to Prevent Poisonings during a Power Outage, North Caroline, December 2002." *Morbidity and Mortality Weekly Report*, 53: 189–192, 2004 (1 table, 1 figure, 10 references)

Abstract: An evaluation was conducted of 124 CO poisoning cases that occurred during nine days of power outage after an ice storm hit North Carolina on 4 December 2002. During the power outage, all electric homes can be just as likely as gas-fueled houses to have sources of CO poisoning. Two exposure cases were due to natural gas fireplace logs. It was determined that 96.2% of severe CO poisonings occurred in homes with no reported functioning CO alarms.

Table: Sources of CO Exposure during Power Outage in Mecklenburg County, North Carolina, December 2002

Source	Number of CO exposures
Grill (gas or propane)	26
Electric generator	15
Propane space heater	3
Stove or oven	3
Wood fire	3
Natural gas fireplace log	2
Kerosene space heater	1
Furnace	1
Candles	1

Source: Adapted from Lavonas et al. 2004.

After receiving these findings, the Mecklenburg Board of County Commissioners amended the county's CO alarm ordinance to require an alarm in every residence, regardless of heating source and appliance type, and a battery back-up in each CO alarm. County officials continue to provide free alarms to low-income homeowners and to target recent immigrants and other no-English-speaking residents with a multilingual program to prevent CO poisonings. Officials in other communities should consider these strategies in adopting CO poisoning prevention programs.

Reference Number: 20408

BARN, P., L. GILES, M-E. HEROUX, AND T. KOSATSKY. "A Review of the Experimental Evidence on the Toxicokinetics of Carbon Monoxide: The Potential Role of Pathophysiology among Susceptible Groups." *Environmental Health*, 17 (1): 13, 2018 (2 tables, 1 figure, 91 references)

Abstract: Health Canada recommends that the indoor air CO concentration not exceed 25 ppm for one hour exposure and 10 ppm for 24 hours, which means that this exposure would not lead to a blood COHb of greater than 2%. In one wing of a long-term care facility in Saskatoon, 24 occupants complained of headache, dizziness, and nausea overnight. The ambient CO concentration was measured at 63 ppm, and the wing was evacuated. Three of the evacuated residents died within one month of the evacuation. Six medical conditions were thought to be at an increased risk of the adverse effects of indoor CO: (1) chronic obstructive lung disease; (2) anemia; (3) cardiovascular disease; (4) heart failure; (5) multiple co-morbidities; and (6) older age (greater than 60 years of age).

However, CO detectors are only intended to prevent acute exposures at levels judged immediately hazardous to health, and therefore it would be advantageous to consider ways to prevent low level exposures through comprehensive CO management programs based on education, prevention, and monitoring. In Canada as elsewhere certified CO detectors are approved to alarm at levels considerably higher than those recommended by indoor air guidelines. Alarms are triggered when a peak value of 400 ppm is reached within 4–15 minutes, 150 ppm is reached within 15–40 minutes and when approximately 70 ppm is reached within 1–4 hours. These values are considered equivalent to an attained 10% COHb level, which is considerably higher than the 2% level used to derive guideline values. Certification and use of more sensitive detectors that both display and alarm at CO concentrations in the lower range would allow for better protection of the health of susceptible occupants.

CHAPTER 3

Types/Effects of Carbon Monoxide Exposure

> This is why I hate Google. You come across one site, match one symptom and all of a sudden you're dying of Carbon Monoxide poisoning or cancer of the big toe.
>
> —K.J. McPike, "Goodreads.com"

> The mind is not clear, except by an effort and one's surroundings seem a little strange. The temper is easily upset, very much as in alcohol intoxication and the judgment is likely to be bad.
>
> — R.M.J. Henderson et al., "Physiological Effects of Automobile Gas" (1921)[1]

This chapter will discuss the numerous and various types of carbon monoxide (CO) exposure such as accidental occult and acute CO poisonings, suicides, and homicides by CO and CO exposures due to fires. Its effect on children and during pregnancy and its frequent delayed and prolonged sequelae and treatment will also be explored. Finally, the surprising beneficial and therapeutic effects of CO will be presented.

Chapter 3

1 Y. Henderson et al., "Physiological Effects of Automobile Gas and Standards of Ventilation for Brief Exposures," *Journal of Industrial Hygiene*, 3: 79, 1921.

3.01 OCCULT/SUBACUTE EXPOSURE

You can't get a vaccination against this winter flu.

—James Wigmore, "You Can't Get a Vaccination against This Winter Flu" (2022)[2]

In one of these cases at any rate, it has been found that there was a small leak of gas in the bedroom of the patient, and her symptoms, which had been previously regarded as neurotic disappeared completely following the repair of this leak.

—G. Slot, "Chronic Carbon Monoxide Poisoning" (1929)[3]

Chronic or occult CO poisoning is sometimes referred to as "winter flu" as the symptoms of low CO poisoning are similar to the flu and tend to occur in the winter months when the doors and windows are closed and heating sources and other potential CO sources are on. An aide-memoire—COMA—is suggested to assess the possibility of CO exposure in the emergency room (ER):

C: Cohabitees/companions; is anyone else affected, including pets?
O: Outdoors; do your symptoms improve outdoors?
M: Maintenance; are your fuel-burning appliances/vents properly maintained?
A: Alarm; do you have a CO alarm?[4]

Occult CO poisoning is frequently misdiagnosed as winter flu (30101–20), although nearly 90% of patients complained of headache (30101). Occult CO poisoning may also cause hearing loss (30124), cardiac failure (30114), and epileptic fits (30114).

Reference Number: 30101

RAJIAH, K., AND E.M. MATHEW. "Clinical Manifestation, Effects, Diagnosis, Monitoring of Carbon Monoxide Poisoning and Toxicity." *African Journal of Pharmacy and Pharmacology*, 5: 259–264, 2011 (3 tables, 20 references)

---

2 James Wigmore, "You Can't Get a Vaccination against This Winter Flu," *LinkedIn*, 22 October 2022.

3 G. Slot, "Chronic Carbon Monoxide Poisoning," *Lancet*, 23 March 1929, 635.

4 F.S. Kar-Purkayastha and V. Murray, "Low Level Exposure to Carbon Monoxide," *British Journal of General Practice* 62: 404, 2012.

Abstract: Carbon monoxide is the product of the incomplete combustion of hydrocarbons with insufficient oxygen to complete oxidation to carbon dioxide (CO_2). In 1776, the French chemist Joseph-Marie-Francois de Lassone produced CO by heating zinc oxide with coke, but he mistakenly thought it was hydrogen as it burned with a blue flame. This view was corrected by the Scottish chemist William Cumberlain Cruikshank in 1800. At atmospheric pressure, the half-life of carboxyhemoglobin (COHb) is four to six hours but is decreased to 40 to 80 minutes by breathing 100% oxygen.

> Carbon monoxide poisoning is a multi-system condition and can cause a confusing constellation of clinical features, precipitating presentation to general practitioners, accident and emergency departments, acute care physicians, general surgeons, neurologists and even psychiatrists. With increasing specialization, within the medical profession the diagnosis may be missed by the specialist who fails to recognize the significance of pathology outside his or her own area of interest. The benefits of prompt diagnosis are threefold. Firstly, recommended therapy in the form of 100% normobaric oxygen in all cases and hyperbaric oxygen in cases of life threatening poisoning can be instigated. Secondly, as illustrated by this case, unnecessary, expensive and painful investigations can be avoided. Thirdly, and perhaps more importantly, the dire consequences of discharging a patient home to, or allowing others access to a potentially fatal environment, can be avoided.

Reference Number: 30102

DOLAN, M.C., T.L. HALTOM, G.H. BARROWS, C.R. SHORT, AND K.M. FERRIELL. "Carboxyhemoglobin Levels in Patients with Flu-Like Symptoms." *Annals of Emergency Medicine*, 16: 782–786, 1987 (5 tables, 1 figure, 34 references)

Abstract: Fifty-five patients attending a university hospital in the inner city of Louisville, Kentucky, with complaints of headache, dizziness, nausea, vomiting, diarrhea, weakness, general malaise, or shortness of breath provided three capillary tubes of blood for COHb determination. Thirteen patients (23.6%) had blood COHb greater or equal to 10% (range 10–21%).

Table: Chief Complaint, Discharge Diagnosis, and COHb Concentration in 13 Patients with COHb Great or Equal to 10%

Chief complaint	Discharge diagnosis	COHb concentration (%)
Nausea, cough	Urinary tract infection	17
General malaise	Viral syndrome	21
Fever, cough	Acute pulmonary tuberculosis	12
Headache	Tension headache	14
Shortness of breath, cough, fever	Bronchitis	15
Headache, abdominal cramping	Drug-induced headache	11
Weak, dizzy, cough	Viral syndrome/bronchitis	14
Shortness of breath, cough	Bronchitis	12
Headache, fever	Urinary tract infection, vasovagal	11
Headache	Headache	11
Abdominal pain	Gastritis	13
Right upper quadrant pain	Biliary colic	10
Weak, tired	Orthostatic hypotension secondary to diuretic use	18

Source: Adapted from Dolan et al. 1987.

The literature is well supplied with reports of patients with subacute CO poisoning who were misdiagnosed as having influenza and sent home with disastrous consequences. Emergency physicians must be aware of the protean presentations of CO poisoning and include it in the differential diagnosis of patients with flu-like illness to prevent the return of patients to hazardous environments.

Reference Number: 30103

BLUMENTHAL, I. "Carbon Monoxide Poisoning." *Journal of the Royal Society of Medicine*, 94: 270–272, 2001 (1 figure, 32 reference)

Abstract: CO is known as the silent killer as it has no odour or smell and is not noxious. CO is produced endogenously in small amounts as a by-product of heme catabolism. Environmental CO is produced by the incomplete combustion of any carbon-containing fuel such as coal, peat, or natural gas. Approximately 50 persons die in Britain and 200 are severely injured by CO poisoning every year. The conversion from coal gas to CO-free natural gas in the UK during the 1960s and 1970s has caused a dramatic drop in CO poisoning. CO has about 210 times the affinity for hemoglobin (Hb) than oxygen and an even greater affinity for myoglobin. Binding to cardiac myoglobin causes myocardial depression, hypertension, and arrhythmias, which can lead to ever greater tissue hypoxia and death. Heavy smokers may have a COHb saturation of 15%. Levels of 5% are found in hemolytic anemia. Symptoms usually appear at levels of 10% or greater. Pulse oximeters are not suitable for the diagnosis of CO poisoning as most use a wavelength that cannot distinguish between oxygen hemoglobin or COHb.

Public education about the danger of carbon monoxide, with emphasis on safety in the home and workplace, is the key to effective prevention. Professional education targeted at community workers is also needed. This could be achieved through a media campaign when risk is greatest during the winter.

Reference Number: 30104

HAMPSON, N.B., AND L.K. WEAVER. "Carbon Monoxide Poisoning: A New Incidence for an Old Disease." *Undersea and Hyperbaric Medecine*, 34: 163–168, 2007 (1 table, 1 figure, 22 references)

Abstract: The vast majority of cases of CO poisoning are non-fatal, but these poisonings may result in significant sequelae. It is estimated in this study that, in 2005, there were 32,413 to 56,037 (mean 46,929) emergency department (ED) visits due to CO poisoning. This is much higher than the commonly cited 10,000 total non-fatal cases in 1974. Thirty-two percent of patients not treated with hyperbaric oxygen (HBO_2) had cognition sequelae one year later compared to 18% of those treated with HBO_2.

While this new estimate of national ED visits for CO poisoning is 3–5 times greater than previous ones, it may still be an underestimate. Because symptoms of CO poisoning can be so nonspecific (e.g., headache, nausea, vomiting, dizziness), the condition is notorious for being underdiagnosed. The current estimated of US ED visits for CO poisoning was made using rates of diagnosed CO poisoning in five different states. Based upon studies examining under-or misdiagnosis of CO poisoning, it is reasonable to assume that many more cases of undiagnosed and the true number of ED visits due to CO poisoning is greater than estimated here.

Reference Number: 30105

HECKERLING, P.S. "Occult Carbon Monoxide Poisoning: A Cause of Winter Headache." *American Journal of Emergency Medicine*, 5: 201–204, 1987 (2 tables, 10 references)

Abstract: Occult CO poisoning may be defined as accidental CO poisoning of which the patient and physician are initially unaware. Approximately 90% of patients with known acute CO poisoning complain of headaches. A study was conducted of 37 patients attending the ER between December 1984 and March 1985 with the chief complaint of a headache. Venous blood samples were collected within one to three hours of arrival, and COHb analysis was conducted using differential spectrometry. Seven patients (19%) had a COHb greater than 10% (range 10.7% to 21.1%). Thirty patients had a COHb less than 10% (range 0.3–7.7%). Three of the seven patients had a headache for one week and used gas space heaters or gas stoves as an alternative source of heat. In six of the seven patients with a blood COHb greater than 10%, gas company investigations showed a toxic CO source.

> In conclusion, occult CO poisoning is an important, frequently overlooked cause of winter headache in urban emergency department patients. Patients who use gas space heaters or stoves for supplemental heat or who have cohabitants with headache may define a subgroup at higher risk for CO-related illness.

Reference Number: 30106

WEAVER, L.K. "Carbon Monoxide Poisoning." *New England Journal of Medicine*, 360: 1217–1225, 2009 (3 figures, 55 references)

At endogenous concentrations, CO acts as a neurotransmitter that can moderate inflammation, apoptosis, and cell proliferation, but a high COHb concentration can result in headache, myalgia, dizziness, and, eventually, confusion, loss of consciousness, and death. A case vignette involved a 39-year-old female executive who had a history of fatigue, headache, and memory lapse for several months. Multiple specialists were consulted, but no diagnosis was made. A friend arrived at her residence and found her semi-comatose. She was transferred to the ER and had a COHb of 18% and was treated with high-flow oxygen. The source of the CO was a faulty furnace.

Patients who have had carbon monoxide poisoning, such as the patient in the vignette, should be treated immediately with normobaric oxygen (with a fraction of inspired oxygen as high as possible), which speeds up the excretion of carbon monoxide. The source of carbon monoxide should be identified and removed.

Reference Number: 30107

TOMASZEWSKI, C. "Carbon Monoxide Poisoning. Early Awareness and Intervention Can Save Lives." *Postgraduate Medicine*, 105: 39–50, 1999 (29 references)

Abstract: CO is a colourless, odourless, non-irritating, poisonous gas that causes vague flu-like symptoms that are often misinterpreted by both physicians and patients.

Table: Symptoms and Degree of CO Poisoning

Degree of Poisoning	Symptoms
Mild	Dizziness, headache, nausea, vomiting
Moderate	Ataxia, blurred vision, disorientation, tachycardia, tachypnea
Severe	Coma, myocardial ischemia, death of muscle fibres, seizures, ventricular dysrhythmias

Source: Adapted from Tomaszewski 1999.

This winter a 22-year-old man presents to your emergency department following a syncopal episode. At home over the weekend, he has had a headache, chills, nausea and dizziness. Your diagnosis is influenza. The patient is given intravenous hydration and is discharged neurologically intact with normal vital signs. Two days later, the patient is readmitted semi comatose, and this time the diagnosis is carbon monoxide poisoning. He is treated with hyperbaric oxygen, but chronic headaches, memory problems and parkinsonian-type tremor eventually develop. Sixteen months later, you receive a letter from the patient's attorney requesting compensation for allegedly failing to recognize carbon monoxide poisoning at the initial visit. The letter claims that your misdiagnosis resulted in reexposure of his client to a faulty furnace, which caused permanent neurologic sequelae.

Reference Number: 30108

DOLAN, M.C. "Carbon Monoxide Poisoning." *Canadian Medical Association Journal*, 133: 392–399, 1985 (1 table, 99 references)

Abstract: Non-lethal CO poisoning is commonly misdiagnosed as a flu-like viral illness or even myocardial infarction. CO poisoning is often listed with syphilis, tuberculosis, and subdural hematoma in its ability to mimic a variety of systemic diseases. Fires contain 0.1–10% CO. Smoking a cigarette is a significant source of CO in which the sidestream smoke contained 2.5 times more CO than mainstream smoke. The natural catabolism of Hb produce a COHb of less than 1% which may increase to 2% due to environmental pollutants in non-smokers. Freeway traffic in Los Angeles, California, can increase blood COHb concentrations to a mean of 5%. Smokers typically have a COHb saturation of 5–9%.

Carbon monoxide poisoning is a significant cause of illness and death. Its protean symptoms probably lead to a gross underestimation of its true incidence. Low levels of carbon monoxide aggravate chronic cardiopulmonary problems, and high levels are associated with cardiac arrhythmia and cerebral edema. Patients who survive acute poisoning are at risk of delayed neurological sequelae. The measurement of carboxyhemoglobin levels does not reveal the tissue levels of carbon monoxide but is useful in determining therapy. Treatment includes the monitoring and management of cardiac arrhythmias and oxygenation. Hyperbaric oxygenation is beneficial but there are currently no definite criteria for use.

Reference Number: 30109

KIRKPATRICK, J.N. "Occult Carbon Monoxide Poisoning." *Western Journal of Medicine*, 146: 52–56, 1987 (3 tables, 19 references)

Abstract: Twenty-six patients who sought medical treatment over a two-year period in a general internal medicine practice, due to fatigue/headache/nausea of at least one-month duration, were found to have occult CO poisoning. The COHb concentrations ranged from 1% to 36.8%. Most CO exposures were due to a faulty gas furnace (58%) and auto exhaust (31%).

Table: Sick Pets in Households with CO Exposure

Number of pets	Pet/Condition
1	Dead canary
4	Dead puppies
1	Dead cat
2	Cats behaving strangely
3	Dogs behaving strangely
1	Dog with diabetes insipidus (polyuria, polydipsia)

Source: Adapted from Kirkpatrick 1987.

The historical findings that should trigger an investigation of possible CO sources are persons whose occupation expose them to vehicle exhaust; households where people and pets are all feeling less well than usual; persons who use solvents containing 1.01 methylene chloride, including paint strippers and carburetor cleaners, and anyone who has several of the symptoms associated with ongoing CO exposure. The problem with CO accumulation in homes may become more frequent as more people insulate their homes. In the Pacific Northwest, where electric heat has long been much less expensive than other heat sources, electric power is now rising in cost. Many homes may be converted to using combustion fuels, with greater potential for CO leaks. Clinicians in other areas where higher priced electricity is seldom used for heating, may already be seeing a significant number of patients with CO exposure not caused by smoking.

Reference Number: 30110

HARPER, A., AND J. CROFT-BAKER. "Carbon Monoxide Poisoning: Undetected by Both Patients and Their Doctors." *Age and Aging*, 33: 105–109, 2004 (3 tables, 44 references)

Abstract: In the United Kingdom, some 250,000 gas appliances are condemned each year, and, if only a small proportion were emitting CO, then the number of CO poisonings could be much greater than those currently detected. In one study of elderly patients being admitted over the winter in southern England, 12 of 314 had abnormally high COHb concentrations that were greater than 2%, and, in three of these cases, the CO source was identified. Average CO levels in houses without gas stoves are 0.5–5 parts per million (ppm) and, with properly adjusted gas stoves, around 5–15 ppm around the cookers. Poorly adjusted gas stoves can emit CO greater than 30 ppm.

The majority of poisonings occur in winter when there is increased use of fossil fuels for heating and ventilation is minimized to conserve heat. Symptoms may be better after a break away from home or worse on weekends when patients spend longer in the homes. They may be worse after cooking, most commonly noted in the evening after supper. With gas heating a yellow flame that leaves black sooty marks on the surrounds is suggestive of incomplete carbon combustion, as opposed to the clearer blue flame. Smoke may accumulate and while CO itself is odourless, other gases in the smoke may be apparent.

Reference Number: 30111

AMIRI, H., S.S. VAHDATI, S. GHAFFARZEDH, N. GHODRATI, P. RAOUFI, AND P. HABIBOLLAHI. "Study About Relation between Carboxy Hemoglobin Levels in the Patient with Headache in the Cold Season." *Emergency Medicine*, 4 (5): 1–3, 2014 (1 table, 2 figures, 16 references)

Abstract: The blood COHb concentrations were determined in 25 men and 25 women (mean age 42.1 years) who visited an ER in Iran in the winter of 2011 with the main complaint of headaches. The COHb was less than 10% in 41 patients (81%), between 10–20% in 7 patients (14%), and between 20% and 30% in 2 patients (4%).

Carbon monoxide has a broad range of symptoms and complications varying from a mild type like mimicking a viral disease to a reverse severe type which the patient is brought to ED in coma and has developed cerebral edema. Headache, nausea, vomiting usually have delayed onset of 1 to 2 hours. These symptoms significantly show a decrease in carbon monoxide level in tissue contents and plasma. Headache induced by carbon monoxide is very similar to migraine type headaches and there is no such pattern that one can differentiate between two. In other words, the carbon monoxide induced headache is a migraine type headache accompanied by nausea, vomiting and malaise as the most important symptoms.

Reference Number: 30112

BENNETTO, L., L. POWTER, AND N.J. SCOLDING. "Accidental Carbon Monoxide Poisoning Presenting without a History of Exposure: A Case Report." *Journal of Medical Case Reports*, 2 (1): 118, 2008 (2 figures, 10 references)

Abstract: A 42-year-old man was found semi-comatose and incontinent on the floor of his trailer. He presented at hospital with amnesia, pyrexia, hypostasis, and a rash on his left leg and buttocks. Other features included rhabdomyolysis, renal failure, dysarthria, parkinsonism, peripheral neuropathy, and cerebral hemorrhage. Over the three weeks of treatment and numerous clinical tests, such as magnetic resonance imaging (MRI), lumbar puncture, electroencephalogram, and muscle biopsy, no diagnosis was determined; after three weeks of hospitalization, he remembered that he had just turned on the gas heating of his trailer. An emergency gas engineer found the patient's gas boiler to be in a dangerous state of repairs with a heavy growth of ivy impeding ventilation. It was removed and replaced.

Carbon monoxide poisoning is a multi-system disease and can cause a confusing constellation of clinical features precipitating presentations to general practitioners, accident, and emergency departments, acute are physicians, general surgeons, neurologists, and even psychiatrists. With increasing specialization within the medical profession, the diagnosis may be missed by the specialist who fails to recognize the significance of pathology outside his or her own area of interest. The benefits of prompt diagnosis are threefold. Firstly, recommended therapy in the form of 100% normobaric oxygen in all cases and hyperbaric oxygen in cases of life threatening poisoning can be instigated. Secondly, as illustrated by this case unnecessary expensive and painful investigations can be avoided. Thirdly, and perhaps most importantly, the dire consequences of discharging a patient home to, or allowing others access to a potentially fatal environment can be avoided.

Reference Number: 30113

BALZAN, M.V., J.M. CACCIOTTOLO, AND S. MIFSUD. "Unstable Angina and Exposure to Carbon Monoxide." *Postgraduate Medical Journal*, 70: 699–702, 1994 (2 tables, 15 references)

Abstract: The blood COHb concentration was determined in 104 cases of unstable angina attending a Malta hospital. Three had elevated COHb concentrations of 15.2%. 15% (back-calculated), and 15% (back-calculated). All had used kerosene heaters in poorly ventilated rooms.

Unstable angina is one form of presentation of carbon monoxide poisoning. Carboxyhemoglobin levels of 15–20% were detected in two patients with underlying coronary artery disease. Patients routinely admitted to coronary care with unstable angina should be questioned about possible exposure to carbon monoxide. Carboxyhemoglobin levels should be determined when this is suspected so as to allow prompt treatment with oxygen in high concentration.

Reference Number: 30114

THEUMA, A., AND M.T. VASSALLO. "Occult CO Poisoning Presenting as Epileptic Fit." *Postgraduate Medical Journal*, 73 (861): 448, 1997 (2 references)

Abstract: A 66-year-old man attended an ER due to having fainted, lost consciousness, and sustained involuntary movements and incontinence. This was witnessed by the patient's brother, who had a weekly routine of starting the car engine in the garage for 30 minutes. The patient's admission blood COHb was 14.8% (back-calculated to 25%). He was administered normobaric oxygen and HBO_2.

This case illustrates the fact that patients presenting to the emergency room with epilepsy may have occult CO poisoning. Furthermore, the presence of chronic metabolic conditions, such as chronic renal failure can prevent physicians from seeking an alternative cause of fits, such as CO poisoning.

Reference Number: 30115

HOPKINSON, J.M., P.J. PEARCE, AND J.S. OLIVER. "Carbon Monoxide Poisoning Mimicking Gastroenteritis." *British Medical Journal*, 19 July 1980, 214–215 (1 figure, 5 references)

Abstract: A medical doctor was called in the early morning to a family of five newly arrived vacationers in a static caravan who complained of nausea, vomiting, and abdominal pain after an evening meal. The doctor diagnosed gastroenteritis. Some 30 hours later, a neighbour summoned help. The two sons and the grandmother died, and their post-mortem COHbs were 52%, 63%, and 50% respectively. The mother and father survived with COHbs of 30% and 27% respectively, four hours later. The only ventilation was a small window by which the father was found unconscious. Beside the window was also a cage containing two healthy birds. The source of the CO was due to an altered flue in the propane stove.

The clinical diagnosis of CO poisoning is not always easy and often depends on being aware of such a possibility. Skin pallor is much more often seen than the classical pink colour of the mucous membranes and skin, which indicates severe poisoning. If those in charge of caravan sites were aware of the manifestations of CO poisoning, similar tragedies might be prevented.

Reference Number: 30116

FISHER, T.L. "Carbon Monoxide Poisoning: Legal Responsibility for Failure to Diagnose." *Canadian Medical Association Journal*, 99: 235–236, 1968

Abstract: During the cold winter months, a doctor visited a family several times as a result of multiple complaints of headache and nausea for three children and the mother and father. The father suggested to the doctor that it was gas poisoning, but the doctor diagnosed stomach flu. A new gas furnace had been installed and was poorly ventilated. The next morning, the children were dead, and the mother and father were seriously ill due to CO poisoning. Legal action against the plumber who had installed the furnace and the gas inspector succeeded, but it did not succeed against the doctor.

The action came to trial about two years later and was dismissed against the doctor with costs. In his judgment the judge dealt briefly with the claim against the doctor. He said, "In all the circumstances which I will not detail, I fail to find default on the part of (the doctor), who, in my view could not have been expected to detect an odourless and deadly gas as the explanation of ailments he was examining and displayed a fair, reasonable, and competent degree of skill. Accordingly, the suit against him should be dismissed with costs."

Reference Number: 30117

NG, P.C.Y., B. LONG, AND A. KOYFMAN. "Clinical Chameleons: An Emergency Medicine Focused Review of Carbon Monoxide Poisoning." *Internal and Emergency Medicine*, 13: 223–229, 2018 (3 tables, 86 references)

Abstract: A general review of occult CO poisoning is presented. Due to the non-specific symptoms of CO poisoning, it is mistaken for other diagnoses such as viral infection. Prompt treatment with oxygen is required in most cases.

CO interferes with oxygen transport and delivery to cells. It has greater affinity for hemoglobin when compared to oxygen (approximately 230–270 times), thus impairing oxygen transport and delivery in the bloodstream leading to cellular hypoxia. COHb is formed when CO binds to hemoglobin. As the concentration of COHb increases, physiologically it causes a leftward shift of the oxyhemoglobin dissociation curve resulting in a decreased oxygen delivery to cells. Thus, even with high partial pressures of oxygen, hemoglobin is unable to effectively deliver oxygen to the cells resulting in an intracellular hypoxia. Additionally, CO increases nitric acid formation that causes vasodilatation as well as microvascular damage that may partially account for long-term morbidity. CO binding to myoglobin also causes further myocardial impairment. The toxin can also directly damage cells through intracellular protein binding, lipid peroxidation, mitochondria oxidation, apoptosis, and inflammation. Morbidity initially involves systems with high oxygen requirements, most commonly the cardiovascular and neurological system. More specifically, CO poisoning can lead to deadly dysrhythmias and permanent neurological sequelae if not promptly recognized.

Reference Number: 30118

KNOBELOCH, L., AND R. JACKSON. "Recognition of Chronic Carbon Monoxide Poisoning." *Wisconsin Medical Journal*, 98 (6): 26–29, 1999 (1 table, 15 references)

Abstract: Between 1995 and 1998, three families in Wisconsin suffered chronic occult CO exposure. The exposures were due to malfunctioning/poorly vented gas furnaces, water heaters, and fireplaces. Various symptoms such as headache, nausea, mental confusion, fatigue, and unconsciousness were reported. On the newly installed gas fireplace, soot stains were noted, and the ambient CO concentration was 112 ppm after only 15 minutes of operation. Delayed symptoms of CO poisoning include mental deterioration, personality changes, mutism, severe memory loss, fecal and urinary incontinence, tremors, visual loss, movement disorders, Tourette's syndrome, and Parkinsonian syndrome.

The cases described in this article serve as reminders of the potential health threat posed by carbon monoxide, even when it is present at sublethal levels. Residential sources of this gas include automotive exhaust from attached garages and improperly vented stoves, fireplaces, furnaces, clothes dryers and water heaters. Virtually all homes are at risk. Although there is no legal requirement that homes or apartments be equipped with CO detectors, health care providers should encourage patients to buy at least one detector and install it in the sleeping area of their home. Health care professionals should be aware of the symptoms of CO exposure and consider CO in their differential diagnosis of patients who present with vague flu-like symptoms.

Reference Number: 30119

HART, I.K., P.G.E. KENNEDY, J.H. ADAMS, AND N.E. CUNNINGHAM. "Neurological Manifestations of Carbon Monoxide Poisoning." *Postgraduate Medical Journal*, 64: 213–216, 1988 (2 figures 16 references)

Abstract: A 24-year-old man who had complaints of a throbbing headache for two weeks, which were also the symptoms experienced by his roommate, was admitted to hospital unconscious and in a deep coma. His neurological condition did not improve, and he was transferred to a more specialized hospital. He was initially treated for herpes encephalitis, and gram-negative septicemia and was administered intravenous acyclovir, dexamethasone, chloramphenicol, and erythromycin. At the referring hospital, it was determined that his blood COHb was 10% (2–3 half-lives or 12 hours after initial contact), which would indicate that his blood COHb would be around 40% to 80%. He was also found in a poorly ventilated room with the remains of a gas fire burning. The most severe effects of CO poisoning are seen in organs with a high basal oxygen consumption.

> The classical structural damage produced in the brain by CO is diffuse hypoxic injury with selectively severe involvement of the pallidum. Pallidal changes were present in our case, but there was no evidence of diffuse hypoxic damage. This may be due to the fact that exposure to CO was prolonged and that it was probably in low concentration. Other types of damage have been described including the delayed type of demyelination and general destruction of myelinated fibres. The latter, the most striking feature in the present case, appears to be a relatively uncommon type of structural damage that is not delayed but occurs as a direct effect of intoxication. This seemingly contradicts the view that CO cannot produce effects by a direct cellular action.

Reference Number: 30120

HECKERLING, P.S., J.B. LEIKIN, A. MATUREN, AND J.T. PERKINS. "Predictors of Occult Carbon Monoxide Poisoning in Patients with Headache and Dizziness." *Annals of Internal Medicine*, 107: 174–176, 1987 (2 tables, 18 references)

Abstract: A study was conducted of the COHb concentration of 63 women and 26 men (mean age was 39.2 years) who attended the ED of the University of Illinois hospital in Chicago between December 1985 and March 1986 with headache or dizziness. The mean COHb was 2.9%. Nine patients (10.1%) had a COHb between 5% and 10%, and four patients (4.5%) had a COHb of less than 10%. Thirty-nine percent of the patients used a gas stove for heat.

> A decision rule for obtaining carboxyhemoglobin levels only on patients with headache or dizziness who used gas stoves for heat or who had similarly affected cohabitants would have correctly identified all patients with carboxyhemoglobin levels greater than 10% and correctly exclude 55% of patients with lower levels.

Reference Number: 30121

FUCILI, G., AND M. BRAUZZI. "Tips for Avoiding Common Mistakes in Out-of-Hospital Diagnosis of Carbon Monoxide Poisoning." *Journal of Anaesthesia, Analgesia and Critical Care*, 2 (1): 14, 2022 (3 tables, 3 figures, 38 references)

Abstract: Acute CO poisoning is the great imitator and is difficult to diagnose without environmental gas detectors used by first responders or CO-oximeters in the ER. Older fingertip blood gas analyzers such as the standard pulse oximeter (at 660 and 990 nanometre [nm] wavelengths) cannot distinguish between COHb and HbO_2.

Table: Common Diagnostic Errors in CO Poisoning and the Percentage of Errors

Diagnostic error	Percent (%)
Food poisoning	38
Psychiatric disorders (hysteria, confusion, depression)	18
Cardiac disorders with angina or syncope	13
Alcohol intoxication or delirium tremens	7
Acute solvent poisoning	7
Headache, migraine	6
Brain ischemic pathologies	4
Cerebral hemorrhage	4
Brain tumour	3

Source: Adapted form Fucili and Brauzzi 2022.

Although carbon monoxide poisoning is a very frequent pathology, especially in the industrialized world, it is still a diagnostic challenge for the clinician today. This undoubtedly depends on the type of presentation of the clinical scenarios which is so varied and non-specific; the complexity of the differential diagnosis with pathologies that have the same symptoms in common; the diversity in instrumental equipment in the various emergency-urgency services; the subtle nature of the clinical course, which often with the removal from the source of carbon monoxide sees improvements in the phase closest to acute but which can evolve into worsening in the long term. Perhaps the most important aspect to underline for the purpose of improving health care is that of gaps in knowledge.

Reference Number: 30122

KANBUROGLU, M.K., M.N. CIZMECI, AND A.Z. AKELMA. "A Rare Cause of Chronic Headache That May Be Misdiagnosed as Migraine: Chronic Carbon Monoxide Poisoning." *Turkish Journal of Emergency Medicine*, 14: 132–134, 2014 (8 references)

Abstract: A 15-year-old girl was admitted to hospital on three occasions as a result of a headache that did not respond to analgesics. The headache was worse in the evenings but gradually disappeared at school. CO exposure was suspected as the headache only occurred at home and at the same time as other family members. A venous blood sample taken one hour after leaving the house showed a COHb of 11.9%. Two hours after oxygen treatment, her symptoms had resolved. Her mother and father had blood COHb of 11.1% and 7.9% respectively. The CO source was a defective domestic burner in the kitchen, which resulted in a measured ambient CO concentration of 78 ppm.

In conclusion, in patients who have the diagnosis of migraine or present with non-specific refractory headaches, chronic carbon monoxide poisoning should be suspected, especially if the attacks occur during the winter months.

Reference Number: 30123

ROTH, D., M. KRAMMEL, W. SCHREIBER, H. HERKNER, C. HAVEL, AND A.N. LAGGNER. "Unrecognized Carbon Monoxide Poisoning Leads to a Multiple-Casualty Incident." *Journal of Emergency Medicine*, 45: 559–561, 2013 (1 figure, 10 references)

Abstract: Emergency medical services (EMS) staff attended a basement where a 71-year-old man was unconscious. While attending to the patient, the EMS staff in the basement began to feel increasingly sick, and so the fire department was called due to suspicion of gas exposure. The fire department used self-contained breathing apparatus and measured an ambient CO concentration of 60 ppm. The man had a blood COHb of 21.9% and was administered normobaric oxygen therapy (NBOT). The initial EMS personnel at the hospital had blood COHb concentrations between 10.3% and 17.2%. The other eight EMS personnel who later worked in the basement were evaluated at hospital and released. A defective gas heating system was the source of the CO, and it was repaired.

This is a report of a case in which unrecognized exposure to CO during the treatment of an unconscious STEMI patient led to the hospitalization of 11 EMS personnel. Hazardous environments must always be considered as potential causes of a patient's altered mental status. Together with the correct use of modern equipment, such as permanently activated CO detectors, this might help to avoid harm to both patients and staff.

Reference Number: 30124

SEALE, B., S. AHANGER, AND C. HARI. "Subacute Carbon Monoxide Poisoning Presenting as Vertigo and Fluctuating Low Frequency Hearing Loss." *Journal of Surgical Case Reports*, 8: 1–3, 2018 (2 figures, 10 references)

Abstract: It is estimated that approximately 25,000 UK residents are exposed to low levels of CO annually. A 38-year-old woman, living alone in a bungalow, presented with a four-month history of vertigo and sensorineural hearing loss (SNHL). She was diagnosed with the possibility of Meniere's disease and administered betahistine (16 milligrams [mg] three times per day). There was no improvement in her condition until a leakage of CO was detected in her gas fireplace during a routine annual safety check, which she had disconnected. Her symptoms no longer recurred.

In unexplained causes of SNHL, CO poisoning should be considered as a diagnostic possibility especially in ENT out-patient settings. This is supported by the fact that missed diagnosis of CO intoxication can occur easily as commonest symptoms are non-specific. Inexpensive and non-invasive tests such as Carboxyhemoglobin (COHb) measurements should be considered when investigating for unexplained hearing loss, especially in combination with other symptoms. In our patient we were unable to check COHb as the diagnosis was made retrospectively. Early recognition of this phenomenon and subsequent management may reduce the morbidity and improve recovery from any consequent hearing loss.

Reference Number: 30125

WEBBER, A.P. "Recurrent Cardiac Failure of Environmental Origin." *Journal of the Royal Society of Medicine*, 96: 458–459, 2003 (2 figures, 4 references)

Abstract: An 82-year-old woman with a history of hypertension and angina was admitted to hospital on three occasions due to shortness of breath and heart palpitations. She was treated for left ventricular failure and released. On the third occasion, the paramedics who transported her to the hospital stated that there was a smell of gas in her apartment. Her blood COHb was 20%. The gas company was notified, and it fixed the faulty ventilation of her apartment and 9 others in the same apartment block, which had caused the elevated ambient CO concentration.

Most cases of CO poisoning present with flu-like symptoms. Other common features are tachycardia and tachypnoea, which are compensatory mechanisms for cellular hypoxia. Nausea and vomiting, presyncope, syncope, and seizures are less frequent manifestations. An increased cardiac output due to cellular hypoxia can result in or exacerbate acute coronary syndromes, pulmonary edema, and arrhythmias. Despite having no acute signs of cerebral impairment, about 15% of patients develop the delayed neuropsychiatric syndrome, which typically occurs between 2 and 28 days after exposure and can include parkinsonism, dementia, incontinence, psychosis, and personality changes. The severity of any of the above symptoms does not correlate well with COHb concentrations; duration of exposure seems more important. What is clear is that one rarely observed the classic findings of cherry-red lips, cyanosis, and retinal hemorrhages.

3.02 ACUTE ACCIDENTAL POISONING

[A]n eminently toxic gas.

—Claude Bernard (1813–78), cited by G.L. Sternbach and J. Varon, "Claude Bernard: On the Origin of Carbon Monoxide Poisoning" (1991) **(40404)**

[T]he 4 D's of accidental carbon monoxide poisoning-the decrepit, the diseased, the drugged, and the drunk.

—F.S. Fiddes, "Accidental Carbon Monoxide Poisoning. Dangers of Inadequate Ventilation" (1956) **(40901)**

There are many overlooked sources of CO such as barbeques, automobiles, furnaces, stoves, and hot water heaters (30201, 30211). Mass CO poisonings have also occurred in schools (one CO poisoning was prevented by the science teacher!) (30204–5) and churches (30209). Fortunately, in most countries, the mortality rate due to non-fire CO poisonings have dropped, due in part to catalytic converters now installed in automobiles (30202–3). The symptoms of acute accidental CO poisonings differ from chronic, occult CO poisonings (30213). Although still somewhat controversial, the main treatment is hyperbaric oxygen therapy (HBOT) (30210, 30212, 30215).

Reference Number: 30201

RALSTON, J.D., AND N.B. HAMPSON. "Incidence of Severe Unintentional Carbon Monoxide Poisoning Differs across Racial/Ethnic Categories." *Public Health Reports*, 115 (1): 46, 2000 (5 tables, 13 references)

Abstract: A study was conducted of the characteristics of 586 Washington State residents who were treated with HBO_2 for severe acute unintentional CO poisoning between 1987 and 1997. A racial/ethnic designation was identified for 530 patients. Charcoal briquettes were the most common source of CO poisoning and occurred in 66.7% of the Asian, 39.5% of the black, and 65.8% of Hispanic white patients. All 39 CO poisonings associated with boats and all space heater poisonings were among non-Hispanic white patients.

Table: Relative Risk of Being Treated with HBOT for Severe CO Poisoning by Racial/Ethnic Category

Racial/ethnic category	Relative risk
Non-Hispanic white	1.0
Hispanic white	3.96
Asian	1.32
Black	2.91
American Indian	0.73
"Other" Race	11.7

Source: Adapted from Ralston and Hampson 2000.

Cultural practices may explain the variation in typical sources of CO poisoning. About two-thirds of CO poisonings among both Asian and Hispanic white patients were due to indoor burning of charcoal briquettes. In an earlier study, present author NBH found that charcoal was a common source of poisoning among Asian and Hispanic white Americans, perhaps due at least in part to a continuation by recent immigrants of practices common in their homeland. Most cases of CO poisoning related to forklifts occurred among non-Hispanic white patients. Forklift cases are typically occupational exposures. All poisonings onboard boats were among non-Hispanic white patients as were all poisonings by space heaters. In summary this study demonstrates a significantly elevated risk for CO poisonings for members of certain minority groups in the state of Washington.

Reference Number: 30202

LAVIGNE, E., S. WEICHENTHAL, J. WONG, M. SMITH-DORION, R. DUGANDZIC, AND T. KOSTASKY. "Mortality and Hospital Admission Rates for Unintentional Nonfire-Related Poisoning across Canada: A Trend Analysis." *Canadian Medical Association Open Access Journal*, 3: E223–E230, 2015 (4 tables, 3 figures, 18 references)

Abstract: In Canada, between 1981 and 2010, there were 1,808 unintended non-fire CO poisoning-related deaths, and 1,984 CO-related hospital admissions between 1995 and 2010. During those time periods, there was an average annual decrease of 3.46% in mortality and 5.83% for hospital admissions. The mortality and hospital admission rates were elevated for men compared to women. The annual decrease in mortality and admission rates were lowest in residential environments. Deaths and admissions to hospital were typically highest between September and April and peaked in December and January.

Table: Changes in Rates of Mortality and Hospital Admissions in Canada of Non-Fire-Related CO Poisonings, 1981–2009

	1981 rate	2009 rate	Average annual rate	Average annual percent change
Mortality	3.58	1.24	2.18	−3.46%
Admission to hospital	7.07	2.57	3.85	−5.83%

Source: Adapted from Lavigne et al. 2015.

Mortality and admission to hospital rates for unintentional nonfire-related carbon monoxide poisoning have decreased over time in Canada. However, continued efforts should focus on reducing the occurrence of carbon monoxide poisoning among males and during the cooler months, and implementing appropriate measures to prevent carbon monoxide poisoning at home, such as supporting the design of heating and cooking products to improve combustion safety and encouraging the use of home carbon monoxide detectors. Public health surveillance of carbon monoxide should consider broader and novel sources of data, such as information from poisoning control centre call records.

Reference Number: 30203

HAMPSON, N.B. "US Mortality Due to Carbon Monoxide Poisoning, 1999–2014 Accidental and Intentional Deaths." *Annals of the American Thoracic Society*, 13: 1768–1774, 2016 (3 tables, 2 figures, 26 references)

Abstract: CO poisoning accounts for hundreds of deaths and thousands of hospital visits to the ER in the United States annually. The deaths due to CO poisoning, either accidentally or intentionally in the United States between 1999 and 2014 were analyzed. The total number of deaths due to CO poisoning decreased from 1,967 in 1999 to 1,319 in 2014. Accidental deaths decreased by 13% and intentional poisonings by 47% over the same time period. As of January 2010, 19 US states required CO alarms installed in residences.

Table: Top Accidental Age-Adjusted Death Rate (ADR) in US States for Accidental and Intentional CO Poisoning, 1999–2014

State	Rank	Accidental ADR	Rank	Intentional ADR
Wyoming	1	0.44	3	0.88
Alaska	2	0.37	2	0.42
Montana	3	0.37	1	1.01
North Dakota	4	0.31	6	0.76
Nebraska	5	0.30	8	0.72
Idaho	6	0.26	7	0.75
New Mexico	7	0.25	19	0.51

Source: Adapted from Hampson 2016.

The continued decline in accidental deaths is likely related in part to ongoing reduction in CO production by motor vehicle. Automobiles manufactured prior to 1975 continue to disappear from the road, and emissions engineering continues to advance, both contributing to reduction in CO production. The influence of motor vehicles on accidental poisoning deaths should level off when further reduction in CO emissions becomes too difficult or expensive.

Reference Number: 30204

CLARKE, S.F.J., A. CROSBY, AND D. KUMAR. "Early Carbon Monoxide Intoxication: Happy to Be Poisoned?" *Emergency Medicine Journal*, 22: 754–755, 2005 (11 references)

Abstract: Emergency services were called to a secondary school for children with special needs (ages 11–18 years) due to a strong odour in the dining room and feelings of headache and euphoria. After 30 minutes, the kitchen boiler was switched off, and everyone was evacuated. Approximately 43 students and 35 staff members were exposed to CO, but all blood COHb concentrations were less than 5%. Eight staff (23%) and 10 children (24%) reported feeling light-headed, happier, or euphoric.

Euphoria tends to be associated with ingestion of certain drugs of misuse and with forms of acute psychosis, and we suggest that the list of potential differential diagnoses should include early carbon monoxide poisoning. It is an easily overlooked diagnosis because many of the clinical features are non-specific, but it is an easy one to exclude as blood levels are readily measurable in most hospitals. In this case series of 78 people with early acute intoxication with carbon monoxide, euphoria was noted to be a relatively common occurrence, a symptom that to our knowledge had not been previously reported.

Reference Number: 30205

BURNEY, R.E., S-C. WU, AND M.J. NEMIROFF. "Mass Carbon Monoxide Poisoning: Clinical Effects and Results of Treatment in 184 Victims." *Annals of Emergency Medicine*, 11: 394–399, 1982 (3 tables, 2 figures, 9 references)

Abstract: A high school furnace emitted CO fumes from 7:30 am to 10:00 am when the science teacher correctly associated the symptoms of the students and staff with CO poisoning and pulled the fire alarm.

The ambient CO concentration was estimated to be 500 ppm and was due to the exhaust chamber of the furnace being inadvertently left ajar. One hundred and eighty-four people were exposed to CO. One hundred and sixty of them (87%) became ill, and 96 (60%) were transported to hospital for treatment. COHb was measured in 60 victims and ranged from 4% to 28% with about half being between 21% and 25%. The age of the victims ranged from 9 to 86 years (mean 20 years) and included 146 students and 38 teachers. The mean time of CO exposure was 73 minutes for those with no symptoms, 87 minutes for those with symptoms but not treated at hospital, and 107 minutes for those treated at hospital.

Table: Symptoms Reported by 159 Victims of CO Exposure

Symptom	Frequency (%)
Headache	90
Dizziness	82
Weakness	53
Nausea	46
Trouble thinking	46
Shortness of breath	40
Trouble with vision	26
Loss of consciousness	6

Source: Adapted from Burney, Wu, and Nemiroff 1982.

Carbon monoxide poisoning is a frequent occurrence in modern society, and it may affect large numbers of persons. On the basis of this clinical and epidemiologic study exposed to potentially toxic levels of CO, we have been able to draw the following conclusions. The apparent severity of the symptoms, such as headache, dizziness and muscle weakness provides a reliable basis for the selection of patients needing hospital treatment. Delay in triage or transport did not reduce the need for hospital evaluation and treatment. COHb levels are mandatory in hospitalized patients in order to confirm the level of intoxication and plan therapy. Intensive and prolonged oxygen therapy, preferably monitored with repeat blood gases, should be provided, and may reduce morbidity. The end point of treatment is a COHb level less than 5%, in addition to complete symptomatic relief. Smokers did not differ from nonsmokers in their symptomatic response to CO poisoning in COHb levels or in the rate of elimination of CO but did feel back to normal sooner than nonsmokers.

Reference Number: 30205

Reference Number: 30206

HAMPSON, N.B. "Cost of Accidental Carbon Monoxide Poisoning: A Preventable Expense." *Preventive Medicine Reports*, 3: 21–24, 2016 (1 figure, 20 references)

Abstract: CO poisoning is common in the United States, accounting for an estimated 15,200 ED visits annually between 2001 and 2003 for non-fatal, non-fire-related CO exposure. It was estimated in 2015 that 6,600 persons sustained long-term sequelae, with a total loss in earnings of $925 million; 334 die from accidental non-fire-related CO poisonings, with an average loss of 26 years of productivity accounting for $355 million, and 2,800 are hospitalized with acute medical care costing $33 million.

> Available data indicate that accidental CO poisonings in the US conservatively cost society over $1.3 billion, resulting from direct hospital costs and lost earnings. Further, it demonstrates a positive cost-benefit ratio for the uniform use of residential CO alarms.

Reference Number: 30207

SATRAN, D., C.R. HENRY, C. ADKINSON, C.I. NICHOLSON, Y. BRACHA, AND T.D. HENRY. "Cardiovascular Manifestations of Moderate to Severe Carbon Monoxide Poisoning." *Journal of the American College of Cardiology*, 45: 1513–1516, 2005 (2 tables, 31 references)

Abstract: The cardiovascular manifestations of 166 male and 64 female patients with moderate to severe CO poisoning, administrated HBO_2 therapy in Hennepin County Medical Centre in Minnesota between 1994 and 2002 were determined. The mean age of the patients was 47.2 years (range 19–91 years), and 56% were active smokers. Electrocardiograms (ECG), cardiac biomarkers, and echocardiograms were measured, and 37% were assessed to have cardiac injury. Prediction of myocardial injury included being male, a Glascow Comma Scale (GCS) of less than 14, and hypertension. Active cigarette smoking tended to lower the relative risk of myocardial injury.

> In conclusion, myocardial injury is common in moderate to severe CO poisoning. A baseline ECG should be performed and serial biomarkers should be followed in all patients. Patients with abnormal cardiac markers should have an echocardiogram. Patients with persistent LV dysfunction, underlying CAD [coronary artery disease], or risk factors of CAD may benefit from further evaluation including angiography and revascularization.

Reference Number: 30208

AKSU, N.M., M. AKKAS, F. COSKUN, E. KARAKILIC, M. GUNALO, H. AKKUCUK, D.K. ATAMAN, H. OZCAN, AND M.M. OZMEN. "Could Vital Signs Predict Carbon Monoxide Intoxication?" *Journal of International Medical Research*, 40: 366–370, 2012 (2 tables, 26 references)

Abstract: The clinical presentations of 476 CO-poisoned patients attending a hospital in Ankara, Turkey, between 2000 and 2010 were determined. The blood COHb concentrations were less than 10% in 11.1% of the patients, 10–20% in 30.2% of the patients, and greater than 20% in 58.6% of the patients. The most common complaints were headache/dizziness (62.6%), nausea/vomiting (16.16%), and syncope (16.0%). A correlation was observed between increased COHb concentration and incidence of syncope. There was no association between vital signs and COHb concentration. Among patients with a COHb greater than 20%, 68.3% had normal vital signs.

Table: Percent of CO Poisoned Patients Showing Clinical Signs That Are Statistically Significant at Various COHb Concentrations

Clinical measure	COHb < 10% (%)	COHb 10–20% (%)	COHb > 20% (%)
Heart rate 60–100 beats per minute (bpm)	87.2	85.8	85.8
Heart rate < 60 bpm	0	0	0.98
Heart rate > 100 bpm	5.1	12.3	24.9
GCS score = 15	100	98.1	94.1
GCS score < 15	5.1	10.4	2.0
Blood pH < 7.35 (Alkalosis)	2.7	0	4.4
Blood pH > 7.45 (acidosis)	7.7	12.3	16.1

Source: Adapted from Aksu et al. 2012.

In the present study 4.45 and 16.6% of the patients with COHb levels > 20% had acidosis and alkalosis, respectively. Metabolic changes may reflect the toxic effects of CO better than any particular COHb level. Mild CO cases may be accompanied by respiratory alkalosis to compensate for the reduction in oxygen-carrying capacity and delivery. Longer exposures with decreased level of consciousness, result in metabolic acidosis from the lactate production that accompanies tissue hypoxia. This indicates that the patients in the present study presented to the emergency department during the early phase of CO intoxication. Hypoxia, hypotension, metabolic acidosis, and a high lactate level are the primary factors responsible for the neurological effects of CO intoxication.

Reference Number: 30209

MCGUFFIE, C., J.P. WYART, G.W. KERR, AND W.S. HISLOP. "Mass Carbon Monoxide Poisoning." *Journal of Accidental and Emergency Medicine*, 17: 38–39, 2000 (1 table, 18 references)

Abstract: An evening memorial service was held in a church hall attended by 60 adults who arrived around 7:00 p.m. During the service, most of the present experienced light-headedness, headache, and nausea, which most thought were due to an emotional response to the service. At 9:15 p.m., a 54-year-old man fell unconscious to the floor and was taken to hospital where his COHb was determined to be 35%. In total, 22 patients presented at hospital, of whom 17 had a COHb greater than 10%. All were treated with high-flow oxygen. Building construction in the hall several months earlier had impaired the ventilation system, and the central heating boiler had been repaired only two days earlier.

Table: Distribution of COHb Percent in Victims of a Mass CO Poisoning Due to Malfunctioning Furnace

COHb range (%)	Number of cases
0–10	5
11–20	12
21–30	4
31–40	1

Source: Adapted from McGuffie et al. 2000.

All of those admitted were invited to attend a follow up clinic three months after the incident. Eleven patients attended; seven had experienced neuropsychiatric problems, including depression, apathy, irritability, and poor concentration. For four of the patients, these episodes were self limiting, lasting only a few weeks, but three patients described persistent symptoms.

Reference Number: 30210

LAMPROPOULOS, C.E., E. SKLAVOU, C. ANASTOGIANNIS, V. PAPANIKOLAOU, D. TSILVARAKIS, V. VASILIOU, AND S. VAKALI. "Acute Heart Failure in Healthy Young Patient after Severe Carbon Monoxide Poisoning." *European Journal of Case Reports in Internal Medicine*, 12: 001340, 2019 (1 figure, 6 references)

Abstract: A 28-year-old man was admitted to hospital with a reduced level of consciousness after having a seizure while using charcoal briquettes for heating indoors all night. His COHb concentration was 29.5%. His condition deteriorated rapidly, and he was intubated. He suffered acute heart failure. His HBOT was cancelled, and he was admitted to intensive care unit (ICU) instead with NBOT. Four hours later, his COHb was 2.5%. He remained in ICU for five days and was discharged one week later.

Cellular hypoxia after CO exposure is caused by disturbances of oxygen delivery. CO reversibly binds hemoglobin (230–270 times more powerfully than oxygen) and cardiac myoglobin with even greater affinity. Additionally, CO causes a leftward shift of the oxyhemoglobin dissociation curve further decreasing oxygen availability. However, impaired oxygen delivery does not sufficiently explain the extent of toxicity. CO may cause direct cellular damage involving immunological or inflammatory mechanisms (peroxynitrites production, reactive oxygen species, lipid peroxidation, mitochondrial oxidative stress, nitric oxide, free radicals, platelet activation).

Reference Number: 30211

DONAHUE, A., K. LANE, AND T. MATTE. "Death from Unintentional Non-fire-Related Carbon Monoxide Poisoning in New York City during the Cold Season, 2005–2013." *Journal of Environmental Health*, 81: 16–22, 2019 (4 tables, 22 references)

Abstract: There were 32 unintentional non-fire-related CO poisoning deaths during the winter months (October to April) in New York City between 2005 and 2013. Most of the victims were men (84%), and 34% were over the age of 65 years. Most of the deaths occurred in Queens (41%). The mean blood COHb of the victims was 61% (range 35–83%). The mean environmental CO levels detected were 372 ppm (range 200–642 ppm). On average, unintentional non-fire-related CO deaths in New York City were 0.4 deaths per million people annually.

Table: Source of Non-Fire-Related Deaths from CO Poisonings in New York City between 2005 and 2013

Source of CO	%
Automobile	47
Generator	16
Oven	13
Space heater	9
Charcoal fire	3
Furnace	3
Gas water heater	3
Lawn mower	3
Unspecified household exposure	3

Source: Adapted from Donahue, Lane, and Matte 2019.

In NYC, the UNFR [unintentional non-fire-related] CO death rate is lower than national rates with the majority of exposures due to car exhaust. Preventive measures such as ensuring home CO detectors, maintaining home heating equipment and proper use of home heating and cooking equipment as well as reducing CO in consumer products such as generators continue to be relevant. Our study shows that the danger of automobile emissions—including automobiles in attached garages—as a source of fatal CO poisoning at home are important in NYC and should be emphasized in public health messaging.

Reference Number: 30212

GRIEB, G., A. GROGER, A. BOZKURT, I. STOFFELS, A. PIATKOWSKI, AND N. PALLUA. "The Diversity of Carbon Monoxide Intoxication: Medical Courses Can Differ Extremely—A Case Report." *Inhalation Toxicology*, 20: 911–915, 2008 (2 figures, 23 references)

Abstract: A married 22-year-old man and 28-year-old woman were found lying unconscious on their bathroom floor after approximately 30 minutes exposure to CO fumes from their faulty gas water heater. The man had an arterial COHb of 34.7% and recovered rapidly after NBOT. The woman had an arterial blood COHb of 38% but had severe lung complications resulting in severe lung edema. She had to be intubated. Both patients were administered HBOT. The man was discharged from hospital after one day, but the woman remained in intensive care for four days.

We can confirm that acute and severe CO intoxication can cause a wide variety of symptoms. These cases show that the spectrums of both expected symptoms of acute CO intoxication and their treatment are large. We recommend critical management of CO intoxications. The severity of CO poisoning should not only be judged by the serum level of COHb, but also by detailed medical history, physical examination, and further supporting diagnostic measures. Moreover, we recommend inpatient treatment in a specialized clinic with the continuous monitoring of vital parameters (heart rate, blood pressure, percutaneous oxygen saturation, blood gas analysis) if CO intoxication is suspected. Patients with respiratory insufficiency should be intubated and ventilated with 100% O_2. The indication for hyperbaric oxygen therapy should be made according to the clinical condition of the patient; however, it is strongly recommended when neurological deficits, respiratory insufficiency, or a COHb of > 15% are present.

Reference Number: 30213

ASHCROFT, J., E. FRASER, S. KRISHNAMOOTHY, AND S. WESTWOOD-RUTTLEDGE. "Carbon Monoxide Poisoning." *British Medical Journal*, 365: (6pp), 2019 (1 table, 1 figure, 23 references)

Abstract: A 23-year-old man with no medical history attended an ER with a three-day history of headache, transient visual disturbances, dizziness, and hypertension. His venous blood COHb was 26%, and he was diagnosed with acute CO poisoning. One of the authors (S. Westwood-Ruttledge) developed chronic CO poisoning over a three-year period after moving into a new house with her six-year-old son. They both developed chronic nausea, headaches, and dizziness. The family doctor said that the son was fine and just wanted to get out of school. She was diagnosed with cocaine abuse. A routine inspection of her water heater revealed a CO leak as a result of a disconnected air duct and resulted in her proper diagnosis.

Table: A Comparison of Symptoms of Acute and Chronic CO Poisoning

Acute	Chronic
Headache	Inattention, memory loss, fatigue
Tiredness	Sleep disturbances
Abdominal aches	Personality and mood changes
Nausea and vomiting	Parkinsonism, ataxia, seizures
Chest or generalized pains	Muscular spasms
Shortness of breath	Tinnitus
Seizure	Photophobia
Coma	Chronic pain syndromes
Death	Cardia ischemia

Source: Adapted from Ashcroft et al. 2019.

The London fire brigade advises members of the public to buy a CO detector for all rooms that contain a fuel burning appliance. CO detectors sound a loud alarm if a harmful level of CO gas is reached within an area. It is the only method to detect CO gas within the household, enabling safe evacuation and preventing harmful CO effects. Detectors should be installed in an open unobstructed area of wall or ceiling at a metre's distance from the appliance.

Reference Number: 30214

SENER, R.N. "Acute Carbon Monoxide Poisoning: Diffusion MR Imaging Findings." *American Journal of Neuroradiology*, 24: 1475–1477, 2003 (3 figures, 10 references)

Abstract: A nine-year-old girl was admitted to hospital in a comatose state due to being in a poorly ventilated room with a coal heater. Diffusion magnetic resonance images showed lesions in the white matter of the brain within 12 hours of exposure, which resolved 16 days later. During the early stages of CO poisoning, white matter may be more sensitive than grey matter to ischemia, and the ischemia can be reversible.

With carbon monoxide poisoning, formation of carboxyhemoglobin in the blood results in hypoxia-anoxia and the direct toxic effects of carbon monoxide on mitochondria interferes with oxidative phosphorylation. These leads to anoxic-ischemic encephalopathy with bilateral lesions. It has been reported that the basal ganglia structures globus pallidus, putamen, and caudate nucleus are most commonly affected by ischemic lesions. Also, thalamus, periventricular and subcortical white matter, corpus callosum, cerebral cortex and hippocampus of the temporal may be involved.

Reference Number: 30215

CLOWER, J.H., N.B. HAMPSON, S. IQBAL, AND F.Y. YIP. "Recipients of Hyperbaric Oxygen Treatment for Carbon Monoxide Poisoning and Exposure Circumstances." *American Journal of Emergency Medicine*, 30: 846–851, 2012 (4 tables, 14 references)

Abstract: An analysis was conducted of 864 patients who received HBOT for non-fire-related CO poisoning in 39 US states between 2008 and 2010. Most of the patients were white men, ages 18 to 44 years old. Only 10% of the patients reported having a CO alarm. Nineteen patients (2%) reported prior CO exposure. About half (55%) of the patients treated were discharged, and 41% were hospitalized. The median COHb was 24.6% (range 1.0–77.0%) for those patients hospitalized and 21.0% (range 0.1–46.0%) for those discharged.

Understanding why certain groups of patients experience higher rates of carbon monoxide poisoning is important for case recognition and prevention efforts. For example, higher proportions of carbon monoxide poisoning among men have been reported in populations with fatal carbon monoxide exposures; this pattern has been attributed to engaging more frequently in high risk behaviour such as using fuel-burning tools. A similar pattern is seen in this study populations given that men account for the large majority of those performing home or auto repair and occupational activities when they were poisoned. Most of the patients who lost consciousness for more than 1 hour were men. Finally, men comprise a larger proportion of patients who were hospitalized after treatment than patients who were discharged home.

3.03 ACUTE INTENTIONAL POISONING (SUICIDES)

Unhappiness is a dangerous thing, like carbon monoxide. You don't taste it, it's formless and colorless, but poisons slowly. It sweeps into every pore in your skin until one day your heart just stops beating.

—Bella Pollen, "goodreads.com"

Two major factors have reduced the suicide rate due to CO: catalytic converters in automobiles and the conversion of domestic coal gas to natural gas in the United Kingdom (30301–3, 30309, 30311–12). The amount of CO in the emission of automobiles having a working catalytic converter is now so low that, even after running the exhaust of his car through a hose into the cabin, where he sat for five hours, the driver was still alive and recovered in hospital (30310). Unfortunately, suicides using charcoal barbeques increased in Japan and Taiwan, due in part to their ready availability and glamorization of this type of suicide in the media (30301–4, 30306–8, 30311–12). The difference between a complex versus a complicated suicide is explained (30313). Dr. Jack Kevorkian used CO in his assisted suicides of ill patients (30314).

Reference Number: 30301

NOUMA, YOUSSEF. "Carbon Monoxide Suicide by Charcoal-Burning: A Case Report and Review of Literature." *PanAfrican Medical Journal*, 40 (190): 35059110, 2021 (4 figures, 20 references)

Abstract: Charcoal burning suicide is an ancient method that has been revived and become more prevalent. In the 1990s, the majority of CO poisonings in Europe were accidental, and only 5% were suicidal. Most of the suicides used domestic coal gas or car exhausts, but, since the introduction of natural gas and catalytic converters respectively, the suicide rate has dropped. A 28-year-old man was found dead inside the locked bathroom of a rented house in which black plastic tape had been used to seal the windows and doors and plug the air inlet. Beside the deceased was a suicide note and three barbecue grills and charcoal. A review of his cell phone telephone browser showed searches for easy suicide. The postmortem blood COHb concentration was 40.05%.

> Charcoal burning in a confined space is a duly old but renewed suicide method. In this paper, we report the first documented case of suicide by CB [charcoal burning] in Tunisia and probably in all African countries. Elsewhere, it is being commonly used as a suicide method in many countries, not only due to the CO lethality but also to the faulty perception, wrought by the media and Internet users of a relatively painless death.

Reference Number: 30302

PAN, Y.-J., S.-C. LIAO, AND M.-B. LEE. "Suicide by Charcoal Burning in Taiwan, 1995–2006." *Journal of Affective Disorders*, 120: 254–257, 2010 (1 table, 1 figure, 23 reference)

Abstract: Between 1995 and 2006, there were 33,757 suicide deaths in Taiwan, and 68% of these victims were men. Suicide by charcoal burning accounted for 33.5% of total suicide deaths in 2006. Its incidence increased from 0.22 per 100,000 people in 1991 to 6.48 per 100,000 people in 2006, a 150% increase. The mean age of persons who committed suicide by charcoal burning was 38.76 years compared to 50.74 years for other suicide methods. Persons who committed suicide by charcoal burning had fewer expressed intents and diagnosed psychiatric disorders.

Romanticizing suicide by charcoal burning by the media and cultural influences may have contributed to the increased suicides. Prohibiting sale of charcoal in convenience stores, enforcing media guideline, and setting up carbon monoxide detectors may help prevent suicide by charcoal burning.

Reference Number: 30303

YOSHIOKA, E., S.J.B. HANLEY, Y. KAWANISHI, AND Y. SAIJO. "Epidemic of Charcoal Burning Suicide in Japan." *British Journal of Psychiatry*, 204: 274–282, 2014 (7 tables, 1 figure, 28 references)

Abstract: Between 1997 and 1998, as a result of an economic recession, the number of suicides in Japan increased by 34.7% to reach more than 30,000 per year. The charcoal-burning suicide epidemic occurred between 2003 and 2007. In men and women aged 15–24 years and 25–44 years, there was a substantial increase in charcoal-burning suicides, without a decrease in suicides by other methods. The percentage of charcoal-burning suicides for all age groups increased from 6.2% (1998–2002) to 13.9% (2003–7) in men and from 1.8% to 5.5% in women. In young men, there is a close correlation between suicide rates and unemployment.

Table: Main Methods of Suicide in Japanese Women during 2007

Method	Percent (%)
Hanging, strangulation, and suffocation	62.2
Jumping from a height	13.0
Drowning and submersion	5.7
Other gases and vapours	4.9
Jumping or laying in front of a moving object	3.1
Legal drugs, sedatives, hypnotics, and so on	2.3
Smoke and fire flames	2.2

Source: Adapted from Yoshioka et al. 2014.

Second, in the Internet era, there is ample reason to anticipate an especially rapid spread if a new method appears comparatively more acceptable to vulnerable individuals than existing methods. Since Internet websites and chat rooms provide detailed information on suicide, it has been argued that the Internet may have a more direct influence on copycat suicides than do the print media. One of the first charcoal burning suicide victims in Taiwan explicitly stated that he learned of the method from a Hong Kong newspaper website. After the first Internet linked suicide pact in Japan in 2003, Internet-linked suicide became more widespread.

Reference Number: 30304

LANDERS, D. "Unsuccessful Suicide by Carbon Monoxide: A Secondary Benefit of Emissions Control." *Western Journal of Medicine*, 135: 360–363, 1981 (3 tables, 3 figures, 12 references)

Abstract: A 35-year-old man was found lying unconscious in a closed garage 12 inches away from the exhaust pipe of a 1980 Chevette that had a catalytic converter. He was pulled from the garage and recovered consciousness by the time the paramedics arrived. At the ER, his COHb concentration was 36% and blood alcohol content (BAC) was 0.09 grams per 100 millilitres. He was administered 100% oxygen at hospital and was transferred to a mental health unit for 10 days before being released without neurological sequelae.

Table: Automobile Manufacturer Exhaust CO Emission Standards in the United States

Model year of vehicle	CO (%)
Pre-1968	8.5
1968–9	7.0
1970–4	5.0
1975–9	3.0
1980 Chevrolets (measured)	0.1

Source: Adapted from Landers 1981.

Reduction in suicides can be shown to be the result of primary prevention—that is, improved knowledge and treatment of psychiatric illness—and secondary prevention—reduction in the lethality of traditional methods of suicides. In this case, secondary prevention of suicide is aptly demonstrated and represents a significant and heretofore unrecognized benefit of emissions control.

Reference Number: 30305

CLARKE, R.V., AND D. LESTER. "Toxicity of Car Exhausts and Opportunity for Suicide: Comparison between Britain and the United States." *Journal of Epidemiology and Community Health*, 41: 114–120, 1987 (1 table, 4 figures, 18 references)

Abstract: Between 1963 and 1971, there was a 35% decrease in the suicide rate in England and Wales due to the progressive reduction of the amount of CO in the domestic gas supply. It was estimated that the detoxification of domestic gas in England and Wales saved approximately 6,700 lives between 1963 and 1970. The rate of suicides due to car exhaust also declined substantially in the United States due to emission controls. But the rate of car exhaust suicides in England and Wales (without emission controls) increased and is now two times that of the United States (with emission controls). It is estimated that emission controls in the United States saved approximately 48,000 lives due to the reduced car exhaust suicide rate between 1968 and 1984.

The results of this study support the idea that emission controls have brought about a decrease in suicide by car exhaust gas in the United States. Because of the particular advantages of exhaust gas as a means of suicide displacement to other methods is unlikely to have been complete and the savings in life may have been substantial. In Britain, where emission controls have yet to be introduced, the rate of exhaust suicides has greatly increased in recent years, possibly as a result of increased knowledge about this method. The British rate as expressed per million of the population is now much higher than the United States though for many years previously, it was much lower.

Reference Number: 30306

CHEN, Y.-Y., O. BENNEWITH, K. HAWTON, S. SIMKIN, J. COOPER, N. KAPUR, AND D. GUNNELL. "Suicide by Burning Barbecue Charcoal in England." *Journal of Public Health*, 35: 223–227, 2012 (1 table, 18 references).

Abstract: The barbeque suicide cases from 12 (11%) of the 107 coroners in England between 2005 and 2007 were analyzed. There were 11 cases of charcoal-burning suicides detected, which could indicate about 100 charcoal-burning suicides in England over that three-year period. Victims of this suicide method were younger (33.4 versus 44 years) and more likely to be unemployed (70.0% versus 30.1%) and unmarried (100% versus 70%) than those using other methods. Alcohol consumption was reported in three cases. Internet involvement and research was found in four cases. In one case of murder suicide, a 30-year-old mother killed her five-year-old son.

> The incidence of charcoal-burning suicide in England and the West is considerably lower than that seen in Hong Kong and Taiwan, where it has been endemic for over a decade. It is unclear why the method has not become popular in the West. A possible explanation is the lack of media publicity given to this method in the West and the absence of high profile cases of suicide using this method. The use of this method in Hong Kong is thought to have followed the high profile reporting of a single case, whereas in Taiwan, greater reporting intensity of charcoal burning suicide compare with other suicide methods has been documented.

Reference Number: 30307

LABERKE, P.J., H. BOCK, V. DITTMANN, AND R. HAUSMANN. "Forensic and Psychiatric Aspects of Joint Suicide with Carbon Monoxide." *Forensic Science Medicine Pathology*, 7: 341–343, 2011 (1 figure, 26 references)

Abstract: A 23-year-old schizophrenic woman and a 30-year-old depressed man were found dead inside a holiday apartment beside a patio barbeque that obviously had been used and now was out with a large amount of ashes. They had first met at a suicide forum on the web. The woman had a post-mortem blood amphetamine concentration of 32 nanograms per millilitres (ng/mL) and COHb of 65%. The man had a post-mortem blood amphetamine concentration of 20 ng/mL and COHb of 70%.

> At first glance the cases appeared to be straightforward accidental CO intoxication in connection with the use of an indoor barbeque. Not until inconsistencies concerning the relations between the deceased emerged and further investigations were initiated were the deaths revealed to be a joint suicide. Date from the computer of the woman showed that she had been gathering information about committing suicide and had become a member of a suicide Internet forum months before. Two months before

the incident she corresponded with her final suicide partner on this platform.

Reference Number: 30308

KATO, K., F. AKAMA, K. YAMADA, M. MAEHARA, K. KIMOTO, Y. TTAKAHASHI, R. SAO, Y. ONISHI, AND H. MATSUMOTO. "Frequency and Clinical Features of Patients Who Attempted Suicide by Charcoal Burning in Japan." *Journal of Affective Disorders*, 145: 133–135, 2013 (2 tables, 12 references)

Abstract: The rate of suicide in Japan (25.8 per 100,000 annually) is one of the highest among developed countries. A study was conducted of 647 consecutive psychiatric patients who tried to commit suicide andwere admitted to a Japanese hospital between 2010 and 2012. Twenty of the 647 patients (3.1%) attempted suicide by charcoal burning.

The rate of men to women in charcoal burning group was significantly higher compared to those in the other methods group. The age group of 20–39 years (50%) in the charcoal burning groups was the highest among the age groups. Moreover, the proportion of patients with a mood disorder was significantly higher in the charcoal burning group than in the other methods group. The occurrence of a psychiatric history in patients in the charcoal burning group was significantly lower than in the other methods group.

Reference Number: 30309

KREITMAN, N., AND S. PLATT. "Suicide, Unemployment, and Domestic Gas Detoxification in Britain." *Journal of Epidemiology and Community Health*, 38: 1–6, 1984 (2 tables, 3 figures, 15 references)

Abstract: The detoxification of the domestic gas supply in the United Kingdom was begun in 1957 when it contained 14% CO. By 1968, it was 12% CO. In 1970, the concentration of CO in the domestic gas supply had decreased to 2% and to less than 1% in 1972. Between 1962 and 1971, the UK suicide rate fell by 34%, while the unemployment rate increased to 50%. The opposite occurred in other Western countries.

Table: Correlation (r) between the Rates for Unemployment and Suicide in Eight Countries

Country	Period	Correlation (r)
United States	1962–76	+0.79
France	1962–74	+0.64
Japan	1962–76	+0.53
Canada	1962–75	+0.51
Sweden	1962–76	+0.40
West Germany	1962–75	+0.39
Italy	1962–74	0.00
Britain	1961–71	−0.59

Source: Adapted from Kreitman and Platt 1984.

It is hypothesized that the British Anomaly is due to the coincident detoxification of domestic gas after 1963. Unemployment and suicide by means other than domestic gas were found to correlate +0.88 between 1955 and 1980 and +0.70 between 1964 and 1974. These findings are discussed in respect to the role of availability of method in the epidemiology of suicide. It is also suggested that the relation between unemployment and suicide is complex and probably largely indirect.

Reference Number: 30310

O'BRIEN, J.T., AND A.F. TARBUCK. "Suicide and Vehicle Exhaust Emissions." *British Medical Journal*, 304 (6838): 1376, 1992 (5 references)

Abstract: A 43-year-old man was admitted to hospital after five hours of exposure to the exhaust of his car (a Saab) running in a garage. He had a hose connecting the tailpipe to the cabin of his car, where he was found semi-conscious. At hospital, he had a COHb concentration of 21%.

Our patient suffered relatively mild carbon monoxide poisoning despite spending five hours inhaling exhaust fumes. Previous reports indicate that when people use ordinary vehicles and a similar method to that reported here, death may be expected within 20–30 minutes of them starting to inhale fumes. At that stage carboxyhemoglobin concentrations would be above 50–60%. We speculate that this suicide failed because the car was fitted with a catalytic converter. To our knowledge, this is the first such report in the United Kingdom.

Reference Number: 30311

STRIFE, B.J., AND L. PAULOZZI. "To Make Further Progress against Carbon Monoxide Poisoning, Focus on Motor Vehicles." *Injury Prevention*, 10: 74–75, 2004 (1 table, 19 references)

Abstract: A short overview of acute motor vehicle CO suicides are presented. Over 90% of the people who kill themselves with motor vehicle exhaust are found inside the passenger cabin of a car or truck.

> Catalytic converters, however, do not remove all CO from exhaust. Motor vehicles emit toxic levels of CO when their emission controls malfunction or after cold starts, and the engine can be fooled into putting out more CO when the vehicle is running in a tightly controlled space. Emission controls therefore have not eliminated the MVCO [motor vehicle CO] poisoning problem, even though essentially all American cars now have them. In 1998, the most recent year for which deaths traceable to MVCO are available in the US, the US motor fleet caused 1,511 MVCO deaths which accounted for 72% of the 2,160 CO deaths from all sources combined.

Reference Number: 30312

FEORELLI, D., G. MANDARELLI, L. SPAGNOLO, F. MISCEO, G. STRISCUILLO, L. POLO, B. SOLARINO, AND A. DELL'ERBA. "An Unusual Suicide by Carbon Monoxide Intoxication: 91% HbCO Saturation in a Sealed Setting." *Emerging Trends in Drugs, Additions, and Health*, 1 (5): 100021, 2021 (1 table, 4 figures, 32 references)

Abstract: CO is one of the most common, lethal poisons and has a worldwide death rate of 2.2 per 100,000 CO poisonings. About 60% of the deaths occur at home. In this case, a 60-year-old man was found dead inside his house with a strong smell of burning inside. The windows and doors were sealed with scotch tape, and a burned charcoal brazier was found inside. A cherry red hypostasis was observed at the rear portion of the body as well as a red colouration of the nails. No drugs or alcohol could be detected, but the post-mortem COHb was 91%.

> The value of 91% HbCO saturation we detected in the present case was the highest among the 55 cases in which HbCO assessment was requested at the Institute of Forensic Medicine in Bari. Among these cases, 16 are those in which the cause of death was eventually identified as a carbon monoxide intoxication with an average HbCO saturation of 57.12%.

Moreover, to confirm the exceptionality of this data we reviewed two different databases (Pub Med, and Google Scholar) that were questioned using the main keyword-carbon monoxide poisoning—that was crossed with—HbCO saturation. The value detected in the present case was one of the highest ever reported.

Reference Number: 30313

SIMONIT, F., U.D. BROI, C. FURIOSO, AND L. DESINAN. "A Burned Body with a Gunshot Wound in the Mouth and a Suicide Note: A Complex or Complicated Suicide?" *Journal of Forensic and Legal Medicine*, 72 (2020): 101958, 2020 (2 tables, 6 figures, 37 references)

Abstract: A suicide involving more than one method has been classified as a complex suicide, whereas a suicide that is caused by an unforeseen accident following a suicide attempt is classified as a complicated suicide. The body of a 53-year-old man was found in the driver's seat of a burning car. The hand brake was on, the ignition was on, and his right foot was on the accelerator. He had been suffering from a paranoid disorder and left a suicide note. There was a gunshot wound to his mouth caused by his handgun. The fire was due to an overheated engine. The post-mortem blood COHb was 40%.

The case is consistent with a complicated suicide involving a gunshot to the head and exposure to fire. After ruling out the homicidal manner, a differential diagnosis between complex and complicated suicide was performed through the analysis of the cause of death and the intention of the victim. Brain lesions did not result in being lethal, whereas COHb levels were consistent with a fatal intoxication. Fire investigation reported that the burning was started in the engine by the prolonged pressure of the accelerator that caused the overheating of the engine. This fact was more consistent with accidental fire than arson. The case was therefore classified as a complicated suicide in which the victim shot himself to the mouth without causing lethal brain injuries and then unintentionally pressed the accelerator pedal for a prolonged time causing the car to catch fire.

Reference Number: 30314

ANONYMOUS. "Kevorkian Cleared Again." *British Medical Journal*, 312: 656, 1996.

Dr. Jack Kevorkian was acquitted a second time on the charge of illegally assisting the suicide of two ill patients with CO in Michigan.

> But a juror interviewed afterwards said that the issue was more fundamental. Noting that suicide is not illegal in Michigan, he asked, "How could we convict someone for helping to do something that is legal?" Kevorkian was accused of helping in the suicides of a 72 year old woman with amyolateral sclerosis and a 61 year old doctor with cancer. Both died by carbon monoxide poisoning.

3.04 HOMICIDES

> A Hong Kong court on Wednesday sentenced a Malaysian professor and anesthesiologist to life in prison for killing his wife and daughter using two yoga balls filled with carbon monoxide, a murder, the court said was premeditated and calculated, the South China Morning Post reports.
>
> —Casey Quackenbush, "Hong Kong Sentences Professor to Life in Prison" (2018)[5]

Compared to suicides, the number of murder cases involving CO are infrequent but do occur (30401–4).

Reference Number: 30401

PRZEPYSZNY, L.M., AND A.J. JENKINS. "The Prevalence of Drugs in Carbon Monoxide-Related Deaths: A Retrospective Study, 2000–2003." *American Journal of Forensic Medicine and Pathology*, 28: 242–248, 2007 (3 tables, 15 references)

Abstract: Post-mortem blood COHb, alcohol, and other drug concentrations were determined in 122 cases involving CO in Cleveland between 2000 and 2003. Eighty-four of the cases (69%) were accidental, 31 (25%) were suicides, and 7 (6%) were homicides. The accidental CO victims were typically white males, ages 40–59 years old residing in the city of Cleveland. The suicide victims were also middle age white males but mainly residing in the suburbs. A positive COHb was detected in 87% of the victims, which typically was due to long survival times before death. Suicide cases (usually by car exhaust) had the highest mean COHb (66%),

5 Casey Quackenbush, "Hong Kong Sentences Professor to Life in Prison for Killing His Wife and Daughter with Gas-Filled Yoga Ball," *Time*, 20 September 2018.

followed by accidental cases (55%), then homicide cases. A positive BAC was detected in 26% of the accidental victims (mean 190 mg per decilitre [dL]), 35% of the suicide cases, but only 1 (14%) of the homicide cases. The homicide cases were all related to fire, and nearly one-half of the victims were under the age of 12 years. Cocaine and cannabis were detected in 12% of the accidental, 10% of the suicidal, and in only one of the homicide cases. Therapeutic drugs were detected in 30% of the victims. The most common drugs detected were antidepressants and antihistamines in suicide victims and pain medication and antihistamines in the accidental cases.

> Ethanol did not appear to be a factor in homicide cases in which the cause of death was related to CO, since ethanol was only detected in 1 case. It is interesting to note that this case was also the only homicide case that involved illicit drugs, specifically cocaine. The young age of the majority of decedents may explain the low incidence of ethanol.

Reference Number: 30402

JANIK, M., M. UBLOVA, S. KUCREROVA, AND P. HEINA. "Carbon Monoxide Related Fatalities: A 60-Year Institution Experience." *Journal of Forensic and Legal Medicine*, 48: 23–29, 2017 (1 table, 5 figures, 46 references)

Abstract: Between 1947 and 2006 in the Czech Republic, a total of 1,233 CO-related deaths were examined. Forty-five percent of the deaths were accidental, 40% were suicidal, 1% were homicidal, and 14% were undetermined. The mean age of the victims was 48 years, and 59% were male. The sources of all poisonings were coal gas (58%), exhaust from cars (15%), fires (12%), faulty gas appliances (11%), and faulty wood stoves (4%). The mean COHb of the deceased was 66%. The COHb concentration was zero in 13 cases due to delayed deaths. Of the 1,017 cases in which the BAC was measured, 45% were positive with an average BAC of 41 mg/dL.

Table: Source of CO in Suicide, Accident, and Homicide Deaths

Source	Suicide (495 victims)	Accident (548 victims)	Homicide (14 victims)
Fire-related smoke	6	116	1
Vehicle exhaust fumes	94	47	0
Coal gas	326	239	13
Faulty gas devices	21	87	0
Faulty wood furnaces	16	21	0

Source: Adapted from Janik et al. 2017.

This study shows that the prevalence of CO-related deaths has decreased significantly in the Czech Republic following the widespread detoxification of the domestic gas in 1990–1995. Our findings suggest that acute ethanol use, poorer socioeconomic position, and inadequate education status about the danger of CO are associated with an increased risk of fatal CO poisoning. Finally, our results demonstrate the continued value of the autopsy monitoring in monitoring global public health security issues and socioeconomic situation. Further, similar large scale studies in different populations are needed to improve the targeting of interventions to the groups with the highest level of risk, and to understand the sources of variation in CO-reduced mortality.

Reference Number: 30403

AKAISHI, S., S. OSHIDA, K. HIRAIWA, I.M. SEBETAN, Y. OHNO, F. KURODA, T. SUZUKI, AND S. KASHIMURA. "Homicidal and Camouflaged Carbon Monoxide Poisoning in Japan." *Zeichschrift für Rechtsmedizin*, 88: 297–304, 1982 (2 tables, 1 figure, 13 references)

Abstract: Between 1969 and 1980, there were 1,322 suicidal, 662 accidental, and 1 homicide cases of CO poisoning in the Tohoku district in Japan. A nationwide search showed four other homicidal cases using CO. A gas stove was used in three cases of homicide, exhaust from a car in one case, and one case involved CO produced by combining sulphuric acid, oxalic acid, and caustic soda. In two cases, the original diagnosis was CO poisoning, but the post-mortem blood COHb was zero.

On January 19, 1970, a man aged 33 was found alone in bed, and his paramour aged 31 with whom he had lived together was missing. The hose of the gas stove was found off, and city gas was being discharged into the room. The constricting metal ring of the hose was found on the floor near the hose. Autopsy revealed the cause of death to be CO poisoning and at first an accidental death was suspected. However, the ring was difficult to come off spontaneously, and his paramour was suspected and arrested. She confessed that he had suggested to her to dissolve their cohabitation. So, while he was sleeping, she took off the hose from the stove, discharged the city gas into the room and ran away. She was sentenced to 8 years' penal servitude.

Reference Number: 30404

CONWAY, K.S., C.J. SCHMIDT, AND T.T. BROWN. "Medical Examiner Review of the Characteristics of Fire-Related Homicides." *Academic Forensic Pathology*, 10 (2): 87–93, 2020 (3 tables, 2 figures, 18 references)

Abstract: Thirty-eight cases of fire-related homicides were identified by the medical examiner in Detroit between 2007 and 2017, of which 30 had post-mortem blood COHb concentrations determined. The mean age of the victims was 41.5. Most victims were black (68%), and most were males (63%). The causes of death were smoke/soot/CO inhalation (n = 24), thermal injuries (n = 20), and/or trauma (n = 9). An accelerant was used to start a fire in four cases (11%). The victims were found dead at the scene or died shortly after in 26 of the cases (68%) and survived a day or more in 12 of the cases (32%). The mean blood COHb concentration was 41%. The median COHb was 46% for inhalation deaths, 16% for thermal injuries, and 24% for trauma.

Table: Postmortem COHb Concentrations in Fire-Related Homicide Victims with Traumatic Injuries

Victim age/gender	Location of body	Soot (Yes/No)	Blood COHb (%)
21-year-old man	House fire	No	Not detected
21-year-old man	Car fire	Yes	56
56-year-old woman	House fire	Yes	2
89-year-old man	House fire	Yes	59
91-year-old man	House fire	Yes	7

Source: Adapted from Conway et al. 2020.

Fire-related homicides present unique diagnostic challenges because multiple insults frequently contribute to the cause of death. Death at the scene and COHb levels about 10% are the most useful factors in establishing smoke and soot inhalation as the cause of death. Some autopsy findings are helpful in establishing or ruling out if smoke and soot inhalation was contributing to or sole cause of death, but an evaluation of the entire circumstances and autopsy findings is necessary.

3.05 ACCIDENTAL CO EXPOSURE DUE TO FIRES

> Wood smoke—in other words, the crude gas driven off from heated wood—is intensely poisonous. The large proportion of oxygen contained in wood and similar material gives rise to the presence of a very large percentage of carbonic oxide [CO]; and it is probable that nearly all the deaths which occur during fires, whether above or below ground are due to carbonic oxide poisoning.
>
> —*British Medical Journal*, 2 July 1889, 33

The main cause of death in house fires is typically due to CO, which after only a few minutes can approach a fatal COHb blood concentration (30510). Other factors such as heat and inhaled soot can contribute to the fire-related death of a victim, as shown by the mean blood COHb concentration of fire-related victims, which was 23.7% compared to 56.2% in non-fire-related CO poisoning victims (30505). A zero blood COHb concentration in fire-related victims may be due to death before the fire started or a flash fire in which death occurred rapidly (30509–10). Children, the elderly, and those with pre-existing medical conditions such as drug/alcohol abuse are especially at risk of death due to CO poisoning in house fires (30502, 30506–7, 30510).

Reference Number: 30501

MURESAN, C.-O., R.E. ZAVOI, R.O. DUMACHE, C.V. PRECUP, V. CIOCAN, O.S. BULZAN, C. FLOROU, AND A. ENACHE. "Co-Morbidities in Multiple Victims of the Silent Killer in Carbon Monoxide Poisoning." *Romanian Journal of Morphology and Embryology*, 60: 125–131, 2019 (14 figures, 45 references)

Abstract: Deaths due to CO poisoning represented about 1.8% of all autopsy cases in Romania between 2000 and 2018. An analysis was conducted of 199 autopsy cases during that time period that involved CO poisoning, including fire deaths. Of the 63 fire victims who died of CO poisoning, soot was found in the upper air passage in 34 cases (54%). Most of the fire victims had a zero BAC (56%), and there was no correlation between COHb and BAC. The red discolouration of the livor mortis is the most consistent indicator of CO poisoning.

Table: Monthly Distribution of Fatal CO Poisoning Cases in Romania between 2000 and 2018

Month	Number of cases
January	33
February	26
March	14
April	11
May	5
June	1
July	2
August	4
September	7
October	9
November	12
December	14

Source: Adapted from Muresan et al. 2019.

Although in most cases (83.36%) death occurred at COHb concentrations over 70%, CO could cause death at concentrations below 50% when pre-existing organic conditions were associated (the autopsy most frequently revealed: myocardial fibrosis, cardiac lipomatosis, pulmonary sclero-emphysema, pulmonary tuberculosis, pulmonary carcinoma, pulmonary fibrosis, pneumonia or hyaline membrane disease—in a 2 week old suckling) Degenerative cardio-pulmonary conditions (myocardial fibrosis and systematic arteriosclerosis) as the most frequent, followed by hepatic dystrophy and steatosis and renal dystrophy. Although multiple deaths cases, because of the similar conditions of death onset, one would expect COHb concentrations to be similar, practice shows that this is not always the case. The COHb blood concentrations of multiple victim incidents can differ with as much as 50% and this can even lead to concluding toward different causes of death within the same incident.

Reference Number: 30502

PURSER, D.A. "Effects of Pre-Fire Age and Health Status on Vulnerability to Incapacitation and Death from Exposure to Carbon Monoxide and Smoke Irritants in Rosepark Fire Incident Victims." *Fire and Materials*, 41 (5): 555–569, 2017 (4 tables, 13 figures, 9 references)

Abstract: Pre-fire health and post-mortem or hospital COHb concentrations were determined in 18 elderly residents (mean age 85 years, range 61–99 years) with a variety of illnesses who had been exposed to a fire at the Rosepark Care Home facility. Ten died in their bedrooms, four died later at hospital, and four who had less exposure to the fire survived. The post-mortem COHb ranged from 55% to 81.8%. COHb concentrations causing a collapse, incapacitation, or unconsciousness is around 30% for active subjects and 40% for resting subjects. Few people survive a COHb of 50% or more, even if treated. At hospital, the COHb for the survivors ranged between 19.6% and 38% and was back-calculated to between 31% and 48% at the time of rescue.

The % COHb levels at incapacitation and at death in elderly and infirm subjects exposed in the Rosepark fire incident were close to those reported for incapacitation in experimental studies of younger, healthy humans and other primates and similar to the averages and range of levels found in younger and elderly human fire fatalities. The results support a rapid increase in probability of incapacitation for all subjects about 25–30% COHb and death above 45–50% COHb. There was an indication that subjects with pre-existing heart disease may have a somewhat increased susceptibility.

Reference Number: 30503

OSHIMA, T., K. YONEMITSU, A. SASAO, M. OHTANI, AND S. MIMASAKA. "Detection of Carbon Monoxide Poisoning That Occurred before a House Fire in Three Cases." *Legal Medicine*, 17: 371–375, 2015 (5 figures, 15 references)

Abstract: A correlation was found between post-mortem blood benzene and COHb concentrations (r^2 = 0.7908) in 38 housefire deaths. Gasoline-involved fire deaths were excluded. In the three house fire cases reported here, the post-mortem blood benzene concentration in the fire victims was much lower than expected, and, therefore, it is likely that death was due to CO poisoning that occurred before the fire. In two cases, a small portable charcoal/kerosene stove was found in the fire debris. The presence of soot in the airways is also a useful sign of exposure to fire before death.

> A high CO-Hb concentration without a hydrocarbon component such as benzene, indicates that the deceased inhaled carbon monoxide that was not related to the smoke from the fire. Comparing volatile hydrocarbons with CO-Hb concentrations can provide more information about the circumstances surrounding fire-related deaths. We are currently convinced that this is the best method to detect if carbon monoxide poisoning occurred before a house fire started.

Reference Number: 30504

ONAL, O., D. CELIK, E. ASLANLAR, A. AVCI, AND J.B. CELIK. "Myocardial Infarction in a Young Patient after Acute Exposure to Carbon Monoxide." *Anesthesia and Pain Medicine*, 6 (3): e33154, 2016 (4 figures, 9 references)

Abstract: A 27-year-old man lost consciousness while fighting a fire in the workplace and went into cardiac arrest. He was resuscitated by first responders and given 100% oxygen. He was unconscious at hospital and intubated. His blood COHb was 22%. He was treated with HBOT and extubated 11 days after admission to ICU.

> This report of a rare case should remind physicians that acute, severe CO intoxication can cause a wide variety of symptoms; thus, cardiovascular investigations, including serial ECG; cardiac necrosis marker measurements, such as troponin; and COHb should be performed in patients with acute CO poisoning in the first hours after CO exposure.

Reference Number: 30505

VREMAN, H.J., R.J. WONG, D.K. STEVENSON, J.E. SMIALEK, D.R. FOWLER, L. LI, R.D. VIGORITO, AND H.R. ZIELKE. "Concentration of Carbon Monoxide (CO) in Postmortem Human Tissues: Effects of Environmental CO Exposure." *Journal of Forensic Sciences*, 51: 1182–1190, 2006 (5 tables, 3 figures, 29 references)

Abstract: Post-mortem CO and COHb concentrations were determined in blood, spleen, lung, kidney, heart, muscle, brain, and adipose tissue after the release of CO using sulfosalicylic acid, in 14 control samples, 13 from victims of fires, and 7 who died due to CO poisoning. Human muscle has five to six times higher myoglobin than rat muscle.

Table: Mean and Range of COHb (%) in Control, Fire Deaths, and CO Poisoning Deaths

Case type	Mean COHb (%)	Range of COHb concentrations (%)
Non-fire, non-CO deaths (controls)	1	0–5
Fire deaths	23.7	1–101.5
CO poisoning deaths	56	16–90

Source: Adapted from Vreman et al. 2006.

> In conclusion, on the basis of the results of the present study, it appears that tissues such as blood, muscle, brain, lung, and kidney are the most suitable index tissues for diagnosing death due to lethal exposure to CO and may be used to differentiate fire only related deaths from those due to CO inhalation before fire exposure. Spleen, heart and adipose appear to be least suitable.

Reference Number: 30506

HOMER, C.D., D.A. ENGELHART, E.S. LAVINS, AND A.J. JENKINS. "Carbon Monoxide-Related Deaths in a Metropolitan County in the USA: An 11-Year Study." *Forensic Science International*, 149: 159–165, 2005 (1 table, 1 figure, 19 references)

Abstract: A study of the post-mortem reports for all deaths at the Cuyahoga County Coroner's Office in Cleveland between 1988 and 1998 showed 209 accidental CO deaths due to fires and 182 suicidal CO deaths. The majority of fire deaths occurred in the city of Cleveland, but the majority of suicides occurred in the suburbs and occurred mainly in the months of December and April respectively. In 1992, there were 4.2 fire deaths per 100,00 people, which decreased to 0.8 per 100,000 people in 1998 after a community-wide effort to provide free smoke detectors.

Table: Comparison of Accident Fire and Suicidal CO Deaths by the Main Category Most Represented

Category	Accidental fire	Suicide
Gender	Male	Male
Race	Black	White
Age (years)	0–9	70–9
Location	City	Suburbs
Month	December	April

Source: Adapted from Homer et al. 2005.

Current research has shown that children and the elderly are particularly at risk of death by CO poisoning due to fires in the home. Surviving a fire depends upon a variety of factors including escaping the home when it is on fire. Those victims with disabilities or those who are dependent upon others for their caregiving have less of an opportunity escaping a fire in the home than those who are independent and in good health. It follows from this that children are at a greater risk of death due to fire in the home due to their dependence on caretakers. Similarly, elderly individuals not only can be highly dependent on caretakers, they may also suffer from various disabilities which could preclude a faster escape from the fire and make them more susceptible to succumbing to CO poisoning.

Reference Number: 30507

DRAGELYTE, G., J. PLENTA, S. CHMIELIAUKAS, A. JASULAITIS, R. RAUDYS, T. JOVAISA, AND R. BADARAS. "Myocardial Rupture Following Carbon Monoxide Poisoning." *Case Reports in Critical Care*, 2014 (1): 281701, 2014 (1 figure, 8 references)

Abstract: An 83-year-old woman was hospitalized four hours after an attempt to extinguish a fire in her house. Her blood COHb was 15.4%, and, due to the low result, she was administered NBOT, intravenous fluids, and antibiotics. The next day, she suffered a sudden cardiovascular collapse, became progressively hypotensive, and suffered an asystolic cardiac arrest. All attempts at resuscitation failed.

Diagnosis following postmortem was as follows: acute poisoning with carbon monoxide and other products of combustion. Complications were myocardial infarction with myocardial rupture and pericardial tamponade. Concomitant disease was first degree burns of the nose and upper airway.

Reference Number: 30508

HIRSCH, C.S., AND L. ADELSON. "Absence of Carboxyhemoglobin in Flash Fire Victims." *Journal of the American Medical Association*, 210: 2279–2280, 1969.

Abstract: Five victims of flash fires or explosions were found to have no significant post-mortem COHb.

> As in explanation for the findings in our cases and those reported by James, it is postulated that initial exposure of the upper respiratory tract to an intense flash of heat can cause impregnable laryngospasm or immediate respiratory arrest. Thus, the victim's first breath of the products of combustion is his last and effective ventilation ceases before a significant amount of carbon monoxide containing atmosphere reaches the lungs. Vagal reflexes could mediate a response of this nature. Additional mechanisms should also be considered. An extremely well-ventilated fire or one occurring out of doors may have its carbon monoxide blown away from the victim. The same reasoning may apply to an individual who is running while his clothing burns.

Reference Number: 30509

SCHWERD, W., AND E. SCHULTZ. "Carboxyhemoglobin and Methemoglobin Findings in Burnt Bodies." *Forensic Science International*, 12: 233–235, 1978 (1 table, 8 references)

Abstract: Since the Vienna Ringtheater fire in 1881, COHb findings are considered to be the most important indicators that burn injuries occurred before death. The post-mortem COHb and methemoglobin concentrations were determined in eight victims of a smoldering bowling alley fire. In another case, a 16-year-old student doused herself with gasoline in a school bathroom and set herself on fire. An immediate explosion occurred, and the bathroom window was blown 20 metres away. Her post-mortem blood COHb was zero.

Table: Survival Time and Post-Mortem COHb and Methemoglobin Concentrations in Eight Victims of a Bowling Alley Fire

Case number	Survival time	COHb (%)	Methemoglobin (%)
1	Minutes	0	0.7
2	Minutes	10	0
3	Minutes	15	5
4	Minutes	32	0
5	1.5 hours	18	37
6	54 hours	–	5
7	57 hours	–	1.4
8	13 days	–	7.3

Source: Adapted from Schwerd and Schultz 1978.

There must have occurred not only sudden cessation of breathing, but also arrest of the circulation. This can be explained in that from the explosion-like outbreak of the fire in both cases, very hot gases influenced the outcome, thereby causing what is normally understood as burn shock. Included in the decidedly reflex occurrence cessation of breathing due to inhalation of very hot gases and heart failure cause by intense vagus irritation. This best explains the absence of carbon monoxide in the blood and the relatively small changes in the upper respiratory tract, as well as the absence of deep soot inhalation.

Reference Number: 30510

SIMONSEN, C., K. THORSTEINSSON, R.N. MORTENSEN, C. TORP-PEDERSEN, B. KIAERGAARD, AND J.J. ANDREASEN. "Carbon Monoxide Poisoning in Denmark with Focus on Mortality and Factors Contributing to Mortality." *PLoS ONE* 14 (1): e0210767, 2019 (1 table, 5 figures, 33 references)

Abstract: Between 1995 and 2015, 22,930 patients suffered from CO poisoning in Denmark, and 21,138 of these patients (92%) were hospitalized. A total of 2,102 victims (9.2%) died within 30 days after exposure to CO. Most of the victims (85%) were found dead at the scene, while the remainder (15%) died during hospitalization. Of the 1,132 deaths due to gas/vapours, 8.3% were accidental, and 91.9% were intentional. Of the 825 deaths due to smoke/fire CO exposure, 87% were accidental, and 7.2%

were intentional. HBOT was not associated with survival benefits compared to those who received no HBOT.

Table: A Comparison of Those Who Died of CO Poisoning at the Scene and Those Who Died at Hospital

Variable	Dead at scene	Died during hospitalization
Mean age	52.6 years	39.3 years
Drug abuse (%)	17.4	10.3
Alcohol abuse (%)	21.8	12.2
Psychiatric disease (%)	12.9	8.3
Stroke (%)	5.1	3.3
Heart failure (%)	3.1	1.9

Source: Adapted from Simonsen et al. 2019.

> Poisoning from smoke and/or CO is a frequent occurrence in Denmark, accounting for numerous contacts with hospitals and numerous deaths every year. Suicide and suicide attempts contribute heavily to these numbers. Both intoxication and mortality are highly associated with co-morbidities, especially conditions that might interfere with cognitive and physical function. treatment with HBO is not associated with an effect on short-term or long-term survival after adjusting for CO-morbidities and age.

3.06 CHILDREN

> A child is more at risk for carbon monoxide poisoning if they live in a house with any of these:
>
> - An appliance powered by oil, wood, gasoline, natural gas, propane or coal
> - A space heater
> - A faulty fireplace and chimney.
>
> —University of Rochester Medical Centre,
> *Carbon Monoxide Poisoning in Children* (2025)

Children are more susceptible to CO poisoning than adults due to their smaller size, faster breathing, and metabolic rates, and their developing bodies are more vulnerable to the effects of oxygen deprivation (30623).

Reference Number: 30510

A common symptom of CO poisoning in children is vomiting (30601, 30613, 30619, 30621). Oxygen therapy and even HBOT has been proven safe for children (30602, 30604–7, 30609, 30616–17, 30619–22). Delayed neurological sequelae occurred in 11–17% of the children poisoned by CO (30601, 30603, 30604, 30606, 30613). Endogenous CO concentrations in infants were not associated with sudden death (30618).

Reference Number: 30601

CHANG, Y.-C., H.-Y. LEE, J.-L. HUANG, C.-H. CHIU, C.-L. CHEN, AND C.-T. WU. "Risk Factors and Outcome Analysis in Children with Carbon Monoxide Poisoning." *Pediatrics and Neonatology*, 58: 171–177, 2017 (4 tables, 1 figure, 32 references)

Abstract: Between 2002 and 2010, 81 children (mean age 6.2 years) were treated at a tertiary medical centre in Taiwan for CO poisoning. The mean COHb was 21% (range 7–37%). The most common sources of exposure were indoor heating systems (54.3%) and improperly vented heater exhaust. The most common presenting symptoms were vomiting (32.1%) and consciousness changes (30.9%). Twenty-one patients (26%) received HBO_2 treatment. Delayed neurological sequelae occurred in 12 patients (15%), and eight patients had permanent neurological sequelae (PNS).

> In conclusion, many effects of CO resemble those associated with various acute illnesses in childhood, making it easy to overlook the possibility of CO intoxication. Although the mortality rate and rate of developing DNS in CO poisoning in the pediatric group are lower than those in the adult group, significant neurological sequelae occur in some children. CO may have a unique effect on the maturing nervous system. Special attention should be given to, and follow-up should be performed on, patients treated in the ICU because of prolonged loss of consciousness and rescue by a ventilator to determine whether DNS [delayed neurological sequelae] or PNS may occur, particularly epilepsy and cognitive deficits.

Reference Number: 30602

SAC, R.U., M.A. TASAR, I. BOSTANCI, Y. SIMSEK, AND Y.B. DALLAR. "Characteristics of Children with Acute Carbon Monoxide Poisoning in Ankara." *Journal of Korean Medical Science*, 30: 1836–1840, 2015 (3 tables, 18 references)

Abstract: Twenty-six girls and 54 boys (ages 1 month to 16 years; mean age, 8.3 years) attended hospital as a result of CO poisoning. Most of the children were hospitalized during the winter months (66.3%) and none in the summer months. The sources of CO were coal or wood stoves (71.2%) and water heaters (28.8%). The stoves and water heaters were used improperly or unsafely. For example, the stove ventilation was closed at night to reduce coal use. The median COHb concentration was 19.5%. The most common complaints were headache and vertigo (59%). One child was treated with HBO_2, the others were administered high-flow, 100% oxygen using a tight-fitting mask. All the children recovered, although seven (8.8%) indicated that their academic marks had declined.

In conclusion, acute CO poisoning in Ankara arises mainly from stoves or hot water heaters used at homes of low economic status. Instructive assistance given to the families after the poisoning was effective to recent recurrent poisoning. This study emphasizes that all people who have risks of CO poisoning should be educated to prevent poisoning. In countries that uncontrolled heating systems like coal stoves and hot water heaters are being used, there is a need to take environmental and policy controls to curb this public health problem.

Reference Number: 30603

CHO, C.-H., N-C. CHIU, C-S. HO, AND C-C. PENG. "Carbon Monoxide Poisoning in Children." *Pediatric Neonatal*, 49: 121–125, 2008 (3 tables, 28 references)

Abstract: A study was conducted of 30 children, ages 2–17 years old (mean 9.5 years), attending a hospital in Taiwan between 1996 and 2007 as a result of CO poisoning. The mean initial COHb was 35.3% (range 7.4–60.3%). Thirteen patients required intensive care, including six who required mechanical ventilation and two who received inotropic agents due to hypotension. The most common sources of the CO poisoning were water heaters (16 patients, 53.3%), house fires (8 patients, 26.7%), and intentional poisoning by the parents using charcoal burning (7 patients, 20%). One child died, and five (16.7%) had delayed neurologic sequelae (DNS), but all recovered within two months. Child CO poisoning victims tend to have less incidence of DNS than adults. Only one patient was administered HBO_2. Children with DNS were more likely to have seizures, severe metabolic acidosis, and hypotension than adult victims.

Table: Characteristics of Children with CO Poisoning from Fires and Non-Fires

Characteristics	House fire (n = 8)	Non-Fire (n = 22)
Mean age (years)	7.8	10.2
Mean initial GCS	9.0	12.2
Mean initial COHb (%)	37.3	34.6
Number who had severe metabolic acidosis	5	1
Consciousness recovery time (hours)	27.4	8.1
Mean hospital stay (days)	8.1	4.0
Mean ICU stay (days)	2.5	0.4

Source: Adapted from Cho et al. 2008.

In conclusion, pediatric CO poisoning in Taiwan is most commonly caused by improperly vented heater exhaust and usually occurs in cool weather. Victims of house fires tend to have more severe clinical courses and need intensive care. Intention al CO poisoning by parents using charcoal burning is an important means of child abuse. Regular monitoring of high-risk groups and prompt interventions are mandatory. DNS in children is more likely to occur in patients with hypotension, seizures, and severe metabolic acidosis. It is usually self-limited and resolves without the use of hyperbaric oxygen.

Reference Number: 30604

GANDINI, C., A.F. CASTOLDI, S.M. CANDURA, S. PRIORI, C. LOCATELLI, R. BUTERA, C. BELLET, AND L. MANZIO. "Cardiac Damage in Pediatric Carbon Monoxide Poisoning." *Clinical Toxicology*, 39: 45–51, 2001 (1 figure, 47 references)

Abstract: A 12-year-old boy was admitted to the ED as a result of a faulty heating system in his home. The rescuers found the father dead (postmortem blood 57% COHb) and his mother with a headache (for one week), nausea, and drowsiness (24.5% COHb). The boy complained of a persistent headache and had no other symptoms but tachycardia and diffuse ST segment elevation. His initial COHb was 13% and was administered 100% oxygen and HBO_2 until his COHb was 0.3%. He suffered from cardiac damage despite the mild symptoms and low COHb. Myocardial and mitral valve dysfunctions were observed suggesting an ischemia-like syndrome. The heart is extremely sensitive to CO-induced hypoxia due to its high oxygen demand. Cardiac damage was completely reversible after one month.

This case report supports that a prolonged carbon monoxide exposure can cause cardiac damage in children even in the absence of specific symptoms, cerebral failure, and high carboxyhemoglobin concentrations.

Reference Number: 30605

DE SUREMAIN N., J. NGO, S. LOSCHI, L. HAEGY-DOEHRING, J. AROULANDOM, AND R. CARBAJAL. "Carbon Monoxide Poisoning from Waterpipe (Narghile) Smoking in a Child." *Archives de Pediatrie*, 26: 44–47, 2019 (1 figure, 19 references)

Abstract: A 13-year-old boy attended hospital as a result of an injury he suffered on his chin while losing consciousness for five seconds. Twenty minutes after arrival, he again lost consciousness and had tonic-clonic movements of his upper limbs. One hour after his arrival, his blood COHb was 23.1%, and he was administered oxygen. The source of the CO was due to smoking a waterpipe or shisha. Shisha is considered to be non-toxic by 22% of smokers. The mouthpiece of the shisha pipe is seldom changed and, therefore, is a risk of transmission of diseases such as tuberculosis.

Shisha smoking should not be trivialized. Surveys have demonstrated that it is very popular among adolescents, including younger adolescents who for the most part are unaware of the risks involved. The risks of CO poisoning are related firstly to high levels of CO in shisha smoker, which are much more substantial than those contained in cigarettes, and secondly to poor ventilation on the premises used for smoking. It is of utmost importance to ensure that public information campaigns are organized to warn young people against those understated health hazards.

Reference Number: 30606

HERMAN, L.Y. "Carbon Monoxide Poisoning Presenting as an Isolated Seizure." *Journal of Emergency Medicine*, 16: 4290–4320, 1998 (20 references)

Abstract: A three-year-old boy was brought to the ER due to a seizure. It was determined that the child lived in the same dwelling in which two adults and one four-year-old child were admitted to the ER with blood COHb concentrations of 27.8, 30.0, and 23.8% respectively, probably due to a recently serviced gas heater. The three-year-old boy had a blood COHb of 24.2% and was treated with HBOT. All patients had no significant sequelae seven months later. Seizures are generally thought to occur at COHb of 40% or greater.

> This case report describes the presentation of a child with an isolated seizure found to have carbon monoxide poisoning at levels not previously thought to be associated with serious toxicity. A review of the literature revealed that the occurrence of seizures cannot be correlated with any particular level of COHb. In conclusion, our best tool for making the diagnosis of carbon monoxide toxicity in seizure patients remains consideration of the possibility.

Reference Number: 30607

ONVLEE-DEKKER, I.M., A.C.H. DE VRIES, AND A.D.J.T. HARKEL. "Carbon Monoxide Poisoning Mimicking Long-QT Induced Syncope." *Archives of Disease in Childhood*, 92: 244–245, 2007 (2 figures, 8 references)

Abstract: A nine-year-old girl became unconscious after sitting on the swim platform of an idling ski boat. She was administered 100% oxygen by facemask before being taken to hospital. At hospital, she had an initial COHb of 25%. She had a prolonged QTc interval according to a 12 lead ECG, which normalized after NBOT.

> Our case history describes a patient with unexplained syncope and reversible increases in QTc and QTc dispersion related to CO intoxication. The CO poisoning resulted from direct exposure to CO in the exhaust of a ski boat. Making the diagnosis without delay had important therapeutic implications, as administration of 100% oxygen showed the COHb formation to decrease rapidly, with concurrent normalization of the ECG.

Reference Number: 30608

VENNING, H., D. ROBERTSON, AND A.D. MILNER. "Carbon Monoxide Poisoning in an Infant." *British Medical Journal*, 284 (6316): 6, 1982 (5 references)

Abstract: A 13-week-old baby girl travelled with her family for six hours in a carry cot placed in the back seat of a car. The child was hypotonic and vomited. At hospital, she had convulsions. Her initial COHb was 60% and, after being administered 100% oxygen, declined to 25% and 10% at 5 and 17 hours after exposure respectively. She was unconscious for about 48 hours and had convulsions off and on for 18 days, which were difficult to control, Six weeks later, all of her symptoms had resolved. A hole in the floor of the car caused by a minor collision allowed exhaust gas to leak into the car. The father (a smoker) had a blood COHb of 15%, and her mother had 10% COHb.

> When unexplained neurological symptoms and signs occur in an infant who has been a passenger in a vehicle the COHb saturation must be measured. Laboratory diagnosis is simple and quick and treatment with 100% oxygen or hyperbaric oxygen will reduce the half life of COHb and improve tissue and cellular oxygenation.

Reference Number: 30609

YARAR, C., A. YAKUT, A. AKIN, B. YILDIZ, AND E.C. DINLEVICI. "Analysis of the Features of Acute Carbon Monoxide Poisoning and Hyperbaric Oxygen Therapy in Children." *Turkish Journal of Pediatrics*, 50: 235–241, 2008 (6 tables, 30 references)

Abstract: A study was conducted of 43 girls and 31 boys (mean age 9.5 years) hospitalized due to CO exposure in Turkey between 2003 and 2005. All patients received normobaric oxygen (NBO_2) until the COHb concentration was close to 2%, and 38 patients were also administered HBOT as well. The median blood COHb was 26.3% at admission (range 6.0–56.0%). Fifty-four percent of CO exposure was due to coal fires and 19% to gas water heaters. One patient had a cardiopulmonary arrest, while six patients had no symptoms of CO intoxication.

> In conclusion, especially in winter because of the poorly functioning heating systems, acute COP has continued as an important health problem in adults and children in our country. The spectrum of COP symptoms may be wide and variable in children. Detailed history, physical examination, and suspicion are important for the diagnosis of COP. We suggest that HBO therapy can be used safely in children with COP. It might also be effective in preventing the delayed neurologic sequelae.

Reference Number: 30610

GEMELLI, F., AND R. CATTANI. "Carbon Monoxide Poisoning in Childhood." *British Medical Journal*, 291: 1197–1198, 1985

Abstract: Poorly maintained gas devices and badly ventilated houses give rise every winter to undiagnosed CO poisoning. Three case reports are presented in which two boys and one girl (ages seven to nine years) who were originally diagnosed with gastroenteritis, and had COHb of 16%, 18%, and 32% respectively. In addition to the children, 12 adults in the same households were found to have COHb concentrations of 23–40%.

> We found that the carboxyhemoglobin concentrations of three children were considerably lower than those of 12 adults (22% v. 35.6%), so that children must be particularly susceptible to carbon monoxide poisoning. These two features—the different symptoms and the earlier presentation—often make it difficult to diagnoses the condition in childhood.

Reference Number: 30611

MENDOZA, J.A., AND N.B. HAMPSON. "Epidemiology of Severe Carbon Monoxide Poisoning in Children." *Undersea and Hyperbaric Medecine*, 33: 439–446, 2006 (7 tables, 2 figures, 16 references)

Abstract: Between 1979 and 1994, 38.2% of poisoning deaths in children ages 10 to 19 years old were due to CO. This detailed study was conducted on 250 consecutive children (ages up to 17 years; mean age 8 years) who received HBOT for CO exposure at a Seattle hospital between 1978 and 2004. Their mean initial blood COHb was 19.0% (range 0.3–57.6%). Most of the poisonings were unintentional (94%), and, of the remaining 6% intentional cases, most involved an adolescent male using a car for the CO source. The highest proportion of CO poisonings occurred between November and January (41%).

Table: CO Sources in Pediatric Poisonings by Percentage in Each Age Group

	0–2 years (%)	3–5 years (%)	6–10 years (%)	11–13 years (%)	14–17 years (%)
Motor vehicle	21.4	31.5	36.2	38.9	38.8
Charcoal	40.4	13.0	30.4	16.7	18.4
Fire	4.8	20.3	10.1	8.3	10.2
Generator	4.8	0	0	8.3	12.2
Furnace	11.9	9.3	8.7	8.3	8.2
Boat	7.1	14.8	5.8	8.3	0
Other	9.5	11.1	8.7	11.1	12.2

Source: Adapted from Mendoza and Hampson 2006.

Severe pediatric CO poisoning demonstrates a seasonal pattern with key sources of exposure varying by gender, age, and race/ethnicity. Minorities are at increased risk for severe CO poisoning. Intentional CO poisoning occurs primarily among adolescent males and by motor vehicle exhaust. Prevention efforts should take these factors into consideration.

Reference Number: 30612

MORI, T., AND K. NAGAI. "Carbon Monoxide Poisoning Presenting as an Afebrile Seizure." *Pediatric Neurology*, 22: 330–331, 2000 (8 references)

Abstract: A one-year-old boy presented at hospital because of an afebrile seizure and loss of consciousness. His father had trouble starting the car as it was surrounded by snow. The child in the back seat was exposed to CO due to a blocked exhaust even though the driver's door was open. The father had no signs of CO poisoning, but the child had a COHb of 25%, metabolic acidosis, and diffuse swelling of the brain. The child was administered HBOT. He recovered uneventfully and showed no neurological deficits eight months later.

The CO-intoxicated patient should be removed from the exposure site as early as possible and 100% oxygen (hyperbaric oxygen chamber at 2–3 atm if possible) should be administered immediately. Hyperbaric oxygen therapy is used to decrease the half-life of CO-Hb. In addition, this therapy has been thought to exert a neuroprotective effect by inducing vasoconstriction and reducing cerebral edema while still enhancing oxygen delivery.

Reference Number: 30613

CROCKER, P.J., AND J.S. WALKER. "Pediatric Carbon Monoxide Toxicity." *Journal of Emergency Medicine*, 3: 443–448, 1985 (2 tables, 11 references)

Abstract: A study was conducted of 28 pediatric patients (< 14 years of age) exposed to CO who were attending the Darnall Army Community Hospital in Fort Hood, Texas, between 1983 and 1984. Blood COHb concentrations were determined with an IL-282-CO-oximeter. Twelve of the patients had COHb less than 15% and were completely asymptomatic. Sixteen patients had a blood COHb of 15% or greater, and every patient in this group experienced nausea, 93% had headache, 75% had nausea, and 69% were lethargic. All patients with a COHb greater than 25%, neurologic findings, acidosis, or syncope were administered HBOT. Eighty-nine percent of the patients had no sequelae. Delayed neurological sequelae occurred in three patients and included chronic headaches, memory deficits, or a decline in school performance.

The ED physician must be alert to the possibility of toxicity. Contrary to expectation, clinical signs such as tachypnea and tachycardia may not be of value in the early identification of those with CO toxicity. Pediatric patients seem to have a lower COHb threshold than adults for both lethargy and syncope. Even minor alternations in the level of consciousness on arrival to the ED may be associated with neuropsychiatric sequelae and should be considered an indicator for HBOT. Patients presenting with CO toxicity under similar circumstances to ours who are treated aggressively with HBOT can be expected to be free from permanent sequelae.

Reference Number: 30614

ESTRELLA, B., R. ESTRELLA, J.K. OVIEDO, X. NARVAEZ, M.T. REYES, M. GUITIERREZ, AND E.N. NAUMOVA. "Acute Respiratory Diseases and Carboxyhemoglobin Status in School Children of Quito, Ecuador." *Environmental Health Perspectives*, 113: 607–611, 2005 (3 tables, 1 figure, 37 references).

Abstract: There is a strong association between acute respiratory infections (ARI) and traffic-related air pollution. The COHb concentration was determined in 295 children (mean age 8.4 years) attending three schools with different traffic densities (low, moderate, and high) in Quito, Ecuador. The mean COHb detected was 2.81%, and a blood COHb concentration greater than the safe 2.5% level was 46.4%.

Table: Mean Blood COHb and Percent of Children with a COHb Greater Than 2.5% in Low, Moderate, and High Traffic Density Schools in Quito, Ecuador

	Low density	Moderate density	High density
Mean Blood COHb (%)	0.7	2.52	5.09
Percent of children with COHb > 2.5%	1	43	92

Source: Adapted from Estrella et al. 2005.

The main finding of the study was that COHb concentrations elevated because of traffic pollution correlate with the occurrence of ARIs in young children. A high COHb level was associated with at least one additional case of ARI in a 12-week period of a 3-fold increase in the annual rate of ARI incidence. These associations remain after adjusting for age, sex, weight, height, BMI [body mass index], hematocrit level, previous history of asthma, presence of smokers, kerosene and/or wood smoke use for cooking and level of crowdedness. Our findings imply that exposure to a high level of CO, the primary reason for increased COHb concentrations may lead to increased susceptibility to ARIs.

Reference Number: 30615

GRANT, M.J.C., AND B. CLAY. "Accidental Carbon Monoxide Poisoning with Severe Cardiorespiratory Compromise in 2 Children." *American Journal of Critical Care*, 11: 128–131, 2002 (30 references)

Abstract: Two three-year-old identical twin girls were exposed to a faulty furnace producing high levels of CO for at least several hours. Their COHb concentrations one to two hours after exposure and oxygen administration were 10.8% and 4.8% respectively. Both girls were in shock with resulting cardiopulmonary collapse and experienced delayed supraventricular tachycardia on day 4 and day 11 respectively. Both patients were treated aggressively with volume resuscitation and alkalinization. Both did not appear to have suffered any long-term neurological sequelae one year later.

In the presented cases, cardiopulmonary compromise and tachyarrhythmias occurred after carbon monoxide poisoning in a pair of 3 year old patients. Carbon monoxide reversibly binds to hemoglobin with an affinity approximately 240 times greater than that of oxygen, thus reducing the total oxygen carrying capacity of the hemoglobin. This competitive binding shifts the oxygen-hemoglobin dissociation curve to the left, resulting in impaired release of oxygen at the tissue level and cellular hypoxia. Myoglobin and mitochondrial cytochrome c oxidase are other heme containing proteins that bind carbon monoxide reducing muscular oxygen stores, possibly leading to poor myocardial function. The affinity of myoglobin for carbon monoxide is 30 to 50 times greater than its affinity for oxygen, whereas cytochrome c oxidase binds carbon monoxide and oxygen equally.

Reference Number: 30616

KLASNER, A.E., S.R. SMITH, M.W. THOMPSON, AND A.J. SCALZO. "Carbon Monoxide Mass Exposure in a Pediatric Population." *Academic Emergency Medicine*, 5: 992–996, 1998 (1 table, 11 references)

Abstract: A mass CO exposure of 504 children (ages 4 to 12 years) occurred over a period of three to four hours on 5 January 1996 at an elementary school in St. Louis, Missouri, and all the children were evacuated. Only 177 children with complaints or showed symptoms of CO exposure were taken to one of three area hospitals and treated with 100% oxygen via a mask. A COHb concentration was determined for 147 children, and the mean COHb was 7.0% (95% confidence interval, 6.6–7.5%). A second COHb result was obtained for 26 children, and the man was 2.7% (95% CI, 2.2–3.2). The mean half-life of CO with 100% oxygen was 44 minutes.

> In this mass CO exposure, we found that children who obtained medical care had symptoms of CO poisoning at lower COHb levels than previously reported. This suggests that children with CO poisoning may have symptoms without corresponding high COHb levels. We found that elimination half-life of COHb in our subjects more rapid than previous studies. All but 1 of these children recovered and went home after O_2 therapy. We believe that children who present to an ED with symptoms of CO poisoning and without significantly elevated COHb levels are likely to respond to O_2 treatment and be released home.

Reference Number: 30617

MEWS, J., K. KROLIKOWSKA, J. SIEWIERA, A. WAWRZYNIAK, AND B. KALICKI. "Assessment of Hyperbaric Oxygen Efficacy in Children with Carbon Monoxide Poisoning." *Pediatria i Medycyna Rodzinna*, 15: 157–163, 2019 (4 tables, 6 figures, 27 references)

Abstract: Nine girls and eight boys (mean age 9 years) attending a Polish hospital as a result of CO poisoning during 2018 were evaluated. The main source of CO poisoning in 14 (82%) of the cases was a faulty gas water heater in the bathroom. The children were divided into eight with mild poisoning symptoms and those with severe poisoning symptoms. The mean COHb concentration before HBOT was 16.4%, and there was no statistically significant difference between the two groups, although the median COHb was 12.4% with mild symptoms and 15.7% in the severe symptoms groups. A significant negative correlation was found between baseline COHb and capillary pH. One session of HBOT at 2.5 atmospheres for all children regardless of symptoms was sufficient to achieve a normal COHb concentration. Children are more susceptible to the negative effects of CO due to their higher respiration and metabolism rates.

Table: Number of Points for Each CO Symptom in Children (Mild 0–3, Moderate 4–6, and Severe > 6)

Symptom	Points
No complaints	0
Headache/dizziness	1
Muscle pain	1
Nausea/vomiting	1
Chest pain	2
Disorientation/confusion	2
Collapse	4
Confusion/loss of consciousness	7

Source: Adapted from Mews et al. 2019.

The results obtained in this study provide a clear indication of the efficacy of hyperbaric oxygen therapy. Apart from high efficacy, this method is characterized by high safety. In patients who underwent HBO therapy, apart from one case complicated with pneumonia, no delayed sequelae of CO poisoning were observed.

Reference Number: 30618

VARIEND, S., AND A.R.W. FORREST. "Carbon Monoxide Concentrations in Infant Deaths." *Archives of Disease in Childhood*, 62 (4): 417–418, 1987 (5 references)

Abstract: The post-mortem COHb concentrations were determined in 50 infants (ages one week to two years of age) who died in South Yorkshire in 1985. There were 12 hospital deaths with ascertainable causes of death (Group A), 11 explained deaths at home (Group B), and 27 unexplained deaths at home (Group C). Carbon monoxide is produced endogenously by the body mainly due to hemoglobin degradation. But any disease that interferes with the diffusion through the lungs can cause elevated COHb concentrations. Two children in hospital had the highest COHb results due to pneumonia and alveolar immaturity respectively.

Table: Mean and Range of Post-Mortem Blood COHb Concentrations in Infants with Known and Unknown Causes of Death

Death	Mean blood COHb (%)	Range of blood COHb (%)
Known hospital	2.9	< 0.5–6.7
Known home	2.2	< 0.5–4
Unknown home	1.9	< 0.5–4/7

Source: Adapted from Variend and Forrest 1987.

The lack of significant difference in carboxyhemoglobin values between the explained and unexplained deaths and between home and hospital deaths argues against carbon monoxide poisoning being a significant factor in the pathogenesis of sudden infant death.

Reference Number: 30619

PALLA, S., J. KUMAR, A. HUSSAIN, AND K. MUKHOPHYAY. "Carbon Monoxide Poisoning in a Neonate." *Indian Journal of Pediatrics*, 86 (8): 754, 2019 (4 references)

Abstract: A 14-day-old boy was brought to an ER due to unresponsiveness, poor feeding, and vomiting for four hours. He and his mother and grandmother were exposed to CO from a coal room heater in a poorly ventilated room. The mother and grandmother were exposed for two to three hours, experienced headaches, and lost consciousness. The neonate was exposed for four to five hours. His initial blood COHb was 10.6%, and he was administered 100% oxygen and intravenous fluids and regained consciousness after three to four hours. At eight hours, his blood COHb was 0.1%.

Carbon monoxide (CO) poisoning in a neonate is an uncommon and unrecognized entity. It poses higher risk to neonates that adults and older children due to non-specific clinical presentation and higher levels of fetal hemoglobin (HbF) in neonates.

Reference Number: 30620

COMFORT, R.J., AND J. DAVELER. "Carbon Monoxide Poisoning in an Elementary School: Report of a Municipal Investigation." *Journal of Environmental Health*, 39: 411–415, 1977 (2 figures, 15 references)

Abstract: On 14 November 1975, 24 grade 6 children were overcome by CO in the school gym. The symptoms included flushed face, drowsiness, and severe headaches, and several students were unconscious. All were promptly evacuated and were administered oxygen. Four had blood samples collected and had COHb between 9% and 16%. The source of the CO was a malfunctioning 20-year-old oil burner heater whose exhaust stack was only three feet higher than the two gravity exhaust vents, which under high wind conditions caused the exhaust to enter the gym. In addition, cars parked near the ground-level intake air duct could have caused CO to enter the gym as well. The ambient CO concentration at the burner stack was 600 ppm, and 9–14 ppm was recorded in the gym.

> It became increasingly apparent to investigators, after all pertinent data had been collected, that factor interactions involving meteorological parameters including high velocity wind conditions and turbulence, the environmental design features of the building, and CO emissions from a malfunctioning oil burner were responsible for the onset of illness in 24 children at the elementary school. Forceful downdrafts of air channeled excessive smoke emissions from the boiler stack through the various ventilation ducts, thus affording heretofore unsuspected means of CO entry into the building. Vehicular CO entry through the gymnasium wall air intake ductwork was thought to be a possibility; however idling motor vehicles were not reported in the vicinity on the day the symptoms occurred.

Reference Number: 30621

GUVEN, D., AND D. SARICI. "Clinical and Laboratory Characteristics Predicting the Severity of Carbon Monoxide Poisoning in Children: A Single-Centre Retrospective Study." *Pediatric Emergency Care*, 39: 207–215, 2023 (4 tables, 2 figures, 54 references)

Abstract: The characteristics of 380 children (mean age 8.9 years) poisoned by CO and attending a hospital in Ankara, Turkey, between 2017 and 2021 were compared to 380 healthy controls. The CO-poisoned children were divided into 189 mild (COHb 10%), 111 moderate (10–25%), and 89 severely (COHb > 25%) poisoned patients. Coal stoves were the most common source of exposure (38.6%), followed by natural gas (28.6%) and natural gas water heaters (10%). A total of 91.3% of the children were treated by HBOT. No deaths or sequelae occurred.

Table: Clinical Symptoms of Severe, Moderate, and Mild CO Poisoned Children as a Percentage

Symptom	Severe CO poisoning (%)	Moderate CO poisoning (%)	Mild CO poisoning (%)
Nausea and vomiting	60.0	55.0	39.7
Syncope	45.0	2.7%	5.8
Vertigo	38.8	31.5	22.2
Confusion	36.3	5.4	0.0
Headache	28.7	36.0	12.2
Dyspnea	8.8	5.4	0.0

Source: Adapted from Guven and Sarici 2023.

In conclusion, COP [CO poisoning] must always be recognized in the diagnosis of headache, changes in behaviour, seizures, or syncope in the absence of an origin, particularly in winter or when there is a history of using gas- or coal-burning devices inside. Our study has shown that COP progressed more severely in children presented with neurological symptoms and elevated RDW [red blood cells distribution width] and MPV [mean platelet volume]. Sometimes, no correlation was found between COHb levels and clinical findings, and normal COHb levels do not exclude COP. From this perspective, the definition of additional laboratory markers such as MPV and RDW in terms of determining the severity of COP may contribute to the literature in establishing a new method for disease scoring and would impact patient care or prognostication. Even in severe COP cases, satisfactory results have been obtained with early and appropriate treatment.

Reference Number: 30622

LIEBELT, E.L. "Hyperbaric Oxygen Therapy in Childhood Carbon Monoxide Poisoning." *Current Opinion in Pediatrics*, 11: 259–264, 1999 (4 tables, 30 references)

Abstract: HBOT was first used to treat CO poisoning in the 1960s. Unfortunately, published studies on the use of HBOT in children and infants is sparse. Acute signs and symptoms of CO poisoning in older children is similar to adults. But infants and children have an increased susceptibility to CO toxicity due to their higher basal rate (that is, an increase in oxygen demand) and their developing nervous systems. In one study of 101 Korean children poisoned by CO, the rate of DNS was

10%. DNS was more likely to occur in children who presented at hospital with coma.

Hyperbaric oxygen therapy should be administered to any infant or child with severe signs and symptoms of CO toxicity. Children may be more susceptible to the detrimental neurologic effects of CO because of their developing central nervous systems and increased metabolic rates. They may have mild or subtle neurologic symptoms that warrant consideration of HBOT to prevent neuropsychological sequelae. Ultimately, the best treatment of CO poisoning is prevention. It is hoped that increasing public awareness about the sources and dangers of CO will decrease the need for HBOT as well as the overall morbidity and mortality caused by CO.

Reference Number: 30623

EL KHASHAB, M., AND F. NEJAT. "Hemorrhagic Cerebral Infarction in Carbon Monoxide Poisoning: A Case Report." *Cases Journal*, 2 (1): 96, 2009 (3 figures, 5 references)

Abstract: Infants and children may be at a higher risk of CO toxicity due to their high-minute rate of ventilation and high metabolic rate. A seven-year-old boy was admitted with a history of persistent vomiting, progressive loss of consciousness, convulsions, and severe right hemiparesis. All of the other family members developed vomiting, headache, and lethargy and recovered after the administration of oxygen, except for the boy. The CO source was a gas stove that had a chimney obstruction. The boy's COHb was 24.5%. A brain computed tomography (CT) performed the next day showed a thick clot in the left middle cerebral artery. The boy showed good recovery except for a global aphasia.

Almost every known central neurological syndrome has been reported as a complication of carbon monoxide poisoning. Hemorrhagic infarct has rarely been considered as an early manifestation of carbon monoxide poisoning. A case of cerebral hemorrhagic infarction is presented. Typical findings, neuropathology and role of vascular injury are discussed.

3.07 PREGNANCY/FETUS

Carbon monoxide (CO) is a colourless and odourless gas that is a by-product of combustion and has the potential to quickly reach hazardous

> concentrations in poorly ventilated areas. In fact, accidental household poisoning is the most frequent cause of poisoning in pregnancy.
>
> —"Winter Danger: Carbon Monoxide Poisoning during Pregnancy," *Contemporary OB/GYN*, 16 December 2015

CO exposure during pregnancy can cause severe injury and death to the fetus and the mother (30703, 30706, 30710–12). CO enters the fetal blood after crossing the placenta and lags behind the maternal COHb. Eventually, the fetal COHb equilibrates at a level 10–15% higher than maternal COHb. Fetal CO elimination also lags behind maternal elimination, and it is suggested that the mother be provided oxygen for five times longer than normal for an adult to effectively clear the fetus of COHb as CO binds to myoglobin as well (30701, 30706). For these reasons, the severity of CO poisoning in the fetus cannot be assessed by the mother's condition alone (30703). HBOT is probably the best treatment in these type of cases (30702, 30705, 30707–8, 30710). A table listing the degree of severity of CO exposure is provided (30712).

Reference Number: 30701

FRIEDMAN, P., X.M. GUO, R.J. STILLER, AND S.A. LAIFER. "Carbon Monoxide Exposure during Pregnancy." *Obstetrical and Gynecological Survey*, 70: 705–712, 2015 (4 tables, 39 references)

Abstract: CO poisoning in pregnancy was first described in 1859 in response to the industrial revolution and the introduction of coal gas for illumination. Fetal hemoglobin has a greater affinity for oxygen than adult hemoglobin, which facilitates the maternal-fetal oxygen exchange. CO enters the fetal blood after crossing the placenta. The fetal COHb lags behind the maternal COHb, but, eventually, the fetal COHb equilibrates at a level 10–15% higher than maternal COHb. Fetal CO elimination also lags behind the maternal, and it is suggested that the mother be provided oxygen for five times longer than normal for an adult to effectively clear the fetus of COHb.

> Carbon monoxide poisoning requires a high degree of suspicion. Diagnosis is made based on initial history and physical evaluation and assessment of environmental CO levels; presenting carboxyhemoglobin levels may be poor indicators of severity of disease. Oxygen therapy should be initiated promptly in all possible cases with consideration of hyperbaric

oxygen therapy in cases of significant maternal exposure. Treatment requires longer duration for fetal CO elimination than in nonpregnant patients. Importantly practitioners should educate pregnant patients on prevention.

Reference Number: 30702

GREINGOR, J.L., J.M. TOSI, S. RUHLMANN, AND M. AUSSEDAT. "Acute Carbon Monoxide Intoxication during Pregnancy: One Case Report and Review of the Literature." *Emergency Medicine Journal*, 18: 399–401, 2001 (1 table, 35 references)

Abstract: A 31-year-old woman, who had been pregnant for 21 weeks, was exposed to CO from a defective water heater. She complained of dizziness and palpitations and was removed from the CO source and administered oxygen via a facemask. A level of 24.6% COHb was measured in her expired breath. High-flow oxygen was started immediately, and HBO_2 was administered for 90 minutes within 30 minutes of CO exposure. She was discharged in good health and delivered a healthy male infant at term weighing 3.8 kilograms (kg).

Acute CO poisoning during pregnancy is comparatively uncommon yet can result in fetal death and functional alterations or anatomical malformations in survivors. The level of maternal CO hemoglobin is a poor indicator of fetal toxicity. Despite maternal well being, fetal morbidity or mortality can still occur. Hyperbaric oxygen seems to be the treatment of choice and all pregnant women suffering CO intoxication should be referred for hyperbaric oxygen therapy. It should be instituted as rapidly as possible, yet its efficacy is still theoretically possible long after initial exposure. In human protocol, the hyperoxic period is short and well tolerated by the fetus.

Reference Number: 30703

DELOMENIE, M., F. SCHNEIDER, J. BEAUDET, R. GABRIEL, N. BEDNAREK, AND O. GRAESSLIN. "Carbon Monoxide Poisoning during Pregnancy: Presentation of a Rare Severe Case with Fetal Bladder Complications." *Case Reports in Obstetrics and Gynecology*, 2015 (1): 687975, 2015.

Abstract: A 22-year-old women at 30 weeks of gestation was found unconscious in her home in cardiorespiratory arrest following a fire. Her initial blood COHb was 52%. After one 90-minute hyperbaric session, her COHb fell to 6.6%. The fetus had no brain damage but presented with bladder lesions not previously described with urinary ascites and severe fetal damage. A 2.4 kg baby was delivered at 35.5 weeks and presented with spastic quadriparesis, inability to suck, and hypoventilation and died 2.5 months later from respiratory distress.

> Monoxide carbon exposure during pregnancy can cause severe damage, including intrauterine hypoxia, serious neurological damage, and even fetal death. CO dissolved in maternal plasma crosses the placenta by passive diffusion and thus combines with fetal hemoglobin. The fetal COHb levels created are 10–15% higher than maternal levels, as fetal hemoglobin has a higher affinity for CO than adult hemoglobin. Moreover, fetal elimination of CO takes more time, for it dissociates much more slowly from fetal than adult hemoglobin. For this reason, the severity of fetal intoxication cannot be assessed solely by maternal condition.

Reference Number: 30704

RUDRA, C.B., M.A. WILLIAMS, L. SHEPPARD, J.Q. KOENIG, M.A. SCHIFF, I.O. FREDERICK, AND R. DILLS. "Relation of Whole Blood Carboxyhemoglobin Concentration to Ambient Carbon Monoxide Exposure Estimated Using Regression." *American Journal of Epidemiology*, 171: 942–951, 2010 (4 tables, 1 figure, 39 references)

Abstract: The blood COHb concentration was measured by headspace gas chromatograph in 7,089 pregnant women (usually at about 13-weeks gestation) in Washington State between 1996 and 2004. Estimated CO exposure was determined for the month and year of the blood draw by using street and population densities and distance to the nearest highway. For every 10% increase in CO exposure, the median COHb increased by 1.29%, excluding smokers. Month and year were the strongest predictors of COHb concentration.

Table: Median Blood COHb Concentrations (and Interquartile Range Twenty-Fifth to Seventy-Fifth Percentile) According to Year and Season of Blood Draw in Pregnant Women

Blood Draw	Median COHb	Range
1996–8	1.09	0.84–1.05
1999–2001	0.88	0.76–1.01
2002–4	0.78	0.66–0.91
Winter	0.88	0.76–1.06
Spring	0.81	0.69–0.95
Summer	0.84	0.71–1.08
Fall	0.86	0.73–0.98

Source: Adapted from Rudra et al. 2010.

Participants were typically 21–34 years old, nulliparous, white, and married. Only 7% reported current smoking. Median carboxyhemoglobin levels were higher among younger women, women without previous livebirths, black or Hispanic women, employed women, unmarried women, current smokers, women reporting secondhand smoke exposure, earlier participants, and those with first trimester blood draws. As expected, given the low prevalence of smoking and the geographic setting, distributions of local ambient carbon monoxide concentrations, estimated carbon monoxide exposures and carboxyhemoglobin concentrations were generally low.

Reference Number: 30705

OZGOK-KANGAL, K. "Long-Term Infant Outcomes after Hyperbaric Oxygen Treatment for Acute Carbon Monoxide Poisoning during Pregnancy." *Diving and Hyperbaric Medicine*, 51: 248–255, 2021 (2 tables, 4 figures, 35 references)

Abstract: A detailed analysis was conducted of 28 pregnant patients who received HBOT for CO poisoning between 2013 and 2016. The mean initial COHb of the patients was 27.9% (range 15.6–55.2%). One fetus was dead before HBOT. Three subsequent adverse events were reported, including abortion, premature birth, and limb malformation of the fetus. The remaining 24 patients delivered healthy-term infants and were healthy a median of 34 months later. Six patients continued to smoke during their entire pregnancy, but none reported drinking.

This is the largest relevant series and longest follow-up to date. Adverse outcomes were likely incidental because the mothers' medical histories revealed alternative aetiologies. There was no definite evidence of fetal morbidity or mortality after HBOT in this study. HBOT may improve long term fetal outcomes after in-utero CO poisoning without complications.

Reference Number: 30706

YILDIZ, H., E. ALDEMIR, E. ALTUNCI, M. CELIK, AND S. KAVUNCUOGLU. "A Rare Cause of Perinatal Asphyxia: Maternal Carbon Monoxide Poisoning." *Archives of Gynecology and Obstetrics*, 281: 251–254, 2010 (1 figure, 20 references)

Abstract: A 29-year-old, 33-week-gestation mother was found half-unconscious in her bathroom due to exposure to CO. At hospital, her initial COHb was 25%, and she was administered 100% oxygen via a mask. No HBOT was available. She was discharged after one day, and, one week later, she complained of decreased fetal movement. An emergency cesarian section was conducted, but the infant died 56 days later. With high levels of CO poisoning, CO is bound to the intracellular myoglobin and cytochrome a3 and P450, which acts as a reservoir and accounts for the persistence of CO poisoning even after oxygen therapy. CO dissolved in the maternal plasma crosses the placenta and combines with fetal hemoglobin and causes COHb to be 10–15% higher in the fetus. In addition, the binding of CO to fetal blood is 2.5–3 times higher than adult COHb. The half-life is also longer in the fetus. Acute CO poisoning of the mother has resulted in a fetal mortality of 36–67%.

If a HBO facility has not been available locally as in our case, the mother with sings of acute CO poisoning should be internalized to intensive care unit, administered 100% oxygen through a tight-fitting face mask and monitored for the fetus well being closely. In our case we could not obtain any information about the fetal signs of compromise during maternal acute CO poisoning and the duration of oxygen treatment. At the time of acute CO poisoning, our case was at the 33rd week of gestation which had a very high chance of survival with less risk of morbidity. After the stabilization of the mother, delivery option should be thought as soon as possible.

Reference Number: 30707

KOSAKI, Y., H. MAEYAMA, T. NOJIMA, T. OBARA, A. MAKAO, AND H. NAITO. "Carbon Monoxide Poisoning during Pregnancy Treated with Hyperbaric Oxygen." *Clinical Case Reports*, 9 (5): e04138, 2021 (18 references)

Abstract: A 30-year-old, 31-week-pregnant woman was transferred to an ER due to nausea and headache. The patient and her husband had been living in a camping tent and had dinner in front of a fireplace before sleeping. Upon admission, her blood COHb was 28.6%. She was treated with HBOT three times over two days. At 41 weeks, she delivered a healthy infant weighing 2.8 kilograms.

Carbon monoxide poisoning is toxic to both the fetus and mother in terms of delayed CNS [central nervous system] sequelae and CNS disorders. Hypoxia appears to be the primary mechanism of the deleterious effects on the fetus associated with CO poisoning during pregnancy. CO dissolves in the maternal blood and crosses the placenta via passive diffusion, thus combining with fetal hemoglobin. COHb levels become 10–15% high in the fetus than the mother, since fetal hemoglobin has a higher affinity for CO than adult hemoglobin. Fetal CO washout is slower, as the dissociation between CO and hemoglobin takes longer than adult hemoglobin. Moreover, the fetus is unable to adequately increase ventilation or tissue blood flow to combat the ischemic insult. Therefore, clinicians should be aware that the fetus is at greater risk than the mother for CO-induced adverse effects. Although the fetal brain has a high threshold to withstand the effects of sustained hypoxia, once a critical level of deoxygenation is surpassed, the result is major brain damage with early death.

Reference Number: 30708

KRESHAK, A.A., S.M. LAWRENCE, S.T. ONTIVEROS, T. CASTELLANO, AND K.B. VAN HOESEN. "Perinatal Carbon Monoxide Poisoning: Treatment of a 2-Hour Neonate with Hyperbaric Oxygen." *American Journal of Perinatology Reports*, 12: e113–e116, 2022 (28 references)

Abstract: A 38-week-pregnant woman presented at labour and delivery with dizziness, headache, and decreased fetal movement, after 12 hours exposure to CO from a charcoal grill used to heat the apartment. Twelve hours after exposure, her blood COHb was 7.5%. The baby was delivered by routine Cesarian section. The arterial cord blood COHb was

11.9%. HBOT was administered to both mother and baby. The infant and mother were discharged in good condition three days after delivery.

HBO treatment was well tolerated in this newborn infant who sustained inadvertent perinatal CO poisoning. HBO is a treatment option for perinatal CO exposures.

Reference Number: 30709

HENNEQUIN, Y., D. BLUM, E. VAMOS, M. STEIPE, J. GOEDSEELS, AND E. CAVATORTA. "In-Utero Carbon Monoxide Poising and Multiple Fetal Abnormalities." *The Lancet*, 341 (8839): 240, 1993 (5 references)

Abstract: A pregnant mother complained of headache and dizziness over seven months of gestation, until her 16-month-old daughter was found comatose at home. At hospital, the mother had 14% COHb, and the daughter had 27.5% COHb. Both were treated with HBOT. The sources of the CO found by the gas company were due to defective kitchen and bathroom water heaters that caused an ambient CO concentration of 120 and 100 ppm respectively. The mother delivered a badly deformed boy at 41 weeks who died shortly after birth.

This fetal morbidity is consistent with events around the end of the second month of gestation, which coincides with the first symptoms in the mother. In further instances of ill-defined congenital syndromes, chronic exposure of CO during fetal life should be checked for.

Reference Number: 30710

CRAMER, C.R. "Fetal Death Due to Accidental Maternal Carbon Monoxide Poisoning." *Journal of Toxicology and Clinical Toxicology*, 19: 297–301, 1983 (20 references)

Abstract: A 20-year-old, 8-month-pregnant woman ignited a natural gas stove to heat the house before retiring for the evening. During the night, several family members awoke with headaches and vomiting. In the morning, the gas valve in the oven was on, but the flame was out. The family was brought to hospital. The expectant mother had a blood concentration of 27.3% COHb and was administered 100% oxygen. The father had 45% COHb and was treated as well. No heartbeat was detected in the fetus. Labour was induced three days later, and a stillborn male infant was delivered. At post-mortem, the dead fetus was found to have a blood COHb of 25% in the right ventricle and 31.5% in the liver and spleen samples.

Treatment of CO poisoning usually involves administration of high concentrations of oxygen at ambient pressure. In selected cases, hyperbaric oxygen therapy is beneficial. In the present case the fetus was dead upon admission of the mother to the hospital, and the mother responded satisfactorily to oxygen administered at ambient pressure, therefore hyperbaric oxygen was not deemed necessary. CO levels remain unaltered for months following fatal CO poisoning in adults. Since fetal blood flow ceases with fetal demise and since CO is firmly bound to the tissues of the fetus, fetal CO levels in cases of fetal CO poisoning are valid for long periods of time following fetal death; in Phillips case, CO could be detected in fetal tissue two months after fetal demise.

Reference Number: 30711

OMALU, B.I., J.L. LINDER, J.K. JANSSEN, U. NNEBE-AGUMADU, AND V. WEEDN. "The Role of Environmental Factors in the Causation of Sudden Death in Infants: Two Cases of Sudden Unexpected Death in Two Unrelated Infants Who Were Cared for by the Same Babysitter." *Journal of Forensic Science*, 52: 1355–1358, 2007 (29 references)

Abstract: A two-month-old female infant died in the apartment of a babysitter after an overnight stay, followed by a four-month-old female infant 39 days later. No cherry red lividity was observed in either infant. The post-mortem blood COHb concentrations were 13% and 14% respectively. No nicotine or cotinine could be detected. The surviving adults in the apartment had a blood COHb of 5%. The source of the CO was a poorly ventilation (back drafting) and defective water heater located across the hallway in which an ambient air concentration of 40 ppm was detected. The cause of death in both infants was determined to be acute accidental CO toxicity.

Fetal hemoglobin comprises 50–95% of a neonate's hemoglobin at birth. These levels decline after 6 months as adult hemoglobin synthesis is activated while fetal hemoglobin synthesis is deactivated. It takes about 2 years for a baby to completely switch over to adult hemoglobin. The adverse effects of CO on mitochondria functioning further accentuate the vulnerability of early infants to CO toxicity because of the high basal metabolic rate (BMR) in infants. BMR is defined as the minimal rate of energy generation and expenditure, which will sustain life. Infants require basal metabolic rates per unit body size which are much higher than older children and adults. Highest values of BMR in humans are observed in infants and per unit of body weight the BMR at age 20 is only just over 30% of its value in the first year. In our two cases the two infants were 2 and 4 months respectively, and among the other individual in the apartment, they were the most naive to CO exposure; and the most vulnerable to the toxic effects of CO intoxication even at lower levels.

Reference Number: 30712

KOREN, G., T. SHARAV, A. PASTUSZAK, L.K. GARRETTSON, K. HILL, I. SAMSON, M. ROREM, A. KING, AND J.E. DOLGIN. "A Multicenter, Prospective Study of Fetal Outcome Following Accidental Carbon Monoxide Poisoning in Pregnancy." *Reproductive Toxicology*, 5: 397–403, 1991 (7 figures, 45 references)

Abstract: An analysis was conducted of 40 cases of CO poisoning treated at various hospitals in Ontario, Alberta, Wisconsin, and several other US states. The main sources of CO were furnaces, hot water heaters, cars, and methylene chloride. Of the five pregnancies with severe CO toxicity two resulted in stillbirths and one resulted in an infant with cerebral palsy with ischemic damage. The blood COHb of the mothers in these three cases were 40.5%, 26%, and 25% respectively, and all had been treated with high-flow oxygen. In the other two cases of severe toxicity treated with HBOT, the babies were normal. All other babies exposed in utero to mild or moderate CO poisoning showed normal physical and neurobehavioural development.

Table: Severity of CO Exposure According to Grade

Grade 1	Alert, orientated Symptoms headache, dizziness, nausea
Grade 1+	As above, but a relative was unconscious
Grade 2	Alert, alterations of mental state Symptoms more pronounced than above
Grade 3	Not alert, disorientation, loss of recent memory, muscle weakness, or incoordination
Grade 4	Disorientated, depressed sensorium, response to simple commands limited and inappropriate
Grade 5	Comatose, responding only to pain or not responding to any stimulus

Source: Adopted from Koren et al. 1991.

It is concluded that while severe CO poisoning poses serious short and long term fetal risks, mild accidental exposure is likely to result in normal fetal outcome. Because fetal accumulation of CO is higher and elimination is slower than in the maternal circulation, hyperbaric oxygen may decrease fetal hypoxia and improve outcome.

3.08 ALCOHOL/OTHER DRUGS

Drinking was a great leveler, not because it made everyone equally drunk, but because it made everyone equally guilty.

—S. Churchwell, *Careless People* (2014)[6]

Alcohol is the most common drug found in CO poisoning victims and especially in fires (30801–6). Alcohol increases the risk of accidental CO poisoning and the inability to rapidly escape (30805), but it does not appear to cause death at a lower COHb (30806).

Reference Number: 30801

SOHN, C.H., J.W. HUH, D.W. SEO, B.J. OH, K.S. LIM, AND W.Y. KIM. "Aspiration Pneumonia in Carbon Monoxide Poisoning Patients with Loss of Consciousness: Prevalence, Outcomes and Risk Factors." *American Journal of Medicine*, 1301465.e21–1465.e26, 2017 (4 tables, 22 references)

Abstract: A study was conducted of 537 CO-exposed patients who lost consciousness and attended a hospital in Seoul, Korea, between 2008 and 2015. Aspiration pneumonia deaths within 48 hours occurred in 103 of the patients (19.2%) and was associated with increased ventilator use (52.4% for aspiration pneumonia compared to 3.2% for non-aspiration pneumonia group). Alcohol was detected in 46% of the patients. The mean CO exposure duration was 7.5 hours in the aspiration pneumonia group compared to 3.0 hours in the non-aspiration group.

In this study we observed that the prevalence of aspiration pneumonia was 19.2% in patients with carbon monoxide poisoning who had transient or persistent loss of consciousness and aspiration pneumonia was associated with poor outcomes. Additionally, altered mental status on emergency department arrival, white blood cell count, and increased

6 S. Churchwell, *Careless People: Murder, Mayhem and the Invention of the Great Gatsby* (Toronto: Random House Canada, 2014).

exposure duration were independently associated with the development of aspiration pneumonia. Carbon monoxide causes tissue hypoxia by forming carboxyhemoglobin and shifting the carboxyhemoglobin dissociation curve to the left. Exposure to carbon monoxide also causes inflammation through multiple pathways independent of the pathways that result in tissue hypoxia, resulting in neurologic and cardiac injury. Therefore, severe exposure to carbon monoxide may cause loss of consciousness and, consequently, a high risk of aspiration pneumonia in patients with carbon monoxide poisoning.

Reference Number: 30802

SILVERS, S.M., AND N.B. HAMPSON. "Carbon Monoxide Poisoning among Recreational Boaters." *Journal of the American Medical Association*, 274: 1614–1616, 1995 (1 table, 12 references)

Abstract: Of the 512 patients treated for CO poisoning in a hyperbaric department in Seattle between 1984 and 1994, 39 (8%) were poisoned in 27 incidents related to recreational boating. The patients were 22 men and 7 women with an average age of 37 years and with a mean COHb of 22.1% (range 0.8–38.4%). Incidents typically occurred in salt water rather than fresh water, and all boats were longer than 22 feet (range 23–47 feet). Alcohol consumption was reported in five incidents (19%); the source of CO was due to engine exhaust in 21 incidents, water heater exhaust in two, space heater exhaust in two, and generator exhaust in two. Loss of consciousness occurred in two-thirds of the patients, and nearly 60% of the incidents occurred while cruising.

Carbon monoxide poisoning is a serious hazard associated with recreational boating. The installation of CO detectors aboard boat types typically associated with this syndrome should be strongly encouraged.

Reference Number: 30803

YOON, S., S.C. MACDONALD, AND G. PARRISH. "Deaths from Unintentional Carbon Monoxide Poisoning and Potential for Prevention with Carbon Monoxide Detectors." *Journal of the American Medical Association*, 279: 685–687, 1998 (1 table, 1 figure, 14 references)

Abstract: Between 1980 and 1995, 695 deaths (1% of all deaths) in New Mexico investigated by the Office of the Medical Investigator were deemed to be due to CO poisoning. Of these deaths, 136 (19.6%) were determined to be unintentional inhalation. The median age of these victims was 39 years (range 1–87 years), and 100 (74%) were males. Fifty-six victims (42%) had a BAC that was less than 10 mg/dL. Thirty-nine (49%) of the 80 residential fatality deaths were most likely asleep compared to 10 of the deaths (18%) occurring in and around motor vehicles. The most common source of residential CO poisoning was natural gas- or propane-fueled supplemental heaters.

Differences exist between deaths due to unintentional CO poisoning that occur in residences and those that occur in or around motor vehicles. Carbon monoxide detectors, whether the electronic or chemical reagent types, may have prevented approximately half of these deaths. The high proportion of decedents with alcohol in their blood indicates that effective public health campaigns must address the role of alcohol in CO poisoning deaths.

Reference Number: 30804

TEIGE, B., J. LUNDEVALL, AND E. FLEISCHER. "Carboxyhemoglobin Concentrations in Fire Victims and in Cases of Fatal Carbon Monoxide Poisoning." *Zeitschrift für Rechtsmedizin*, 80: 17–21, 1977 (1 table, 3 figures, 3 references)

Abstract: The post-mortem blood COHb concentrations were determined in 54 victims of CO poisoning and 87 fire victims (72 had severe burning and 15 had no burns) over a period of 11 years in Oslo. In most cases of CO poisoning, there were no significant BACs detected, whereas a positive BAC was detected in 45% of the fire victims. As the age increased in the CO poisoning victims, the lower the mean fatal COHb detected, the greater its variation. In the 87 fire victims, 32% had a COHb that was less than 45%.

Table: Mean Post-Mortem Blood COHb in Victims of Non-Fire CO Poisoning

Age (n)	Mean COHb (%) (standard deviation)
< 40 years (11 victims)	74 (5)
40–50 years (12 victims)	74 (7)
51–60 years (18 victims)	70 (11)
> 60 years (13 victims)	65 (12)

Source: Adapted from Teige, Lundevall, and Fleischer 1977.

1. Carbon monoxide poisoning is probably not the cause of death when postmortem carboxyhemoglobin concentration is below 45%. 2. Burns rather than carbon monoxide poisoning is the main cause of death in at least thirty percent of the fire victims. 3. Alcohol intoxication does not seem to influence the fatal HbCO concentration. 4. A low post mortem carboxyhemoglobin concentration is often found in victims suffering from disease.

Reference Number: 30805

GIRMAN, J.P., Y.-U. CHANG, Y.-U., S.B. HAYWARD, AND K.-S. LIU. "Causes of Unintentional Deaths from Carbon Monoxide Poisonings in California." *Western Journal of Medicine*, 168: 158–165, 1998 (4 tables, 2 figures, 9 references)

Abstract: A detailed analysis was conducted of 444 cases of unintentional fatal CO poisonings in California between 1979 and 1988. The annual number of fatalities ranged from 27 to 58 over the decade analyzed. The victims were mainly male (72.5%), and the fatal CO poisonings were more likely to occur in ages 20 to 39 years and older than 50 years. The main source of CO were motor vehicles (30.6%), heating and cooking appliances (40.1%), and charcoal grills (13.1%). At least 137 (31%) of the 444 unintended CO deaths involved alcohol.

Table: Percentage of Alcohol-Related Unintentional CO Deaths by CO Source

CO source	Percent alcohol related (%)
Automobiles	43.4
Heating/Cooking appliance	18.5
Charcoal Grill/Hibachi	37.9
Small engine	33.3
Camping equipment	10.0

Source: Adapted from Girman et al. 1998.

The public is well informed that one of the most important causes of traffic accidents is drinking and driving. It is not well known, however, that drinking even without driving (for instance simply parking a car and allowing the engine to run) can also be dangerous. Alcohol was involved in 43% (59 cases) of the 136 vehicular deaths. Not only was parking in a car in a garage without stopping the engine, highly risky, CO deaths also occurred in cars parked and running in the open.

Reference Number: 30806

MOON, J.M., B.J. CHUN, Y.S. CHO, AND J.G. MUN. "Does Alcohol Play the Role of Confounder or Neuroprotective Agent in Acute Carbon Monoxide Poisoning?" *Clinical Toxicology*, 58: 161–170, 2020 (3 tables, 2 figures, 44 references)

Abstract: The COHb and serum alcohol concentrations (SAC) were determined in 158 patients who presented as a result of CO exposure at a South Korean hospital between 2017 and 2018. The mean blood COHb was 8.2% and ranged between 3.2% and 25.6%. The mean SAC was 56.8 mg/dL and ranged up to 408 mg/dL. The patients were divided into low (< 2 mg/dL), moderate (2–100 mg/dL), and high (> 100 mg/dL) SAC. The mean duration of hospital stay was three days for all SAC groups.

Table: Various Variables Associated in CO-Poisoned Patients at Low, Moderate, and High SAC

Variable	Low SAC	Moderate SAC	High SAC
Age	55 years (30–74)	40 years (34–56)	37 years (32–50)
Initial GCS	15 (14.3–15)	15 (10–15)	14 (6.8–15)
Duration of exposure to CO (minutes)	150 (30–525)	110 (30–300)	30 (20–60)
Poor neurological outcome at 30 days (%)	15.4	26.5	13.2
In hospital deaths (%)	0	2.9	2.6

Source: Adapted from Moon et al. 2020.

Predicting the severity of CO poisoning should never be taken based on the serum lactate alone without adjustment for BAC, particularly those cases with both acute CO poisoning and alcohol consumption. Initial GCS score can be used as a predictor for the severity of CO poisoning and neurological outcome 30 days after acute CO poisoning in the clinical field without considering a history of acute alcohol consumption. Alcohol, regardless of the amount consumed does not have a neuroprotective benefit in acute CO poisoning.

3.09 DELAYED/PROLONGED SEQUELAE

> The following characteristics of carbon monoxide should be borne in mind: the gas is odorless, it does not affect a candle's flame, its presence in the atmosphere at a strength of 0.1 per cent may be fatal, and its action is cumulative.
>
> —C.E. Sundell, "The Clinical Manifestations of Gas in Military Mines" (1916) **(41304)**

DNS unfortunately occurs in many CO-poisoning victims and can first appear days or even weeks after the initial exposure (30901). CO causes more damage to organs with a high oxygen demand such as the heart, brain, and kidneys (30904). Hence, the features of DNS include dementia (30902, 30911), memory loss (30912, 30925), brain injury (30910), hearing and vision loss (30913, 30914), Parkinson Disease (PD) (30907), acute kidney injury (30905), and cardiovascular events (30906, 30908). The longer the exposure time to CO and the more prolonged the coma, the more likely the victim will develop DNS (30915, 30922). HBOT will not prevent the development of DNS (30916, 30919). DNS may persist for 29 to 35 years after the initial CO poisoning (30903, 30924).

Reference Number: 30901

LAI, C.-Y., Y.-W. HUANG, C.-H. TSENG, C.-L. LIN, F.-C. SUNG, AND C.-H. KAO. "Patients with Carbon Monoxide Poisoning and Subsequent Dementia: A Population-Based Cohort Study." *Medicine*, 95: e2418 (7pp), 2016 (4 tables, 1 figure, 50 references)

Abstract: The hazard ratio (HR) of developing dementia was determined in 9,401 adults with recent CO poisoning compared to 36,160 adults without CO poisoning in Taiwan between 2000 and 2011. Those who received HBOT had a greater HR of dementia (but not statistically significant) compared to those who did not receive HBOT.

Table: Adjusted Hazard Ratio of Dementia in Patients with CO Poisoning Compared to Those without CO Poisoning

Outcome	Adjusted HR
All	1.50 times
Male	1.83 times
Ages 20–49 years	2.62 times
Ages 50–64 years	1.99 times
≥65 years	1.05 times

Source: Adapted from Lai et al. 2016.

In conclusion, this study suggests that CO poisoned patients may be associated with an increased risk of dementia. It is important to note there is a large number of CO poisoning events, which occur to mainly young populations. Further data showed the cumulative mortality over 12 years (2000–2011) was approximately 10.2% (922/9041) in the CO poisoning cohort and 32.3% (20/62) in those with dementia developed. People suffering from depression may need more medical attention, particularly for the young population. Further study needs to differentiate between late neuropsychiatric effects of CO poisoning and dementia.

Reference Number: 30902

MAZO, J., E. MUKHTAR, Y. MAZO, N. ABHIRAM, AND M.T. MANTELLO. "Delayed Brain Injury Post Carbon Monoxide Poisoning." *Radiology Case Reports*, 15: 1845–1848, 2020 (3 figures, 16 references)

Abstract: The brain is particularly sensitive to CO toxicity, and poisoning can result in white matter demyelination, the necrosis of the globus pallidus, and reversible anoxic-ischemic encephalopathy or irreversible dementia. A rare, delayed post hypoxic leukoencephalopathy can also occur, as shown in this case. A 66-year-old man was admitted to the ER after being found unresponsive in an auto body shop, laying adjacent to a car with its engine on. His COHb concentration was greater than 27%, but he was not administered HBOT. Four months after his hospital stay, he was readmitted due to COVID-19, and his mental status had progressively declined, becoming mute, non-verbal, and quadriplegic and requiring mechanical ventilation and tube feeding.

Improved awareness of the neurological sequelae associated with carbon monoxide toxicity and its rare progressive disease is paramount. Delayed posthypoxic leukoencephalopathy is under-recognized due to its late manifestation. The employment of repeat MRI CT and Magnetic Resonance spectroscopy provides a diagnostic approach that can defend clinical symptomology and allow for accurate prognosis.

Reference Number: 30903

MISHIMA, T., F. MIURA, AND Y. TSUBOI. "MRI 35 Years after Carbon Monoxide Intoxication." *Internal Medicine*, 51: 3307–3308, 2012 (1 figure, 2 references)

Abstract: A 28-year-old man suffered severe CO poisoning. He was in a semi-coma and was only able to talk after 30 days and walk four months later. He suffered from apathy, disinhibition, and executive dysfunction, which lasted until he died of colon cancer at the age of 73 years. An MRI, 35 years after his CO poisoning showed diffuse cerebral atrophy and white matter hyperintensity of the parietal and occipital lobes on T2 weighed images.

The sequelae and MRI abnormalities of CO intoxication persist for long periods.

Reference Number: 30904

DOGAN, F.S., O. GUNEYSEL, E. GOKDAG, M. GUNES, AND S.G. SUMEN. "Demographic Characteristics and Delayed Neurological Sequelae Risk Factors in Carbon Monoxide Poisoning." *American Journal of Emergency Medicine*, 28: 2552–2556, 2020 (3 tables, 1 figure, 13 references)

Abstract: A study was conducted of 36 male and 36 female CO-poisoned patients (mean age 33 years) with an initial COHb of 10% or more. Initial physical examination and neurological examination were normal in 70 (97.2%) and 65 (88.9%) of the patients respectively. The DNS rate was 7%. The mean COHb in patients with DNS was 30% compared to 25% of those who did not develop DNS. A patient with an initial pathological neurological examination was 18.6 times more likely to develop DNS. Only two patients (2.8%) attended hospital due to CO poisoning in the summer compared to 39 (54.2%) in the winter.

It induces more damage to the organ with high oxygen demand such as heart and brain. Carboxyhemoglobin levels differ according to oxygen amount in the surrounding air, carbon monoxide level, exposure time to carbon monoxide, age, and minute ventilation. Due to the direct toxic effects on cells, CO interferes with cellular oxygen usage and causes tissue hypoxia. It is thought that dissolved CO directly attaches the cytochrome a3 (cytochrome c oxidase) enzyme, which is a terminal enzyme for oxygen electron transport chain and affecting the cellular respiration. In the central nervous system (CNS), reoxygenation damage occurs after tissue hypoxia. Oxygen radicals are formed due to hyperoxygenation and reperfusion injury eventually occurs. Besides that, CO causes lipid peroxidation and leads to reversible demyelination of CNS lipids. Most of the symptoms occur due to the toxic effects of CO on CNS. In patients with COP, the most common cause of morbidity is DNS. After full recovery of COP patients, DNS can be seen within 3–240 days.

Reference Number: 30905

HUANG, T.-L., M.-C. TUNG, C.-L. LIN, AND K.-H. CHANG. "Risk of Acute Kidney Injury among Patients with Carbon Monoxide Poisoning." *Medicine*, 100 (38): e27239, 2021 (5 tables, 1 figure, 37 references)

Abstract: A comparison was conducted between the incidence of acute kidney injury (AKI) in 8,593 patients with CO poisoning and 8,593 control patients without CO poisoning in Taiwan between 2000 and 2011. The adjusted hazard ratio of AKI was 2.27 times in patients with lower-severity CO poisoning and 3.77 times in patients with high-severity CO poisoning compared to patients with no CO exposure.

AKI has rarely been documented in the evaluation of the severity of CO poisoning. In this study, the cumulative incidence of AKI was higher in the CO poisoning cohort than in the comparison cohort. Further data analysis also showed that patients with severe CO poisoning who had longer hospital stays had a higher HR for developing AKI. Traditionally, we thought that rhabdomyolysis was not common among patients with CO poisoning. CO itself may act as a nephrotoxin via multiple pathways, and we suggest that AKI can be used as an indicator of the estimated severity of CO poisoning and the need for HBOT.

Reference Number: 30906

WONG, C.-S., Y.-C. LIN, L.-C. SUNG, T.-T. CHEN, H.-P. MA, Y.-H. HSU, S.-H. TSAI, Y.-F. LIN, AND M.-Y. WU. "Increased Long-Term Risk of Major Adverse

Cardiovascular Events in Patients with Carbon Monoxide Poisoning: A Population-Based Study in Taiwan." *PLoS ONE* 12: e0176465, 2017 (2 tables, 3 figures, 15 references)

Abstract: A comparison was conducted of the risk of a major adverse cardiovascular event (MACE) in 13,939 patients with CO poisoning in Taiwan between 2005 and 2010 and in 55,757 matched controls without CO poisoning. After adjusting for the known confounding factors, the HR was two for CO-poisoned patients and 2.98 for patients aged 18 to 44 years for MACE.

> CO poisoning remains a major cause of accidental and intentional injuries worldwide. Several studies have indicated that CO poisoning is associated with a risk of toxicity to the central nervous system and heart. The present population-based cohort study conducted using Taiwan's NHIRD [National Health Insurance Research Database] revealed that patients with CO poisoning have a higher risk of MACE than do controls. The study also revealed a significant risk of MACE in a long-term follow-up after adjustment for age, sex and comorbidity.

Reference Number: 30907

LAI, C.-Y., M.-C. CHOU, C.-L. LIN, AND C.-H. KAO. "Increased Risk of Parkinson Disease in Patients with Carbon Monoxide Intoxication." *Medicine*, 94 (19): e869, 2015 (4 tables, 1 figure, 50 references)

Abstract: The incidence of PD was determined in 9,012 adults with CO intoxication compared to 36,408 adults without CO intoxication (age, sex, and year hospitalized matched) in Taiwan hospitals between 2001 and 2011. Most patients were less than 34 years of age. The mean age of the CO and non-CO patients were 39.8 and 39.7 years respectively. The overall incidence of PD was 27.4 per 10,000 person years for CO-exposed patients and 2.53 for the non-CO-exposed patients. After adjusting for age, gender, and comorbidities was 9.08 times increased risk for PD. CO patients who received HBOT were 14.3 times more likely to develop PD compared to 7.97 times for CO patients who did not receive HBOT.

Table: Adjusted HR of PD between Patients with and without CO Exposure

Outcome	Mean Adjusted HR
Male	7.48 times
Female	14.1 times
20–49 years of age	39.4 times
50–64 years of age	5.75 times
65 +years	3.55 times
No comorbidity	15.8 times
Comorbidity	4.15 times

Source: Adapted from Lai et al. 2015.

HBO treatment did not appear to be an effective preventative measure for PD. In addition, HBO treatment seemed unable to prevent reperfusion injury for neurodegeneration. There may be a severe degree of COI [carbon monoxide intoxication] in patients who received HBO treatment, leading to less favorable results. The treatments with hyperbaric oxygen were received to the patients with more severe symptoms of COI. Therefore, we found influence of the severity of COI in the development of PD. Certain aspects of COI possibly cause irreversible oxidative damage and neurodegeneration.

Reference Number: 30908

HENRY, C.R., D. SATRAN, B. LINDGREN, C. ADKINSON, C.I. NICHOLSON, AND T.D. HENRY. "Myocardial Injury and Long-Term Mortality Following Moderate to Severe Carbon Monoxide Poisoning." *Journal of the American Medical Association*, 295: 398–402, 2006 (1 table, 1 figure, 41 references)

Abstract: In total, 230 consecutive moderate-to-severe CO-poisoned patients treated with HBOT between 1994 and 2002 and with follow-up to 2005 were studied. The mean age of the patients was 47.2 years, 72% were male, and 56% were active smokers. Only 52 (22.6%) had hypertension and 16 (7%) had diabetes. Myocardial injury, as determined by blood markers or an abnormal ECG, was present in 85 of the patients (37%). Thirty-eight percent of the patients with cardiac injury died by the median follow-up of 7.6 years compared to 15% of the CO-poisoned patients without cardiac injury.

In conclusion, myocardial injury is a common consequence of moderate to severe CO poisoning and identifies patients at increased risk of mortality. Patients with suspected exposure to CO should be screened for myocardial injury and further cardiovascular risk stratification should be considered in all patients with confirmed myocardial injury. Although CO poisoning is the most common cause of accidental poisoning in adults in the United States, the focus had been on acute outcome, in particular the neurological manifestations. Our data indicate these patients have poorer long-term outcomes, a topic that deserves further study.

Reference Number: 30909

CHAMBERS, C.A., R.O. HOPKINS, L.K. WEAVER, AND C. KEY. "Cognitive and Affective Outcomes of More Severe Compared to Less Severe Carbon Monoxide Poisoning." *Brain Injury*, 22: 387–395, 2008 (2 tables, 66 references)

Abstract: The prevalence of cognitive sequalae, depression, and anxiety was determined in 55 patients with less-severe, and 201 patients with more-severe, CO poisoning at 6 weeks, 6 months, and 12 months between 1992 and 1999 was determined. A most-severe CO poisoning was defined as a COHb that was greater than 15% or loss of consciousness. The mean initial COHb was 25.2% for the more-severe group and 6.8% for the less-severe group. Thirty-five percent of the more severe group received HBO_2 compared to 17% of the less-severe group.

Cognitive sequelae occurred frequently in both more and less CO-poisoned patients and there were no differences in the rate of cognitive sequelae between the two groups. The lack of a difference in the rate of cognitive sequelae from 6 weeks to 12 months suggest there is little recovery in brain injury in 12 months. Previous investigations indicate that healthy normal individuals exposed to less severe CO exposure develop cognitive impairments that are similar to those observed following acute or more severe CO poisoning. There are several possible explanations for why less severely poisoned patients develop affective sequelae at approximately the same rate as those with more severe poisoning. Carbon monoxide poisoning causes brain inflammation. This inflammation may be independent of hypoxia, suggested by the lack of association between COHb levels and outcomes. Clearly hypoxia can cause brain injury also, but even in the absence of brain hypoxia, CO-induced inflammation may be damaging. In addition, low to moderate COHb levels can trigger multiple biochemical cascades resulting in brain injury and morbidity.

Reference Number: 30910

LO, C.-P., S.-Y. CHEN, K.-W. LEE, W.-L. CHEN, C.-Y. CHEN, C.-J. HSUEH, AND G.-S. HUANG. "Brain Injury after Acute Carbon Monoxide Poisoning: Early and Late Complications." *American Journal of Roentgenology*, 189: W205–W211, 2007 (9 figures, 15 references)

Abstract: CO poisoning can produce different patterns of brain injury and delayed sequelae. Neurons of the central nervous system are the most vulnerable to CO poisoning as they have the highest oxygen and glucose demands. Some of the acute brain injuries include diffuse hypoxic-ischemia, encephalopathy, focal cortical injury, necrosis of the globus pallidus, thalamus, brainstem, and cerebellum. Diffuse brain atrophy and cerebral white matter demyelination may occur later.

It is the most common lethal poison worldwide, and neurolgic sequelae are the most frequent form of morbidity. The pathophysiologic mechanisms of CO toxicity can be divided into hypoxia and cellular theories. The affinity of CO for heme protein is approximately that of oxygen and the formation of carboxyhemoglobin reduces the oxygen-carrying capacity of blood causing tissue hypoxia. CO inhibits the mitochondrial electron transport enzyme system and activates polymorphonuclear leukocytes, which undergo diapedesis and cause brain lipid peroxidation, leading to the delayed effects of CO poisoning. The clinical presentations and imaging features of CO poisoning are diverse.

Reference Number: 30911

HU, H., X. PAN, Y. WAN, Q. ZHANG, AND W. LINAG. "Factors Affecting the Prognosis of Patients with Delayed Encephalopathy after Acute Carbon Monoxide Poisoning." *American Journal of Emergency Medicine*, 29: 261–264, 2011 (1 table, 10 references)

Abstract: Delayed encephalopathy after acute carbon monoxide poisoning (DECMP) can occur days or weeks after the disappearance of symptoms of acute CO poisoning. Twenty-eight female and 18 male DEACMP patients (ages 20 to 84 years) were administered hyperbaric treatment. Initially, 27 patients had a high degree of dementia. Twenty-three patients (50%) showed improvement after treatment. The patients that showed improvement were of an older age (59.5 years versus 48.7 years) and had other complications and a higher activity-of-daily-living (ADL) score during the most serious conditions of DEACMP (that is, danger ADL). The patients who were not improved had an ADL score less than 40 points, were incapable of verbal communication, and had completely dependent daily life.

> Hyperbaric oxygen treatment appears useful in treating DEACMP patients, and patients with greater age and more complications, but shorter lucid interval and less danger activities of daily living scores, are more likely to have poor prognosis.

Reference Number: 30912

SMITH, J.S., AND S. BRANDON. "Morbidity from Acute Carbon Monoxide Poisoning at Three-Year Follow-Up." *British Medical Journal*, 1: 318–321, 1973 (25 references)

Abstract: Seventy-four survivors of acute CO poisoning were followed for three years after exposure. Three patients had committed suicide, and eight had died from other causes during that time. Of the 63 patients at follow-up, eight (12.7%) showed an improvement, and 21 (33.3%) showed a deterioration of personality. A total of 27 patients (43%) reported memory impairment. Eight patients (11%) suffered neuropsychiatric sequelae.

> Deterioration of personality and memory impairment were highly correlated. The level of consciousness on admission to hospital in the acute phase of poisoning correlated significantly with the development of gross neuropsychiatric sequelae. These findings emphasize the importance of prompt and efficient treatment of carbon monoxide poisoning and the need to follow-up all cases in the anticipation of a relapsing course or the development of sequelae.

Reference Number: 30913

MEHRPARVAR, A.H., M.H. DAVARI, A. MOLLASADEGHI., M.R. VAHIDI, M. MOSTAGHACI, M. BAHALOO, AND P. SHOKOUH. "Hearing Loss Due to Carbon Monoxide Poisoning." *Case Reports in Otolaryngology*, 2013 (1): 940187, 2013 (1 table, 2 figures, 23 references)

Abstract: A 22-year-old man who worked in a smelting shop was found unconscious on the factory floor. At hospital, he was still unconscious and was admitted to the ICU. He was diagnosed with CO poisoning and administered NBOT. His CT scan showed mild brain edema. He regained consciousness after four days and complained of hearing impairment without vertigo or tinnitus. Three months later, there was no improvement in his hearing impairment.

> Acute or chronic exposure to CO is one of the known but rare causes of hearing loss. This kind of impairment is caused by the effect of CO on blood oxygen carrying capacity or free radical production. It has been evidenced that auditory cortex, inferior colliculus, and cochlea are very sensitive to anoxia, but central auditory pathways are more sensitive than the cochlea. Hearing impairment due to anoxia could be either reversible or irreversible. This type of hearing loss is usually bilateral and variable in severity.

Reference Number: 30914

ERSANLI, D., S. YILDIZ, E. TOGROL, H. AY, AND T. QYRDEDI. "Visual Loss as a Late Complication of Carbon Monoxide Poisoning and Its Successful Treatment with Hyperbaric Oxygen Therapy." *Swiss Medical Weekly*, 134: 650–655, 2004 (2 figures, 18 references)

Abstract: A 34-year-old woman was found unconscious by her husband in the bathroom. She was administered NBO_2 by mask and regained consciousness. The oxygen was administered intermittently for two days. On the third day, she complained of memory problems and blurry vision. On the eighth day, she was administered HBOT, which continued for 25 sessions until some improvement in her vision was noted. In another case, a 32-year-old woman was also found unconscious in the bathroom and was administered NBOT and recovered. But she developed visual problems and, on the sixth day, was administered HBOT, and her vision recovered. In both cases, HBOT should have been administered immediately as both victims were found unconscious.

> It seems evident that the adverse effects of CO poisoning continue to progress, at least in some cases during the late period. We believe that HBO treatment in this period may still be effective and may prevent some of the neurological, neuropsychological, and visual sequelae from becoming persistent. We also believe that it is better to apply HBO treatment, which is recommended, relatively safe and cheap, rather than taking any risk of a permanent disability.

Reference Number: 30915

PEPE, G., M. CASTELLI, P. NAZERIAN, S. VANNI, M. DEL PANTA, F. GAMBASSI, P. BOTTI, A. MISSANELLI, AND S. GRIFONI. "Delayed Neuropsychological Sequelae after Carbon Monoxide Poisoning: Predictive Risk Factors in the Emergency Department—A Retrospective Study." *Scandinavian Journal of Trauma, Resuscitation and Emergency Medicine*, 19 (1): 16, 2011 (5 tables, 20 references)

Abstract: A study was conducted of 347 patients admitted to an ER in Italy due to CO poisoning between 1992 and 2007. One month after hospital discharge, 141 patients participated in a follow-up visit, and 34 (24.1%) were diagnosed with DNS. The mean blood COHb was 26% for the non-DNS group and 26.4 for the DNS group. There was a significant correlation between CO exposure greater than six hours, GCS greater than nine, seizure, systolic blood pressure less than 90 millimetres (mm) of mercury (Hg), elevated creatine phosphokinase (CPK), and leukocytes and risk of DNS. There was no significant correlation of DNS and age, sex, transient loss of consciousness, and blood COHb concentration. The cherry pink colour of the skin was found in only 1% of the patients. Only two patients died (0.57%), and both were elderly.

Table: Comparison of Variables in Patients without DNS and with DNS One Month after Discharge

Patient variable	Without DNS	With DNS
Duration of CO exposure > 6 hours (%)	33.6	52.9
White blood cell count (per microlitre)	9.1	11.9
CPK (units/litre)	159	572
GCS < 9 (%)	5.6	26.4
Systolic blood pressure < 90 mm Hg (%)	0.9	8.8
Seizures (%)	4.6	26.4
Transient loss of consciousness (%)	49.5	64.7

Source: Adapted from Pepe et al. 2011.

Acute CO poisoning is still a challenge in emergency medicine: correct diagnosis, optimal treatment and DNS prevention are fundamental goals. There are no established criteria, if any, for determining the risk for DNS. Appropriate information about DNS should be given to every patient at the moment of hospital discharge to allow prompt recognition of sequelae. In our study, sequelae were common after recovery from acute

CO poisoning. We identified several potential predictive risk factors for DNS development. Early risk-stratification of patients in the Emergency Department might permit an improvement in quality of care, in terms of DNS prevention. No clear evidences are available whether or not HBO therapy is beneficial in preventing DNS, and indications for this treatment in the acute setting remain controversial. Further research is needed to better define the optimal management of CO poisoning.

Reference Number: 30916

ABBOTT, D.F. "Slow Recovery from Carbon Monoxide Poisoning." *Postgraduate Medical Journal*, 48: 639–642, 1972 (2 figures, 7 references)

Abstract: An 18-year-old undergraduate student was found deeply unconscious, and another undergraduate was found dead in their student residence. At hospital, the patient was not treated for CO poisoning as it was thought to be a drug overdose. The patient remained deeply comatose and had spasms 15 hours later. It was determined that the student residence where the patient was found was without ventilation, the windows were closed, and a gas fire was fully on. It is estimated that the gas fire was burning for 25 hours and could have resulted in a room CO concentration of 1%, which could have caused a blood COHb of 70%. The patient's blood COHb concentration in a blood sample collected four hours after the patient was discovered showed a blood COHb of 20%. He was administered HBOT and NBOT over the next several days, and he slowly recovered, but it was not until 38 days in hospital that the patient could hold a reasonable conversation. The other student had a fatal COHb post-mortem blood concentration.

It is suggested that as very slow recovery occurred, treatment of severe carbon monoxide poisoning in young patients should be optimistic. It is also suggested that more rigorous efforts should be made to ensure the patency of the flues of gas burning appliances by regular inspection.

Reference Number: 30917

HUANG, C.-C.-., M.-H. CHUNG, S.-F. WENG, C.C. CHIEN, S.-J. LIN, H.-J. LIN, H.-R. GUO, S.-B. SU, C.-C. HSU, AND C.-W. JUAN. "Long-Term Prognosis of Patients with Carbon Monoxide Poisonings: A Nationwide Cohort Study." *PLoS ONE*, 9: e105503, 2014 (3 tables, 1 figure, 13 references)

Abstract: A retrospective study was conducted on the long-term effects of CO poisoning in 441 CO-poisoning patients and 8,820 matched, control patients between 1999 and 2010 in Taiwan. During that period, 37 of the CO-poisoning patients (8.4%) died compared to 142 (1.6%) of the controls. Older age (> 30 years of age), male gender, diabetes mellitus, hypertension, and low income were independent mortality predictors. After adjusting for these conditions, CO-poisoning patients still had a 4.1 times greater mortality rate than controls.

> In summary, our study shows that COP+ patients had a significantly higher risk of death than did COP controls, especially during the 12 months immediately after an acute COP event. Early follow-up and secondary prevention of death are needed for patients with COP.

Reference Number: 30918

TSUDA, S., J. UCHIDA, T. IWAI, C. MOROOKA, K. KABEI, T. MINO, N. KUWABARA, AND T. NAKATANI. "Delayed Neurologic Sequelae Caused by Carbon Monoxide Poisoning in a Kidney Transplant Recipient." *Experimental and Clinical Transplantations*, 6: 849–851, 2019 (3 figures, 6 references)

Abstract: A 65-year-old man received a kidney transplant from his daughter in 2014. Forty-seven months later, he attended hospital due to losing consciousness. One month later, he developed higher-brain dysfunction. It was determined that he warmed himself with a traditional, Korean wood-fired floor heating system called an Ondol. He was finally diagnosed as suffering from CO poisoning and was treated with HBOT.

> In our case of delayed neurological sequelae after carbon monoxide poisoning in a kidney transplant patient, it was difficult to make a definitive diagnosis due to the numerous differential diagnoses. However, diagnosis was confirmed and treatment was conducted after a follow-up cranial MRI was made and after detailed interview of the patient's family. It may be necessary to consider this condition in kidney transplant recipients who develop higher brain dysfunction.

Reference Number: 30919

KIM, H.-H., S. CHOI, Y.S. JUNG, Y.-G. MIN, D. YOON, AND S.E. LEE. "Stroke Incidence in Survivors of Carbon Monoxide Poisoning in South Korea: A Population-Based Longitudinal Study." *Medical Science Monitor* 26 (2020): e926116, 2020 (3 tables, 4 figures, 27 references)

Abstract: Of the 29,301 patients diagnosed with CO poisoning between 2007 and 2018 in South Korea, 984 (3.36%) were diagnosed with stroke, of which nearly 50% occurred within one year of CO poisoning. The overall standard incident ratio was 19.49 during the first year, which decreased to 5.64 during the second year of recovery. It was equal to the general population after seven years. Cumulative HRs did not differ between patients who received and did not receive HBOT.

Vessel injuries in the central nervous system may result from a disruption of microvascular endothelial integrity caused by excessive increases in superoxide, nitric oxide and peroxynitrites. These increases may be caused by ischemia/reperfusion injury in patients with CO poisoning resulting in the development of ICH [intracranial hemorrhagic] stroke. However, because stroke occurs soon after CO poisoning followed by a rapid decrease of risk during the second year, the development of SAH [subarachnoid hemorrhagic] may result from events during the initial stage of CO poisoning process, such as an epinephrine-induced catecholamine surge of CO induced dysregulation of cerebral blood control. These phenomena may increase intracranial pressure, affecting existing cerebral aneurysms, or may directly induce aneurysmal changes in cerebral arteries. If these result aneurysms do not remain stable, they can eventually lead to SAH stroke after CO poisoning. The secondary peaks in the incidence patterns of ischemic stroke and ICH stroke are thought to be consequences of the long-term effects of endothelial dysfunction and disruption, causing subtle changes in the cerebral micro vessels if they do not undergo spontaneous repair.

Reference Number: 30920

CHUNG, W.-S., C.-L. LIN, AND C.-H. KAO. "Carbon Monoxide Poisoning and Risk of Deep Vein Thrombosis and Pulmonary Embolism: A Nationwide Retrospective Study." *Journal Epidemiological and Community Health*, 69: 557–562, 2015 (3 tables, 1 figure, 33 references)

Abstract: The incidence of deep vein thrombosis (DVT) and pulmonary embolism (PE) was determined in 8,316 patients with CO poisoning and 33,624 controls without CO poisoning in Taiwan between 2000 and 2011. The adjusted HR was 3.85 times for CO patients developing DVT and 1.66 times (not statistically significant) for CO patients developing PE.

The epidemiological causes of increased risk of DVT in patients with CO poisoning are unclear. Exogenous exposure to high amounts of CO can result in toxic effects, causing hypoxia and inflammation. CO enters the blood to induce hypoxia through formation of carboxyhemoglobin, which causes a leftward shift in the oxyhemoglobin dissociation curve and binds to protein to impair mitochondrial function. CO also causes inflammation by increasing the levels of cytosolic and oxygenase-1-protein, and inducing the release of myeloperoxidase, protease and reactive oxygen species which results in oxidative stress. The pathogenesis of VTE [venous thromboembolism] is reportedly associated with inflammation and oxidative stress.

Reference Number: 30921

CHOI, S., S. HAN, S. NAH, Y.H. LEE, Y.S. CHO, H. LIM, M.S. KIM, AND G.W. KIM. "Effect of Ethanol in Carbon Monoxide Poisoning and Delayed Neurologic Sequelae: A Prospective Observational Study." *PLoS ONE*, 16: e0245265, 2021 (3 tables, 2 figures, 29 references)

Abstract: The BAC of 171 patients attending the ER in Korea between 2016 and 2019 was determined as well as the development of DNS (at two weeks, one month, and three months). On average, CO exposure was of 190 minutes duration for non-DNS patients and 375 minutes for the DNS patients. Intended suicides accounted for 82% of the patients, and DNS occurred in 28 patients (16%). The mean BAC was 11 mg/dL (95% range, 0–81 mg/dL) for non-DNS patients and 1 mg/dL (95% range, 0–6 mg/dL) for DNS patients.

Table: Characteristics and BAC of Non-DNS and DNS Patients

Characteristics	Non-DNS patients (n = 143)	DNS patients (n = 28)
Age	42 years	47 years
Male gender (%)	71	54
GCS score	15	10
Lactate (mmol/L)	3.84	2.34
BAC (mg/100mL)	11.0	1.5

Source: Adapted from Choi et al. 2021.

Higher initial ethanol levels and higher initial GCS scores were associated with lower incidence of DNS. Ethanol could have a neuroprotective effect on the occurrence of DNS in CO-poisoning patients.

Reference Number: 30922

GAN, Y., H. GU, L. YANG, Z. LI, AND J. ZHANG. "Prognosis of Patients in Prolonged Coma after Severe Carbon Monoxide Poisoning." *Human and Experimental Toxicology*, 40: 1355–1361, 2021 (2 tables, 1 figure, 21 references)

Abstract: Eight female and five male patients (ages 23 to 71 years) poisoned by CO due to coal heating and poor ventilation, who had a coma for more than seven days were evaluated. Their initial blood COHb concentrations ranged from 2% to 40% (mean 26.3%). The patients were treated with HBOT daily for six days. Only two of the 13 patients (15.4%) regained consciousness. One year later, two patients had died, nine were in a persistent vegetative state, one was moderately disabled, and one (age 23 years) had a good recovery.

In this study, most patients with prolonged coma after severe CO poisoning had a poor prognosis, and the younger patients had a better prognosis. Severe respiratory failure, hypotension, and renal failure during the course of the disease were associated with a poor prognosis. The prognosis of patients with injuries in two sites in early CT was poor. Multiple lesions (> 3) on MRI and extensive white matter damage (Fazekas grade (PVH or DEMH = 3) on MRI of chronic phage were also associated with a poor prognosis.

Reference Number: 30923

BHAGWAT, S., AND G. BRUXNER. "Not Quite Out of the Woods: Potential for Misdiagnosis of Delayed Neurologic Syndrome of Carbon Monoxide Poisoning as Relapse of Mental Illness." *Australasia Psychiatry*, 25: 494–496, 2017 (1 figure, 12 references)

Abstract: The incidence of DNS in acute CO poisonings range from 3% to 40%. DNS occurs between two and six weeks post-exposure in 50% of the victims. The clinical sequelae of DNS are very variable and include mild cognitive impairment, personality change, Parkinsonism and movement disorders, hearing and visual loss, subcortical and cortical dementia, depression, and psychosis. A 58-year-old man presented in hospital following a suicide attempt by burning charcoal in a closed room. He was treated with oxygen therapy and, 22 days later, had poor sustained attention and short-term memory impairment. He developed delirium and was unable to walk, which resembled a relapse of his psychiatric illness rather the DNS.

Delayed neuropsychiatric presentation of CO poisoning are not uncommon and are likely to occur in a psychiatric treatment setting. DNS should always be considered with a sudden deterioration in mental state after apparent recovery from CO poisoning.

Reference Number: 30924

ROOHI, F., R.W. KULA, AND N. MEHTA. "Twenty-Nine Years after Carbon Monoxide Intoxication." *Clinical Neurology and Neurosurgery*, 103: 92–95, 2001 (1 figure, 27 references)

Abstract: A 40-year-old woman was admitted to a Long Island College Hospital in 1997 with the diagnosis of acute gastrointestinal bleed, anemia, and multiple sclerosis. She had suffered a severe accidental CO poisoning in 1968 at the age of 11 years. She was unconscious and, after a prolonged convalescence period, made an apparently complete physical and mental recovery, except for a very mild right spastic hemiparesis. She was medically stable until 1985 (age 28 years), when she noticed very slowly progressive walking and coordination difficulties, which worsened until she had to quit her job. She developed depression, insomnia, and weight gain and was diagnosed as having multiple sclerosis. The diagnosis of multiple sclerosiswas based mainly on the brain white matter changes detected by MRI, but the same changes could be due to CO poisoning. Delayed and progressive nervous system dysfunction also occur post-radiation and post-polio trauma.

We conclude that a post-CO poisoning syndrome similar to the post-poliomyelitis and post-radiation motor neuron syndromes may occur many months or years after severe acute CO intoxication. The syndrome that we describe is most likely under-recognized because, as in our patient, CO intoxication long before the onset of symptoms could be dismissed as non-contributory. Consequently, careful history taking and knowledge of this syndrome may prevent misdiagnosis and inappropriate treatments in patients with progressive neuropsychiatry disorders and appropriate history of CO poisoning.

Reference Number: 30925

BRUNO, A., W. WAGNER, AND W.W. ORRISON. "Clinical Outcome and Brain MRI Four Years after Carbon Monoxide Intoxication." *Acta Neurologica Scandinavica*, 87: 205–209, 1993 (3 figures, 18 references)

Abstract: A 41-year-old woman, her 43-year-old husband, and 14-month-old girl granddaughter were found unconscious in their home after being exposed to CO from a natural gas wall heater for 36 hours. The vent pipe of the heater was full of soot, which released toxic amounts of CO under winter conditions. Their COHb concentrations at hospital were all less than 20% after they had received 100% oxygen from the start of their discovery. They all recovered and were released from hospital. Four years later, the MRI of the wife showed a left globus pallidus lesion but no clinical signs except sensorineural hearing loss. The MRI of the granddaughter had parieto-occipital periventricular lesions but not cognitive deficits. The husband suffered from depression, and his MRI showed lesions in the left thalamus medial temporal lobe and cerebellum.

> One reason for poor clinical-MRI correlation in our patients may be that some lesions caused relatively mild and transient symptoms, which were no longer apparent on examination four years later. Another reason for poor clinical-MRI correlation may be that MRI is so sensitive that it demonstrates partially damaged, but still functioning brain regions. It is likely that a variety of brain lesions caused by CO intoxication will continue to be identified with MRI. The clinical significance of these lesions, differing in location and size remains to be determined.

3.10 TREATMENT

> As regards treatment, Mr. Miller lays most stress on perfect rest in the recumbent position, hypodermic injections of ether, and external application of warmth.
>
> —Anonymous, "Recommended Treatment of CO Poisoning," *British Medical Journal* (1898)

> Tissue hypoxia is the major outcome of CO intoxication, therefore based on chemical and pathophysiological data, O_2 is the natural antidote.
>
> —L.D. Prockop and R.I. Chichkova, *Journal of Neurological Science* (2007)[7]

Oxygen is the natural antidote to CO poisoning. The COHb must be removed from the body as quickly as possible by dissociation with high

7 L.D. Prockop and R.I. Chichkova, "Carbon Monoxide Intoxication: An Updated Review," *Journal of the Neurological Sciences*, 262 (1–2): 122–130, 2007.

doses of oxygen. The main controversy is how to deliver the oxygen, normobaric (NBOT) or hyperbaric (HBOT) (30101–21). HBOT offers more risks (baro-trauma, oxygen toxicity, and so on) than NBOT, and there can be delays in treatment in order to access a hyperbaric chamber (31002, 31007, 31009, 31012, 31020). The addition of 5–10% CO_2 to the oxygen stream (carbogen) has been found to increase the ventilation rate and may reduce the half-life of CO to the same rate as obtained with HBOT (31008–9). High serum lactate concentrations may indicate a more severe CO poisoning and, hence, the use of HBOT (31014–16).

Reference Number: 31001

ANNANE, D., K. CHADDA, P. GAIDOS, M.-C. JARS-GUINCESTRE, S. CHEVRET, AND J.-C. RAPHAEL. "Hyperbaric Oxygen Therapy for Acute Domestic Carbon Monoxide Poisoning: Two Randomized Controlled Studies." *Intensive Care Medicine*, 37: 486–492, 2011 (1 table, 3 figures, 20 references)

Abstract: Patients with a transient loss of consciousness due to CO poisoning were assigned at random to either six hours of NBOT (n = 169) or four hours of NBOT and one HBOT session. The one-month complete recovery rate was 58% with HBOT and 61% without HBOT. In total, 206 patients with an initial coma due to CO poisoning were also randomized to either four hours of NBOT and one HBOT session or four hours of NBOT and two sessions of HBOT. The complete recovery rate was lower in the two-session HBOT group (47%) compared to 68% with one HBOT session.

This study did not confirm that hyperbaric oxygen therapy improves recovery from pure CO poisoning. In addition, in comatose patients, repeating hyperbaric oxygen therapy resulted in worse outcomes compared to one session.

Reference Number: 31002

BALZAN, M.V., G. AGIUS, AND A.G. DEBONO. "Carbon Monoxide Poisoning: Easy to Treat but Difficult to Recognize." *Postgraduate Medical Journal* 72: 470–473, 1996 (4 tables, 58 references)

Abstract: A short, general review about CO poisoning, the difficulty of diagnosing, treatments, and outcomes is presented.

Hyperbaric therapy has occasionally been shown to cause baro trauma to the tympanic membrane. It should be avoided after cardiopulmonary resuscitation with external chest compression, pulmonary emphysema, and asthma as life-threatening pneumothorax may develop. Probably frail, elderly patients are more safely treated with atmospheric pressure. Supportive therapy is very important in acute intoxication. Severely intoxicated patients may need intubation, mechanical ventilation, and intensive care. However, the majority of patients do well with hyperbaric oxygen.

Reference Number: 31003

HAMPSON, N.B., AND N.M. HAUFF. "Carboxyhemoglobin Levels in Carbon Monoxide Poisoning: Do They Correlate with the Clinical Picture?" *American Journal of Emergency Medicine*, 26: 665–669, 2008 (3 tables, 1 figure, 9 references)

Abstract: The COHb saturation was measured in 922 male and 485 female patients treated at a HBO_2 centre for CO poisoning between 1978 and 2005. Statistically higher COHb measurements were associated with male gender, adults greater than 18 years of age, and loss of consciousness, but the variability of the COHb measurements was high due to delays and oxygen administration after exposure. Death occurred in one patient with a 3% COHb saturation, and one patient survived with a 72.3% COHb saturation. But the mean COHb results were 32.1% in those who died compared to 23.1% in those who survived. The mean COHb in patients who experienced a loss of consciousness was 24.3% compared to 22.3% in those without a loss of consciousness.

It is probable that a major reasons for this is the delay from termination of CO exposure to the time blood is drawn for COHb measurement, as well as intervening administration of oxygen. Both allow the COHb level to fall before blood is obtained. Almost all patients had COHb measured at an outside hospital before referral, but information was lacking as to the time lapsed from removal from CO exposure to obtaining blood. Although venous blood drawn into a heparinized sample tube at the scene and transported with the patient for subsequent analysis will remain stable even at room temperature, this is not commonly done.

Reference Number: 31004

HAMPSON, N.B., AND N.M. HAUFF. "Risk Factors for Short-Term Mortality from Carbon Monoxide Treated with Hyperbaric Oxygen." *Critical Care Medicine*, 36: 2523–2527, 2008 (2 tables, 2 figure, 22 references)

Abstract: The mortality rate of 1,473 patients (mean age 35 years) treated with HBOT was determined between 1978 and 2005 at the Virginia Medical Centre. The CO exposure was accidental in 1,024 patients (70%) and intentional in 441 patients (30%). Patients exposed to CO from a fire source have a lower arterial blood potential of hydrogen (pH) than those exposed to CO alone.

Table: Characteristics of Survivors and Non-Survivors of CO Poisoning

	Survivors (%)	Non-Survivors (%)
Gender	66 male	53 male
Accidental	70	70
Intentional	30	30
Fire CO exposure	11	53
CO only	89	47
Loss of consciousness	52	100
COHb %	22.6	31.1
Initial arterial pH	7.39	7.20
Intubation during HBOT	16%	76%

Source: Adapted from Hampson and Hauff 2008.

To summarize, the mortality rate of CO poisoning receiving hospital based medical treatment including hyperbaric oxygen is approximately 3%. Characteristics significantly associated with mortality include fire as a source of CO, loss of consciousness, carboxyhemoglobin level, arterial pH, and presence of endotracheal intubation during hyperbaric treatment. Those designed future prospective studies should consider measurement of cyanide levels in patients exposed to CO from fires to determine its contribution to metabolic acidosis as a predictor of mortality.

Reference Number: 31005

ROSE, J.J., M. NOURAIE, M.C. GAUTIER, A.F. PIZON, M.I. SAUL, M.P. DONAHOE, AND M.T. GLADWIN. "Clinical Outcomes and Mortality Impact of Hyperbaric Oxygen Therapy in Patients with Carbon Monoxide Poisoning." *Critical Care Medicine*, 46: e649–e655, 2018 (3 tables, 1 figure, 28 references)

Abstract: CO poisoning affects approximately 50,000 patients annually in the United States. CO poisoning is typically treated with NBOT and/or HBOT and supportive care. There are risks with HBOT in the critically ill, including ear barotrauma, pulmonary barotrauma, and oxygen toxicity. Between 2000 and 2014, 1,899 patients attended the University of Pittsburgh Medical Centre due to CO poisoning. Eighty percent of the patients had a COHb of 3% or more. The most common symptom at presentation was headache (55%). About 26% of the patients received HBOT, and, of these, 39% had a COHb of 19% or more and were mainly older age men. Ninety-one percent of the HBOT patients had at least one high-risk feature (that is, cardiac complications, COHb of 25%+, pH < 7.20, fire exposure, or syncope) compared to 21% of those who did not receive HBOT. A total of 1.3% of the patients died during the initial CO-poisoning admission, 2.9% by year 1 and 4.4% by year 2. In the group of CO-poisoning patients exposed to fire, HBOT was associated with lower acute mortality but not with one-year mortality.

Table: Odds Ratio Risk Factors of Acute CO Poisoning Death

Factor	Odds ratio of acute and one-year mortality
Fire exposure	6.37 times
Respiratory distress	31.75 times
Syncope	41.26 times
Cardiac complication	114.75 times
Low pH (< 7.20)	204.57 times
High COHb (29% or greater)	5.95 times

Source: Adapted from Rose et al. 2018.

In this study, we assess a database of over 1,000 poisoned adults. Because there is a low overall acute mortality rate a randomized control trial would be difficult to power a test for mortality benefit. A study such as ours may be the only way to associate a mortality benefit to HBO_2. We found HBO_2 treatment is associated with improved survival in both the acute and long-term setting.

Reference Number: 31006

LIM, M.L., S.J. KIM, M.K. TAN, K.H. LIM, AND H.G. SEE. "Provision of Emergency Hyperbaric Oxygen Treatment for a Patient during the COVID-19

Pandemic." *Diving and Hyperbaric Medicine*, 51: 78–81, 2021 (1 figure, 12 references)

Abstract: In March 2020, the World Health Organization declared COVID-19 a global pandemic. On 19 June 2020, a 21-year-old woman was found in an enclosed room with burning charcoal after a suicide attempt. At hospital, she had third degree burns to 5% of her total body surface on her limbs. Her COHb was 23%, and she was drowsy. She was intubated and treated with HBOT using strict COVID procedures in the HBO_2 chamber. The fire safety of polypropylene gowns needed to be addressed due to the 100% oxygen and the risk of static electricity. The patient was kept intubated to limit the risk of aerosol generation.

> The patient underwent two HBOT sessions within 24 hours, and the first HBOT session was started within six hours of presentation. Two samples of endotracheal aspirates, each collected 24 hours apart, were sent for COVID-19 PCR [polymerase chain reaction] testing. Both samples tested negative for COVID-19. The patient subsequently received surgical treatment for her burns and she recovered well.

Reference Number: 31007

CHO, K., Y. OKUNO, Y. KAKUDA, T. TSUTSUMI, T. KOGAME, S. OHRTSURU, N. SATO, AND K. KOIKE. "Convulsive Seizure and Pulmonary Edema during Hyperbaric Oxygen Therapy: A Case Report." *Journal of Medical Investigation*, 65: 286–288, 2018 (3 figures, 9 references)

Abstract: A 31-year-old man was taken to hospital after a house fire and had no apparent external burns but had a lot of soot in his nasal cavity and a COHb of 27.7%. He was diagnosed with CO poisoning and inhalation injury and, after intubation, was administered HBOT. He was administered HBOT a second time and, during decompression, developed a convulsive seizure and pulmonary edema. The convulsive seizure was attributed to oxygen toxicity in the central nervous system. He was not administered any more HBOT and was discharged from ICU on day 6.

> We present a rare case of simultaneous convulsive seizure and pulmonary edema during HBOT. Through this case, two significant clinical suggests were obtained. First HBOT for acute indications should be performed in multiplace chambers, so that accompanying physicians can treat the various expected and unexpected adverse events should they occur. Second, medical equipment like monitoring devices or syringe pumps

available under hyperbaric conditions should be introduced in Japan. As CO poisoning is one of the leading causes of poison-associated deaths, prompt improvement in the Japanese HBOT environment is required.

Reference Number: 31008

FISHER, J.A., S. ISCOE, L. FEDORKO, AND J.K. DUFFIN. "Rapid Elimination of CO through the Lungs: Coming Full Circle 100 Years On." *Experimental Physiology*, 96: 1262–1269, 2011 (6 figures, 58 references)

Abstract: At the beginning of the twentieth century, the preferred treatment for CO poisoning was carbogen (5–10% CO_2 to increase ventilation and oxygen), and it continued until the 1960s when NBOT and HBOT became the treatment of choice. All three treatments have different problems. Oxygen treatment can result in hypocapnia (a decrease in alveolar and blood carbon dioxide levels below the normal reference range of 35 mm Hg) and can result in a reduction of blood flow to the tissues. It was determined that it was not the CO_2 in the carbogen treatment that increased the removal of CO from the body but the increase in the ventilation rate. Increases in the ventilation rate can result in an elimination half-life of CO almost as short at HBOT. Perhaps a better treatment would be increasing alveolar ventilation while maintaining normocapnia.

Now, half a century later, there is accumulating evidence that hyperbaric O_2 is not efficacious, most probably because of delays in initiating treatment. We now also know that increases in pulmonary ventilation with O_2 enriched gas can clear CO from the blood as fast, or very nearly as fast, as hyperbaric O_2. Compared with hyperbaric O_2, the technology for accelerating pulmonary clearance of CO with hyperoxic gas is not only portable and inexpensive, but also may be far more effective because treatment can be initiated sooner. In addition, the technology can be distributed more widely, especially in developing countries where the prevalence of CO poisoning is the highest. Finally, early pulmonary CO clearance does not delay or preclude any other treatment including subsequent treatment with hyperbaric O_2.

Reference Number: 31009

SEIN ANAND, J., D. SCHETZ, W. WALDMAN, AND M. WISNIEWSKI. "Hyperventilation with Maintenance of Isocapnia: An Old New Method in Carbon Monoxide Intoxication." *PLoS ONE*, 12 (1): e0170621, 2017 (2 figures, 14 references)

Abstract: Since the air-blood gradient of CO is much higher than the blood-air gradient, the uptake of CO from the ambient air is much faster than its elimination. From the 1920s to the 1960s, a method of increasing the rate of elimination of CO from the patient was by using carbogen—a mixture of 6% CO_2 and 94% oxygen—which induced hyperventilation by hypercapnia (elevated CO_2 level). This treatment, however, could cause an elevation in CO_2 levels and subsequent potentially fatal acidosis. This treatment was replaced by HBOT and NBOT. A new method (CleanMate) adjusts the CO_2 added to the breathing mixture to maintain a uniform CO_2 concentration (isocapnia). Eight female and five male chronic smokers (ages 25 to 64 years) were administered either 100% oxygen (non-isocapnia hyperventilation [NIH] and isocapnic hyperventilation [IH]) on two separate occasions. The mean initial capillary blood COHb was 5.0% and 5.1% respectively. The mean COHb after 20 minutes of treatment was 3.6% for NIH and 2.9% for IH.

Table: A Comparison of NIH and IH Treatments in Chronic Smokers

	NIH treatment	IH treatment
Mean $t_{1/2}$ of CO	47.3 minutes	29.6 minutes
Mean COHb after 20-minute treatment (%)	3.6	2.9
Mean pH after 10-minute treatment	7.55	7.41
Percentage of subjects with alkalosis (pH > 7.45) after 10-minute treatment (%)	84.6	0

Source: Adapted from Anand et al. 2017.

The elimination t1/2 of CO during IH was comparable with the values reported during HOT and lower than can be achieved with NOT or NIH. No serious adverse effects were reported during IH procedures. Further studies, especially direct comparisons with NOT and HOT are necessary to evaluate the effectiveness of IH in the treatment of acute CO poisoning.

Reference Number: 31010

OCAK, T., E. TTEKIN, M. BASTURK, A. DURAN, M. SERINKEN, AND M. EMET. "Treatment in Carbon Monoxide Poisoning Patients with Headache: A Prospective Multicenter, Double-Blind, Controlled Clinical Trial." *American Journal of Emergency Medicine*, 34: 2140–2145, 2016 (2 tables, 4 figures, 22 references)

Abstract: A total of 117 CO-poisoned patients with headache were randomized into three treatment groups (NBOT alone; NBOT + metoclopramide; NBOT + metamizole sodium) to determine their effect on the relief of headache. Patients reported improvement rates of headache at baseline, 30 minutes, 90 minutes, and four hours later using visual analogue scale (VAS). During the observations, there was no difference in headache improvement between the three treatment groups.

Table: Improvement in Headache Symptoms after Treatment with NBOT, NBOT + Metoclopramide, and NBOT + Metamizole

	NBOT	NBOT + metoclopramide	NBOT + metamizole
Mean baseline COHb	23.7	19.5	19.7
Mean COHb 4 hours later	5.9	5.7	4.4
Mean baseline VAS score for headache	6.0	6.4	7.2
Mean VAS at 30 minutes	4.5	4.1	5.2
Mean VAS at 90 minutes	2.4	2.1	3.4
Mean VAS at 4 hours	1.4	1.0	2.0

Source: Adapted from Ocak et al. 2016.

The primary outcome of this study was the finding that there was no significant statistical or clinical difference between patients treated with high-flow oxygen alone and patients treated with oxygen plus medication. The additional use of analgesic medicine does not seem to add any comfort to CO-intoxicated patients with headache.

Reference Number: 31011

CAGLAR, R., S. SERIN, G. YILMAZ, A. TORUN, AND I. PARLAK. "The Impact of Treatment with Continuous Positive Airway Pressure on Acute Carbon Monoxide Poisoning." *Prehospital and Disaster Medicine*, 34: 588–591, 2019 (5 tables, 2 figures, 6 references)

Abstract: The blood COHb concentration was determined using the fingertip Masimo CO-oximeter Rad-7 at the initiation and 30 minutes after treatment of 37 CO-poisoned patients with the typical NBOT and 40 CO-poisoned patients with continuous positive airway pressure (CPAP) treatment. CPAP was 1.0 fraction of inspired oxygen (FiO_2) at 12 centimetres of water pressure with non-invasive mechanical ventilation with an oronasal mask. The initial median COHb for the NBOT group was 19%, which decreased to 14% 30 minutes later. The initial median COHb for the CPAP group was 22%, which decreased to 9% 30 minutes later. Headache symptoms declined two units for the NBOT and five units for the CPAP treatment. Three patients could not tolerate the CPAP treatment.

> The use of CPAP was determined to more rapidly reduce COHb level as opposed to high-flow oxygen therapy. It is also thought that it may enable earlier discharge by reducing the duration of the emergency follow-up since it provides a faster improvement in the symptoms of patients.

Reference Number: 31012

OLSON, K.R., AND D. SEGER. "Hyperbaric Oxygen for Carbon Monoxide Poisoning: Does It Really Work?" *Annals of Emergency Medicine*, 25: 535–537, 1995 (21 references)

Abstract: The use of HBOT to treat CO poisoning is still controversial as it is expensive, carries a small risk of pneumothorax, tympanic membrane ruptures, and seizures. The evidence that HBOT is superior to NBOT is largely published in isolated case reports and small case series.

> One hundred years ago, Haldane reported that oxygen at 2 atm [atmospheres] prevented carbon monoxide (CO) poisoning in mice. Since 1962, when Smith described the use of hyperbaric oxygen (HBO) to treat accidentally poisoned patients, reports of the successful use of this intervention have steadily entered the literature. Three decades after Smith's report and a century after Haldane's, many physicians and clinical toxicologists consider HBO the standard of care for CO poisoning. However, most scientific evidence does not support its benefit and no widespread agreement exists on selection of patients for treatment with HBO versus that with 100% oxygen at ambient pressure (normobaric oxygen or NBO).

Reference Number: 31013

LANE, T.R.A., W.J. WILLIAMSON, AND J.M. BROSTOFF. "Carbon Monoxide Poisoning in a Patient with Carbon Monoxide Retention: A Therapeutic Challenge." *Cases Journal*, 1: 102–103, 2008 (4 references)

Abstract: A 70-year-old man and his wife attended hospital as a result of a house fire. At hospital, the man appeared drowsy and not responsive, even when he was administered high-flow oxygen. There was no evidence of external burns, face burns, or singed nostril hairs. With high-flow oxygen, his pH dropped to 7.32, and the partial pressure of CO_2 in his arterial blood increased to 7.80 kilopascal. Upon commencement of controlled flow oxygen at 0.24 FiO_2, the patient recovered. His initial COHb was 9%, and the COHb of his wife, who showed no symptoms, was 11%. The man suffered from COPD.

Our difficulty in the above case was that the patient almost certainly had undiagnosed COPD and hence when high-flow oxygen was given—the preferred initial management strategy—the patient retained CO_2 and became increasingly obtunded, necessitating a reduction in the FiO_2, and hence prolonging the clearance of carbon monoxide.

Reference Number: 31014

ICME, F., N. KOZACI, M.O. AY, A. AVCI, U. GUMUSAY, M. YILMAZ, AND S. SATAR. "The Relationship between Blood Lactate, Carboxyhemoglobin and Clinical Status in CO Poisoning." *European Review of Medical and Pharmacological Sciences*, 18: 393–397, 2014 (4 tables, 10 references)

Abstract: The reduction in oxygen delivery, especially in the brain, muscle, intestine, and erythrocytes with a high rate of glycolysis produced increased lactate formation. Blood lactate levels are an indicator of compromised oxygen delivering and correlated with short-term prognosis in critically ill patients. A detailed clinical and biochemical analysis was conducted in 35 CO-poisoned patients who received HBOT and 166 CO-poisoned patients who received NBOT. There were no significant differences between the two groups in CO_2, oxygen, or bicarbonate ion (HCO_3) upon admission.

Table: Mean COHb and Lactate Concentrations According to GCS and Cardiac Enzyme Levels

Clinical condition	Mean COHb concentration (%)	Mean lactate concentration
GCS < 15	31.44	50.2
GCS = 15	19.52	18.77
Elevated cardiac enzyme levels	18	45
Normal cardiac enzyme levels	22	19

Source: Adapted from Icme et al. 2014.

Blood lactate and COHb levels were both correlated with the changes of consciousness in CO poisoning. Blood lactate levels together with COHb in defining indications for HBO treatment might be suggested.

Reference Number: 31015

CERVELLIN, G., I. CORNELLI, G. RASTRELLI, A. PICANZA, AND G. LIPPI. "Initial Blood Lactate Correlates with Carboxyhemoglobin and Clinical Severity in Carbon Monoxide Poisoned Patients." *Clinical Biochemistry*, 47: 298–301, 2014 (2 tables, 21 references)

Abstract: COHb and blood lactate concentrations were determined in arterial blood in 38 CO-poisoned patients as well as serum troponin levels. Twenty-three patients (61%) underwent HBOT. The median COHb was 8.5% for patients with NBOT and 18.9% for those with HBOT. Blood lactate concentrations was the only parameter associated with hospital admission. The median blood lactate concentration was 1.2 mmol/L in patients not admitted and 1.8 mmol/L in patients admitted to hospital.

In conclusion, the results of this retrospective investigation on 38 patients presenting to the ED with acute CO poisoning suggest that initial blood lactate level may be an additional and useful parameter for risk stratification of patients, especially for predicting the need for hospitalization.

Reference Number: 31016

ALTINTOP, I., M.E. AKCIN, M. TATLI, AND M.S. ILBASMIS. "Factors That Influence the Decision for Hyperbaric Oxygen Therapy (HBOT) in Cases of Carbon Monoxide Poisoning: A Retrospective Study." *Annals of Burns and Fire Disasters*, 31: 168–173, 2018 (1 table, 6 figures, 22 references)

Abstract: A study was conducted of 31 patients with a COHb of 30% or less and 69 patients with a COHb greater than 30%. Approximately 90% of all patients were administered a single session of HBOT.

Table: Patient Characteristics of Those with a Blood COHb Less Than or Equal to 30% and Those with a COHb Greater Than 30%

Variables	COHb ≤ 30%	COHb > 30%
Suicide attempt (%)	3.2	1.4
GCS < 14 (%)	9.7	10.1
Mean Lactate (mmol/L)	2.89	5.10
Mean pH	7.37	7.32
Seizures (%)	9.7	30.4
Syncope (%)	9.7	30.4
Speech impairment (%)	9.7	8.7

Source: Adapted from Altintop et al. 2018.

HBO therapy should certainly be administered to patients with obvious symptoms, low pH, high lactate levels and evident neurological findings. Multiple sessions of HBOT should be planned for patients with persistent symptoms that do not clear. Although it is recommended that patients are followed up to at least 24 hours, a follow-up 3 weeks later is also advised. The results of the study showed no significant differences between the COHb level groups and it must be kept in mind that in COP there is no difference between age groups. The treatment of COP is a race against time. Patients needing HBOT must be carefully selected within the first few hours and kept under close observation.

Reference Number: 31017

BRANDEBURG, J. "Inhalational Injury: Carbon Monoxide Poisoning." *American Journal of Nursing*, 80: 98–100, 1980 (2 tables, 1 figure, 11 references)

Abstract: A general review was conducted regarding domestic CO poisoning. Every exposure to CO should be treated as an emergency. The only treatment of CO poisoning that is not controversial is the administration of oxygen, which was first proposed by J.B.S. Haldane in 1895. Other treatments such as hypothermia and exchange blood transfusion are not recommended. Some of the effects of CO poisoning include personality changes of an aggressive or psychotic mature; memory, vision, hearing, and speech may be impaired as well as movement disorders and ataxia.

Because carbon monoxide is an odorless, colorless gas, it is easy to forget that when a victim is discovered s/he must be removed from the source environment lest the rescuer also become a victim of carbon monoxide poisoning. Until sophisticated equipment arrives, CPR [cardiopulmonary resuscitation] may be the only help a rescuer can offer the unresponsive victim. For the carbon monoxide poisoning victim, CPR unfortunately may not be highly successful. Mouth-to-mouth ventilation delivers approximately 16 percent oxygen to the victim. This concentration of oxygen is insufficient to displace the carbon monoxide from hemoglobin or to increase the amount of dissolved oxygen in the blood. CPR should be used on victims of carbon monoxide poisoning. But the rescuer ought to know that unsuccessful resuscitation attempts are not necessarily the result of poor technique of lack of knowledge. The mechanism of carbon monoxide poisoning itself may prevent the rescuer from successfully resuscitating the victims.

Reference Number: 31018

HUSTON, C.J. "Carbon Monoxide Poisoning." *American Journal of Nursing*, 96 (1): 48–49, 1996

Abstract: A hypothetical case of CO poisoning in a 32-year-old man named Steve is presented. He had a headache, slurred speech, and reddish face and lips and felt like he had the flu. It is recommended that Steve is kept calm and motionless as exertion increases metabolism and oxygen consumption and can worsen hypoxia. Paramedics administered 100% oxygen in a tightly sealed rebreathing mask. At hospital, various tests were conducted. A hypothermia blanket may minimize metabolism activity, and he was kept on strict bed rest. His blood COHb was 40%, and so he was sent to a HBO_2 chamber.

When he's ready to go home, Steve will be told to report any personality or consciousness changes to his physician immediately, since neurologic and psychological impairment may not occur until two to three weeks after exposure to CO. He'll be advised to contact the local utilities company to assess the CO level in his home and warned that this experience has made him more susceptible to CO poisoning in the future. The nurse will recommend that he install a CO detector in his home to prevent future poisoning.

Reference Number: 31019

AYDIN, G. "Hyperbaric Oxygen Treatment Resistant Carbon Monoxide Poisoning." *Journal of Clinical Anaesthesia*, 50: 24–25, 2018 (1 figure, 4 references)

Abstract: A 54-year-old female patient was brought to an ER in an unconscious state. Her blood COHb was 16%. She was administered HBOT, and her COHb decreased to 2%. She did not respond and so was administered five more sessions of HBOT. She still did not respond and was administered a prolonged mechanical tracheostomy was performed, and she was admitted to ICU. Fifty days later, she was able to begin oral feeding and was transferred to the physiotherapy and rehabilitation department.

> In conclusion, clinical severity of COI sometimes may not be related to blood COHb levels, and furthermore, oxygen replacement and/or HBOT sometimes could not be effective to improve the clinical condition of the patient likewise this patient.

Reference Number: 31020

JUURLINK, D.N. "Hyperbaric Oxygen Should Not Be Used Routinely for Carbon Monoxide Poisoning." *British Journal of Clinical Pharmacology*, 89: 942–945, 2023 (1 figure, 19 references)

Abstract: An evaluation of HBOT for the routine treatment of CO poisoning was conducted. CO not only displaces oxygen from Hb and shifts the oxygen-Hb dissociation curve to the left, it also generates reactive oxygen species, inhibits cellular respiration by binding to the mitochondrial enzymes, and directly peroxidases the brain lipids leading to neuroinflammation.

> However, because clinicians tend to anchor their diagnostic and mechanistic thinking around carboxyhemoglobin, it can be tempting to infer that HBO must be beneficial because it promotes carboxyhemoglobin elimination. This reasoning oversimplifies CO poisoning, reducing it to a functional anemia akin to methemoglobinemia and discounting entirely its other mechanisms of toxicity.

Reference Number: 31021

HELGESON, S.A., M.E. WILSON, AND P.K. GURU. "Normal Carboxyhemoglobin Level in Carbon Monoxide Poisoning Treated with Hyperbaric Oxygen Therapy." *British Medical Journal Case Reports*, 2017: bcr-2017, 2017 (2 figures, 9 references)

Abstract: A 22-year-old woman was found unconscious and covered with vomitus at the home of her parents who were found dead. At hospital, her blood COHb was reported to be 2.5%. In spite of her normal blood COHb concentration, she was presumed to have CO poisoning and was administered HBOT. The home had a defective furnace, which the father had attempted to fix. Several weeks later, the post-mortem blood COHbs were found to be 49% in the mother and 35% in the father.

> The case highlights the therapeutic benefit of HBO therapy in suspected CO poisoning even when initial CO levels are within normal limits. The toxic CO gas is colorless, odorless, a non-irritant and easily absorbed through the lungs. The most common sources of poisoning in humans includes exposure to motor vehicle exhaust, poorly functioning heating system (like in the current case) and inhaled smoke. The pathophysiology of CO poisoning-related organ injury is incompletely understood. Because of its high affinity for hemoglobin, CO hampers the body's ability to carry and deliver oxygen. CO also interferes with the cellular oxidative phosphorylation process by binding to proteins and cytochromes. The combination of tissue hypoxia, ischemia reperfusion injury and subsequent inflammatory cascade is primarily responsible for the end-organ damage.

3.11 THERAPEUTICS

> All things are poison, and nothing is without poison, only the dose make a thing a poison or not.
>
> —Paracelsus, Swiss physician (1493–1541)[8]

> It is estimated that 10% of carbon throughout the cosmos is in the form of carbon monoxide (CO). Earth's earliest prebiotic atmosphere included the trinity of gasotransmitters CO, nitric oxide (NO) and hydrogen sulfide (H_2S), for which all of life has co-evolved with.
>
> —C.P. Hopper et al., "A Brief History of Carbon Monoxide" (2021) **(31101)**

CO is not only a poisoning gas but is also formed naturally in the body by the breakdown of heme (30102). It has been found to be a neurotransmitter, known as a gasotransmitter, which is important for the health of humans as well as its role as a therapeutic agent (31101–6). CO has also been used for many years as a single-breath-hold test for lung diffusion

8 "Die dritte Defension wegen des Schreibens der neuen Rezepte," *Septem Defensiones,* 1538 (2): 510, 1965.

capacity (31107). But, as always, CO is a will o' the wisp, as one such test caused the patient to develop CO poisoning (31108).

Reference Number: 31101

HOPPER, C.P., P.N. ZAMBRANA, U. GOEBEL, AND J. WOLLBORN. "A Brief History of Carbon Monoxide and Its Therapeutic Origins." *Nitric Oxide*, 111: 45–63, 2021 (5 tables, 1 figure, 450 references)

Abstract: CO is not only a poisonous gas that has plagued humanity for tens of thousands of years but is also a neurotransmitter referred to as a gasotransmitter. CO is produced endogenously in the body by the enzymatic catabolism of heme, mainly by heme oxygenase. CO also has cytoprotective roles and has therapeutic potential as an anti-inflammatory, anti-malarial, or anti-microbial drug similar to nitric oxide (NO) and hydrogen sulfide (H_2S).

> It is estimated that 10% of carbon throughout the cosmos is in the form of carbon monoxide (CO). Earth's earliest prebiotic atmosphere included the trinity of gasotransmitters CO, nitric oxide (NO) and hydrogen sulfide (H_2S), for which all of life has co-evolved with. The history of CO can be loosely traced to mythological and prehistoric origins with rudimentary understanding emerging in the middle ages. Ancient literature is focused on CO's deadly toxicity which is understandable in the context of our primitive relationship with coal and fire. Scientific enquiry into CO appears to have emerged throughout the 1700s followed by chemical and toxicological profiling throughout the 1800s. Despite CO's ghastly reputation, several of the 18th and 19th century scientists suggested a therapeutic application of CO. Since 2000, the fundamental understanding of CO as a deadly nuisance has undergone a paradigm shift such that CO is now recognized as a neurotransmitter and viable pharmaceutical candidate.

Reference Number: 31102

KIM, H.-H., AND S. CHOI. "Therapeutic Aspects of Carbon Monoxide in Cardiovascular Disease." *International Journal of Molecular Sciences*, 19 (8): 2381, 2018 (2 figures, 83 references)

Abstract: CO is a poison but is also a gasotransmitter that is produced endogenously by heme oxygenase and CO-releasing molecules show anti-apoptotic, anti-inflammatory, antioxidant, and vasodilatory effects.

An increase in reactive oxygen species production by the mitochondria after exposure to CO is also considered one of the underlying mechanisms of cardio protection but, at higher doses that occur in CO poisoning, causes toxic mitochondria inhibition.

It is astounding that CO, which is considered completely toxic, has emerged as a new therapeutic substance. However, in medical practices, many clinicians are currently concentrating on reducing the morbidity and mortality of patients suffering from carbon monoxide poisoning, i.e., to reduce the toxic effects of carbon monoxide. Studies regarding the complications and sequelae of CO poisoning, such as delayed neurological sequelae (DNS), permanent brain injury and cardiovascular abnormalities in CO poisoning are still being actively conducted. Therefore, the application of CO to treat clinical disease might be counterintuitive and lead to hesitancy in the application of CO as a remedy for patients. In addition, although many experimental and clinical studies have demonstrated the vasorelaxant, anti-apoptotic, and anti-inflammatory properties of CO, there are still limitations in the clinical application of CO.

Reference Number: 31103

YANG. X., W. LU, C.P. HOPPER, B. KE, AND B. WANG. "Nature's Marvels Endowed in Gaseous Molecules I: Carbon Monoxide and Its Physiological and Therapeutic Roles." *Acta Pharmaceutica Sinica B.*, 11: 1434–1445, 2021 (2 figures, 237 references)

Abstract: The three gases that are known as endogenous-signaling molecules are NO, H_2S, and CO and are also known as gasotransmitters. CO is produced endogenously by heme degradation mainly in the red blood cell turnover and can diffuse throughout the body, although the proportion that diffuses into the tissues is estimated to be less than 20% as most of the CO is in the form of COHb. CO is considered essential for life and has anti-inflammatory, cytoprotective, and organ protective effects.

CO's therapeutic effect at low doses does not diminish its health risk or lethality at high levels and vice versa. Further simply because a drug is useful in treating a specific pathological condition does not mean that it is good for a healthy person to take as a preventative measure. This is the message that I (Wang) always tell students, just because taxol in a good drug to treat certain forms of cancer does not mean that a healthy person should take taxol to prevent the same cancer because all drugs are inherently toxic is abused. The same is true for CO as well as medicinal plants, herbal supplements, and natural compounds. Prophylactic usage is very different from treatment and should only be based on proven science. Just because CO has the potential to be used as a therapeutic agent for treating certain conditions such as inflammation, it does not mean that a healthy person can self medicate in any form or fashion. All therapeutic agents should be used only based on clinically proven science and under the supervision of appropriately trained medical professionals. Otherwise, they could be harmful or lethal consequences.

Reference Number: 31104

HESS, D.R. "Inhaled Carbon Monoxide: From Toxin to Therapy." *Respiratory Case*, 62: 1333–1342, 2017 (4 figures, 97 references)

Abstract: CO is a colourless, odourless, and tasteless non-irritating gas that is slightly less dense than air. It consists of one carbon and one oxygen atom connected by a triple covalent bond and has a molecular weight of 28 grams. CO is a highly toxic gas but has been used for many years in the pulmonary function laboratory to assess diffusing capacity of the lungs. To measure CO-diffusing capacity, the test gas contains 0.3% (3,000 ppm) CO and inhaled for 10 seconds. Monitoring of breath CO concentration may be of value in smoking cessation programs.

In patients with CO poisoning, administration of 100% O_2 is standard practice. If available, hyperbaric O_2 can be used, although this is controversial. Measurement of exhaled CO might be useful to estimate COHb, such as in smoking cessation programs, but assessment of COHb using pulse oximetry can be misleading. Endogenous CO is generated as the result of heme oxygenase activity. It is becoming increasingly recognized that the results of heme oxygenase activity, specifically CO production, might have important physiologic functions. These include effects on vascular function, inflammation, apoptosis, cell proliferation and signaling pathways. Given the abundance of basic science supporting a therapeutic role for CO, clinical trials are exploring this potential.

Reference Number: 31105

MORSE, D. "Carbon Monoxide: A Modern Pharmakon for Sepsis." *American Journal of Respiratory and Critical Care Medicine*, 185: 800–801, 2012 (1 figure, 16 references)

Abstract: Pharmakon is a word used by the ancient Greeks to describe both a poison and its remedy. CO is made by nearly every cell in the human body, mainly due to the heme-oxygenase enzyme, and stress substantially increases its production.

> Preclinical studies have demonstrated beneficial effects of CO on pathological processes suspect to contribute to poor outcome in sepsis. These include inflammation, oxidative injury, apoptosis, repression of fibrinolysis and ventilator induced lung injury. Despite advances in our understanding of CO biology, a large gap remains in our grasp of the mechanisms by which CO confers protection.

Reference Number: 31106

BUCOLO, C., AND F. DRAGO. "Carbon Monoxide and the Eye: Implications for Glaucoma Therapy." *Pharmacology and Therapeutics*, 130: 191–201, 2011 (1 table, 4 figures, 117 references)

Abstract: Glaucoma is a progressive optic neuropathy and is the leading cause of blindness in the world. It accounts for about 12% of all global blindness. NO was the first identified gasotransmitter (that is, a gaseous neurotransmitter) followed by hydrogen cyanide and CO. CO has been found to be a fine moderator of intraocular pressure and may provide additional therapies for its treatment.

> The role of CO as a key molecule in the ocular system has emerged slowly, due primarily to its negative connotation as a toxic gas. Working strongly against this negative paradigm are highly evolutionarily conserved enzymatic systems that continually generate CO, NO and H_2S. There systems are elaborately interrelated and are able to influence each other's function. Based on the large amount of evidence highlighted in this review, CO has a beneficial effect in different in vitro and in vivo models and it is important to recognize that the toxic connotation of this molecule should not detract from its potential in ocular therapeutic applications. All agents are toxic at a given concentration, and this is also true for CO; however, low concentrations can be helpful in various ocular conditions. CO is a very intriguing molecule, and, therefore, a better understanding of its regulation and interactions with other gases is important to extend our knowledge of the physiological and pathophysiological processes in the eye, in general, and in glaucoma, in particular.

Reference Number: 31107

CASTILLO, A., J. LIAPUR, T. MARTINEZ, K. KISLING, T. WILLIAMS-NKOMO, C. COATES, AND R.S. TEPPER. "Measurement of Single Breath-Hold Carbon Monoxide Diffusing Capacity in Heathy Infants and Toddlers." *Pediatric Pulmonology*, 41: 544–550, 2006 (1 table, 4 figures, 26 references)

Abstract: The measurement of CO-diffusing capacity (DLco) is used to assess lung growth and parenchymal lung disease. Six male and six female children (ages 6–22 months) had their DLco measured using a facemask while sedated with chloral hydrate. The test gas contained 0.3% carbon-18 monoxide, 4% helium, 21% oxygen, and 75% nitrogen. Helium and CO concentrations were measured with a respiratory gas MS system. DLco increased with increasing body length.

> In conclusion, we describe a methodology to obtain a single breath-hold measurement of alveolar volume in sleeping healthy infants. These in vivo physiologic measurements can be used to evaluate the growth and development of the lung parenchyma early in life.

Reference Number: 31108

ERGOZEN, S., A. DEMIR, AND E. ACAR. "A Carbon Monoxide Poisoning Case Due to Lung Diffusion Test." *American Journal of Emergency Medicine*, 38 (5): 1047-e1, 2020 (8 references)

Abstract: The DLCO is used to detect alveolar membrane diffusion capacity. The standard gas mixture used in the DLCO is 0.3% CO, 0.3% methanol, 10% helium, or 0.5 neon, 17–21% oxygen, and the remainder is nitrogen. A 68-year-old male patient present at the ED with a complaint of unconsciousness, fatigue, and a throbbing headache after a DLCO test. His COHb was 25%, and he was administered nasal oxygen. He recovered uneventfully.

> Although not encountered frequently in the literature, high pressure gas cylinders used for the storage of breathing gases may be totally refilled with the wrong gas or wrong percentages of gas mixtures or may be contaminated by some other toxic gases. Labels on gas cylinders must be checked before their use after refilling and technicians must be sure that the gas content is correct. In case of a syncope during DLCO, the above-mentioned factors must be remembered and inhalation of the gas

by the patient must be prevented, blood gas analysis should be checked for CO poisoning, and the gas cylinder must be secured and analyzed for contents. If the patient is diagnosed as acute CO poisoning, treatment must be administered according to guidelines.

3.12 MISCELLANEOUS

In autopsy examinations, the appearance of carbon monoxide poisoning is notable because unembalmed dead persons are normally bluish and pale, whereas dead carbon monoxide poisoned people may appear unusually lifelike in coloration. The colorant effect of carbon monoxide is such post-mortem circumstances is thus analogous to its use as a red colorant in the commercial meat packing industry.

—"Carbon Monoxide Poisoning," *Wikipedia*[9]

CO has a molecular weight of 28 and is lighter than air. It is an odourless gas except at high concentrations greater than 75% where it has a garlic-like odour (31201). Two cats who died of CO poisoning had post-mortem blood COHb concentrations of 57% and 41% respectively (31203). The more likes a YouTube video obtains, the more likely it contains misinformation about CO (31204). Between 1997 and 2017, the risk of dying from CO poisoning globally declined by 36%, yet the incidence of CO poisoning has remained stable (31206). The organs of donors who die from CO poisoning are better preserved as a lot of oxygen has been removed from the tissues and blood (31205). In Chicago, a higher ambient CO level on the weekdays was associated with an increase in hospital admissions for congestive heart failure (31207).

Reference Number: 31201

DIVISION OF INDUSTRIAL HYGIENE, NATIONAL INSTITUTES OF HEALTH. "Carbon Monoxide: Its Toxicity and Potential Dangers." *Public Health Reports*, 56: 421–433, 1941 (5 tables, 20 references)

Abstract: CO has a molecular weight of 28, a specific gravity of 0.9671 (air = 1) and is a colourless and odourless gas, except at high concentrations of 75–100%, it has a garlic-like odour. CO melts at –207 degrees Celsius and boils at –192 degrees Celsius. It burns with a blue flame at

9 "Carbon Monoxide Poisoning," *Wikipedia*, https://en.wikipedia.org/wiki/Carbon_monoxide_poisoning.

650 degrees Celsius, which is extinguished when the oxygen level drops below 13.4%. The maximum permissible exposure CO concentration is 0.01% for exposure not exceeding eight hours a day. Numerous workers can be exposed to CO, including blast furnace workers, blacksmiths, plumbers, solderers, miners, kiln workers and brick burners, fire fighters, cooks, and bakers.

Table: CO Concentration Generated under Different Working Conditions

Type and source	CO in air (%)
Mine explosion immediately after dust explosion	8
Mine fire	1
TNT explosion	60
Bessemer furnace gas	25
Cupola gas	17
Coke oven gas	6
Coal gas	16
Gas range (natural gas) improperly constructed and maintained	0.1
Room heater (natural gas) improperly constructed and maintained	0.5
City fire, black smoke from burning building	0.1
Insulation burning in electric arc	0.5

Source: Adopted from Division of Industrial Hygiene 1941.

It is obvious that the best prevention of carbon monoxide poisoning would be the prevention of any pollution of the air with carbon monoxide. This can be accomplished to a very high degree by proper engineering methods and adequate ventilation. For continued exposure, the concentrations should be kept at or below 100 parts per million by volume (0.01 volume percent). It appears absolutely necessary that the public, and especially those persons who may have occupational exposure to carbon monoxide, be instructed regarding its toxicity and potential dangers. Before entering any space not frequently used and suspected of containing carbon monoxide, air samples should be taken to determine the amount of carbon monoxide present.

Reference Number: 31202

FUJIHARA, J., M. HASEGAWA, T. KATO, M. MIURA, K. IIDA, H. KINOSHITA, N. TANAKA, AND H. TAKESHITA. "A Case of Drowning Lacking Typical

Autopsy Findings of Carbon Monoxide Poisoning Despite the High CO Concentration." *Forensic Toxicology*, 31: 180–182, 2013 (1 table, 6 references)

Abstract: A woman in her fifties was found dead in the bathroom of her house sitting in a bathtub with her face underwater. An autopsy was conducted 12 hours after her death. Her post-mortem blood COHb was 57.2% in right and 55.9% in left cardiac blood using a AOXimeter 4000 screener. Using conventional double wavelength spectrometry, the COHb concentrations were 55.7% and 55.7% respectively. The victim had been using an old country-style bath in which wood is used to heat the water. A gap was found between the bathtub, tile, and fireplace. A reconstruction was conducted, but the victim may have been exposed to higher ambient CO concentrations as she took the bath several hours after the fire was started. The CO concentration was 1,000 ppm before the fireplace was used, 3,000 ppm 50–90 minutes later, and 1,000 ppm after 120 minutes. The victim probably drowned after losing unconsciousness after inhaling CO. These findings helped prevent a repeat accident.

Table: Ambient CO Concentrations Found around the Victim's Face Area and the Gap between the Bathtub and Tile with the Bathroom Door Closed

Time	Face area	Gap
5 minutes	17 ppm	15 ppm
30 minutes	40 ppm	34 ppm
60 minutes	57 ppm	51 ppm
90 minutes	86 ppm	79 ppm
120 minutes	111 ppm	104 ppm

Source: Adapted from Fujihara et al. 2013.

In cases of CO poisoning, blood containing a high CO-Hb level appears bright red and cherry-pink lividity is observed. However, these characteristics are often different between individuals and gross pathology is subjective. Routine CO-Hb analysis in all autopsies can be useful in determining the circumstances contributing to the cause of death. Therefore, we recommend CO-Hb analysis in all forensic autopsy cases to reduce the risk of overlooking the circumstances surrounding the cause of death.

Reference Number: 31202

Reference Number: 31203

SOBHAKUMARI, A., R.H. POPPENGA, J.B. PESAVENTO, AND F.A. UZAL. "Pathology of Carbon Monoxide Poisoning in Two Cats." *Biomedical Central Veterinary Research*, 14 (1): 67, 2018 (4 figures, 21 references)

Abstract: Two cats were found dead in a household with their two deceased owners in the same household. Both cats presented with bright red discoloration on their skin and edematous lungs. The blood COHb concentrations were 57% and 41% respectively. No cyanide could be detected.

> There are no established reference values for lethal COHb concentration in cats. The COHb% values detected in this case which fell within the lethal range reported for other species along with the gross lesions and unique histological findings in the heart suggest a helpful criteria for diagnosis of CO intoxication associated death in cats. This demonstrates that since pets share the same environment as human beings and often are part of their activities, they can be useful adjuncts in potential forensic investigations to help solve human cases.

Reference Number: 31204

KRAKOWIAK, M., M. RAK, P. KRAKOWIAK, K. RACISZ, P. SLONIEWSKI, T. ILCZAK, AND A. KRAKOWIAK. "YouTube as a Source of Information on Carbon Monoxide Poisoning: A Content Quality Analysis." *International Journal of Occupational Medicine and Environmental Health,* 35: 285–295, 2022 (8 tables, 2 figures, 30 references)

Abstract: An analysis was conducted on 8 December 2020 on the accuracy of the information provided about CO in 95 YouTube videos. Six videos contained misinformation that could potentially be life-threatening. Only 4.3% of the videos mentioned the increased danger of CO poisoning in pregnancy. The more likes the video got, the more likely the video contained misinformation.

> Our study results indicate that the overall quality of YT [Youtube] videos concerning CO is poor, the information accessible to the audience can be misleading and does not provide an expected amount of information for a non-medical viewer to fully understand the threat of CO intoxication.

Reference Number: 31205

FUJISAKI, N., A. NAKAO, T. OSAKO, T. NISHIMURA, T. YAMADA, K. KOHAMA, H. SAKATA, M. ISHIKAWA-AOYAMA, AND J. KOTANI. "Can Carbon Monoxide-Poisoned Victims Be Organ Donors?" *Medical Gas Research*, 4 (1): 13, 2014 (1 table, 33 references)

Abstract: Less than 1% of all donated organs in Western Europe and the United States are obtained from CO-poisoned donors. CO potentially protects against cellular injury, relaxes blood vessels, and reduces inflammatory response and ischemia/reperfusion injury. In animals, CO has been found to be beneficial to the survival of organ grafts such as the heart, kidney, liver, and lung.

> In general, little data regarding long-term survival are available, but individual reported cases suggest that the survival of patients transplanted with organs procured from carefully selected CO-poisoned donors may be comparable to that of patients transplanted from non-poisoned donors. While these reports do not support the use of CO therapeutically during transplantation per se, they certainly do not provide any evidence that contraindicates it.

Reference Number: 31206

MATTIUZZI, C., AND G. LIPPI. "Worldwide Epidemiology of Carbon Monoxide Poisoning." *Human and Experimental Toxicology*, 39: 387–392, 2020 (2 tables, 2 figures, 18 references)

Abstract: The worldwide cumulative incidence and morbidity of CO poisoning in 2017 were 137 cases per million and 4.6 deaths per million people respectively. Between 1992 and 2017, the risk of dying from CO poisoning has declined by 36%, but the incidence of CO poisoning has remained virtually stable. The risk of dying is nearly two times that in infants and 26 times higher in the elderly, 80 years of age and older. The risk of dying is also twice as great for men than for women.

Table: Effects of COHb Concentrations

COHB concentration	Effect
0.5–1.5%	Typical levels in non-smokers
10–20%	Nausea, fatigue, tachypnea, emotionality, confusion, clumsiness
21–30%	Headache, exertional dyspnea, angina, visual impairment, decreased sensory perception
31–50%	Dizziness, fainting, confusion, nausea, vomiting, visual impairment, problematic decision-making
> 51%	Come, severe acidosis, death

Source: Adapted from Mattiuzzi and Lippi 2020.

In conclusion, while these data suggest that the worldwide burden of CO poisoning remains stable, and the number of fatal outcomes and percentage of patients who die have both consistently declined during the last 25 years, the unreliability of the primary data sources in many countries with respect to accurate diagnosis of CO poisoning means that caution is required, and that field studies, particularly in poorer countries are required.

Reference Number: 31207

MORRIS, R.D., AND E.N. MAUMOVA. "Carbon Monoxide and Hospital Admissions for Congestive Heart Failure: Evidence of an Increased Effect at Low Temperatures." *Environmental Health Perspectives*, 106: 649–653, 1998 (2 tables, 3 figures, 39 references)

Abstract: The daily hospital admission for congestive heart failure (CHF) and temperate pollutants (including CO) were determined in Chicago over a four-year period. CO had the lowest levels on weekends (mean 1.96 ppm) and highest on Fridays (2.76 ppm). CO was the only pollutant with an elevated relative risk of CHF admissions. Low ambient temperatures (< 4 degrees Celsius) were associated with an increased relative risk of CHF admissions compared to warmer temperatures.

The impact of CO on CHF patients has not been well studied, but in coronary artery disease patients, this effect has been clearly shown to decrease the time to angina at COHb levels as low as 3.0% and perhaps even lower. It is possible that this threshold for adverse effects may be even lower among persons with CHF than persons with coronary artery disease. Our data indicate that exposure to cold air may reduce this threshold still further. In this way, susceptibility, thermal stress, and relatively low levels of ambient CO may combine to induce acute heart failure at an individual level.

CHAPTER 4

Locations and/or Sources of Carbon Monoxide Poisoning

Carbon monoxide (CO) is a true will o' the wisp and appears in unexpected places. One unusual source of CO that is often overlooked is the production of CO as a result of the detonation of explosives in mines or by the military or the firing of firearms (4.13, 4.14). The following equation shows the production of CO from the nitrocellulose (gun cotton), which replaced gunpowder. The detonation of one kilogram (kg) of nitrocellulose produces about 859 litres (L) of gas and about 25% of the gases produced is CO. This percentage can increase to 40% if the charge burns slowly.

$$C_{12}H_{14}O_4(NO_2)_6 = 7CO_2 + 5CO + 4H_2 + 3H_2O + 3N_2$$

It can also poison unsuspecting people in tents (4.19), motels (4.11), indoor skating arenas (4.12), and coffee-roasting establishments (4.06). It even causes "warehouse workers headaches" (4.21) and can accumulate in patients undergoing anaesthesia during an operation (4.01).

4.01 ANAESTHESIA

Closed system anaesthesia techniques allow the accumulation of endogenous and exogenous carbon monoxide which causes binding of hemoglobin and impaired oxygen transport when present in even minute concentrations. Accumulation can be prevented by intermittently opening and flushing the anaesthetic system.

—V. Middleton et al., "Carbon Monoxide Accumulation in Closed Circle Anaesthesia Systems" (1965) **(40105)**

CO accumulates in a closed anaesthetic system due to its endogenous production and the interaction of some anaesthetic gases with a carbon dioxide (CO_2) absorbent to produce CO (40101–7)

Reference Number: 40101

SHERMAN, W.O., C.M. SWINDLER, AND W.S. MCELLROY. "Carbon Monoxide Poisoning Following Ethylene Anaesthesia." *Journal of the American Medical Association*, 86: 1765–1766, 1926

Abstract: Three cases of CO poisoning are reported due to ethylene anaesthesia in hospital. One man died during an operation that was originally attributed to a fat embolism. His post-mortem blood carboxyhemoglobin (COHb) concentration was 50%. The other two patients survived, and the blood COHb of one was found to be 60%. In all three cases, the signs of collapse did not occur until near the end of the operation. Two tanks of ethylene anaesthesia were found to contain CO.

Recently, serious complications following ethylene anaesthesia were encountered with at least one fatality; another case which, in the light of subsequent events could be ascribed to the same cause, and a third case in which the patient was dangerously asphyxiated. The clinical and laboratory findings point to carbon monoxide as the causative factor. We are not aware that any previous reference has been made to carbon monoxide as a source of difficulty in the use of ethylene in anaesthesia. Therefore, this preliminary report is made for the purpose of calling attention to this grave possibility and the necessity for more careful control of the purity of ethylene used.

Reference Number: 40102

BERRY, P.D., D.I. SESSLER, AND M.D. LARSON. "Severe Carbon Monoxide Poisoning during Desflurane Anaesthesia." *Anesthesiology*, 90: 613–616, 1999 (17 references)

Abstract: Anaesthetic gases such as enflurane and sevoflurane interact with desiccated CO_2 absorbers (soda lime) in a closed-breathing circuit and produce a high blood COHb concentration in the patient. A 24-year-old woman was administered general anaesthesia for a clinical study. After 15 minutes of anaesthesia, the desiccated CO absorbent was suspected of causing abnormal readings and replaced with fresh baralyme.

After 30 minutes, her COHb was 36%. The procedure was aborted, and the subject was ventilated with 100% oxygen. Fifteen minutes later, her COHb was 21% and, one hour later, was 12%. The subject reported no adverse sequelae six months later.

> Diagnosis of carbon monoxide intoxication during anaesthesia is difficult because the main clinical features of toxicity are masked by anaesthesia. Furthermore, there is no routinely available means to reliably identify the presence of carbon monoxide within the breathing circuit, nor to detect when carbon dioxide absorbent has been desiccated. In our case, early diagnosis of carbon monoxide production was facilitated by the presence of a clearly erroneous gas analyzer reading (it is not possible to accidentally fill a Tec 6 desflurane vaporizer with enflurane). This false reading probably resulted from trifluoromethane, which is produced along with carbon monoxide by volatile agent degradation. Because it has an infrared absorption profile similar to einflurane, it is identified as such by some gas analyzers, including the Datex Capnomac Ultima. Our case, therefore, confirms the potential role of unexpected gas analyzer readings in the clinical detection of carbon monoxide production. Many gas analyzers, however, do not show clearly erroneous readings, which suggests that the routine incorporation of a carbon monoxide detection module into gas analyzers may be justified.

Reference Number: 40103

WOEHICK, H.J., M. DUNNING, AND L.A. CONNOLLY. "Reduction in the Incidence of Carbon Monoxide Exposures in Humans Undergoing General Anaesthesia." *Anesthesiology*, 87: 228–234, 1997 (1 table, 2 figures, 11 references)

Abstract: CO is produced during anaesthesia when isoflurane, enflurane, and desflurane are passed over dried CO_2 absorbents (soda lime) such as baralyme. Fresh CO_2 absorbents contain 10–15% by weight water but can dehydrate when left exposed to dry gases. It takes about 24–48 hours to dry a cannister of CO_2 absorbents. As such, most of the CO exposures during anaesthesia occur on Monday mornings because the oxygen has dehydrated the absorbents over the weekend. The incidence of CO exposure was 0.46% initially, which decreased to 0.051% when new standards and education about CO production due to dehydrated baralyme was provided.

These interventions reduced patient exposure to carbon monoxide. Monitoring for carbon monoxide exposures during general anaesthesia may be necessary to recognize and end patient exposures that occur despite preventative measures.

Reference Number: 40104

BAXTER, P.J., AND E.D. KHARASCH. "Rehydration of Desiccated Baralyme Prevents Carbon Monoxide Formation from Desflurane in an Anaesthesia Machine." *Anesthesiology*, 86: 1061–1065, 1997 (2 figures, 13 references)

Abstract: Tests were conducted on the formation of CO from desflurane due to desiccated baralyme (a CO_2 absorbent) in an anaesthesia machine. Up to 2,500 ppm CO was detected with an average concentration of 520 ppm.

Desflurane is degraded by desiccated Baralyme in an anaesthesia machine, resulting in CO formation. Adding water to dried Baralyme is an effective means of reducing CO formation and the risk of intraoperative CO poisoning. Although demonstrated specifically for desflurane and Baralyme, rehydration is also applicable to enflurane and isoflurane and to soda lime.

Reference Number: 40105

MIDDLETON, V., A. VAN PAZNAK, J.F. ARTUSIO JR, AND S.M. SMITH. "Carbon Monoxide Accumulation in Closed Circle Anaesthesia Systems." *Anesthesiology*, 26: 715–719, 1965 (2 tables, 2 figures, 11 references)

Abstract: Breath CO concentrations were determined in 22 adult surgical patients under closed system anaesthesia with cyclopropane or halothane. Fifteen patients showed increases in breath CO concentration, which should not affect normal healthy subjects but may affect seriously ill patients.

Closed system anaesthesia techniques allow the accumulation of endogenous and exogenous carbon monoxide which causes binding of hemoglobin and impaired oxygen transport when present in even minute concentrations. Accumulation can be prevented by intermittently opening and flushing the anaesthetic system.

Reference Number: 40106

LEVY, R.J. "Anaesthesia-Related Carbon Monoxide Exposure: Toxicity and Potential Therapy." *Anesthesia and Analgesia*, 123: 670–691, 2016 (1 table, 4 figures, 141 references)

Abstract: Exposure to CO during anaesthesia can result from a volatile anaesthetic gas degradation by dehydrated CO absorbents as well as a build-up of endogenously CO during low-flow anaesthesia. The first reported detection of a build-up of CO during a closed-circuit anaesthesia system in 1965 and reported levels as high as 810 parts per million (ppm). Typically, children can inspire up to 20 ppm of CO and adult smokers 145 ppm using low-flow anaesthesia. The greatest exogenous production of CO occurs with the use of desflurane or enflurane, followed by isoflurane, halothane, and sevoflurane. Higher temperatures of desiccated baralyme causes a greater production of CO.

Anesthesiologists should also understand that the practice of low-flow general endotracheal anaesthesia may result in low concentration CO exposure. Although the effects of such subtoxic are not well understood, animal studies indicate that inspiring low dose CO might be beneficial from a cytoprotective but might also have pathologic consequences depending on the context of exposure. Because of the lack of clinical information and formal study regarding anaesthesia-related CO exposures in patients, further investigation is necessary and anesthesiologist should not change their practice. Instead, they should be aware of ongoing clinical trials and await the results of studies aimed at determining the safety and therapeutic efficacy of low concentration CO.

Reference Number: 40107

INGRAM, C.T., W.R. ELSEA, R.J. FINTON, J.D. SMITH, J.A. WILBER, E.A. BRUNNER, B.J. FRANCIS, R.E. MOON, AND J.N. MACCORMACK. "Elevated Intraoperative Blood Carboxyhemoglobin Levels in Surgical Patients—Georgia and North Carolina." *Morbidity and Mortality Weekly Report*, 40: 248–249, 1991 (6 references)

Abstract: The intraoperative COHb concentration was determined in 16,000 surgical patients using CO-oximetry. In 15 cases (0.1%), the COHb was 8% or greater for smokers and 20% or greater for non-smokers. It was suspected that certain anaesthetic gases (for example, trichloroethylene, which is no longer used) interacted with the CO_2 absorbent to produce CO.

Until the exact mechanism of CO production is identified, the three hospitals have initiated the following actions to minimize the risk of intraoperative anaesthesia associated CO exposure; (1) flushing CO_2 cannisters with high-flow (about 30 L per minute) oxygen for at least 60 seconds before each anaesthetic procedure; (2) replacing CO_2 absorbent canisters that have not been used within 24 hours; and 3) using high fresh gas flows (> 5 L per minute) during anaesthesia. Since implementation of these measures, no further cases have been detected.

4.02 AVIATION

The adverse effects would be more pronounced at elevations during flight than at ground level because of the interactive effects of altitude hypoxia and CO-induced anemic hypoxia.

—A.K. Chaturvedi et al., "Blood Carbon Monoxide and Hydrogen Cyanide Concentrations in the Fatalities of Fire and Non-Fire Associated Civil Aviation Accidents, 1991–1998" (2001) (**40201**)

A combination of high altitude and CO from an airplane engine and heater combine to cause impairment of the pilot and result in several crashes (40201–5).

Reference Number: 40201

CHATURVEDI, A.K., D.R. SMITH, AND D.V. CANFIELD. "Blood Carbon Monoxide and Hydrogen Cyanide Concentrations in the Fatalities of Fire and Non-Fire Associated Civil Aviation Accidents, 1991–1998." *Forensic Science International*, 121: 183–188, 2001 (3 tables, 1 figure, 27 references)

Abstract: Post-mortem blood COHb and cyanide concentrations were determined in 2,837 civil aviation accidents that occurred between 1991 and 1998 in the United States. The presence of both gases can indicate that the victim was alive and inhaled smoke from a fire, whereas the presence of CO alone could indicate CO contamination of the interior. Forty-one percent of the fatalities were associated with fires, and 49% with non-fire-related causes. There were 15 non-fire accidents, which included 17 fatalities in which only an elevated blood COHb was determined (10–69%).

Table: Post-Mortem Blood COHb Concentrations in Non-Fire-Related Aviation Accidents between 1991 and 1998 in Which No Cyanide Was Detected

COHb (%)	Aircraft type	Probable cause of accident
10	Bell 206-B	In-flight structural failure
20	Aeronca 7-AC	Physical impairment of pilot
10	Piper 28–140	Undetermined
16 (smoker)	Gulfstream AA-5A	Physical impairment of pilot
10 (smoker)	Gulfstream 690-B	Severe turbulence
11	Bell 214-B	Structural failure
17	Cessna 177-RG	Pilot error
14 (smoker)	Beach 95-B95	Fuel starvation
12 (smoker)	Cessna 150-L	Pilot error
11	Cessna 150-M	Pilot error
41	Mankovich Sonerai	Possible inadequate sealing of engine's firewall—CO-induced incapacitation
43/ 69	Piper 28–236	Faulty exhaust—two deaths
26 (smoker)	Bell 206-B	CO-induced incapacitation
12	Windex 1200 C	Pilot error
24/ 12	Piper 28–160	Suspected faulty exhaust

Source: Adapted from Chaturvedi et al. 2001.

The most probable scenario for CO poisoning to take place is the leakage of gases into the cockpit/cabin from faulty heaters, and/or exhaust systems. A high COHb (> 21%) values clearly suggest the occurrence of CO-induced dizziness, incapacitation, coma or death. Obviously, there would be relatively mild adverse physiological effects at the low COHb (10–20%) levels. At higher altitudes, there is also a probability of the coexistence of altitude associated hypoxic hypoxia. Therefore, the effects of CO would be further exacerbated in association with hypoxia, thereby hastening a performance impairment that may contribute to an accident. The adverse effects would be more pronounced at elevations during flight than at ground level because of the interactive effects of altitude hypoxia and CO-induced anemic hypoxia.

Reference Number: 40202

STEVENS, P.J. "The Search for the Cause of an Accident." *Proceedings of the Royal Society of Medicine*, 61: 1076–1079, 1968 (12 references)

Abstract: Six fatal aircraft crashes (both commercial and military) are described. Various causes for such accidents are discussed including medical, CO, alcohol, and drugs. In one fatal crash, three pilots whose aircraft veered off course and crashed into high ground, had post-mortem COHb concentrations of 18% (captain), 7% (second officer), and 13% (first officer). The type of cockpit heater used in the aircraft could have leaked products of combustion into the cockpit. The captain was closest to the heater, and the second officer was the furthest.

> Carbon monoxide intoxication of as little as 5% can affect performance in certain tests of higher cerebral function, while 15–20% carboxyhemoglobin can cause poor performance in simple choice discrimination, arithmetic, plural noun underlining and t-crossing tests. It is likely that complex cerebral functions involving judgments, situational decisions and response may well be affected with 20% carboxyhemoglobin or less at sea level.

Reference Number: 40203

KAMINSKI-MORROW, D. "Carbon Monoxide Poisoning Clue Emerges in Fatal DHC-2 Crash Probe." *Flight Global*, 3 July 2020 (3 figures)

Abstract: A De Havilland Canada DHC-2 floatplane carrying five passengers crashed into the waters of Jerusalem Bay, Australia, on 31 December 2017. The pilot and two passengers had elevated COHb concentrations.

> After scrutinizing the validity of the findings, the bureau states that the levels of carbon monoxide were likely to have adversely affected the pilot's ability to control the aircraft. All three victims were probably exposed within the aircraft cabin, it says, and a further examination of the aircraft has determined that a crack in the engine exhaust collector could have leaked exhaust into the engine bay. The exhaust would not have been contained in the engine bay because three of the eight bolts used to secure magneto access panels beneath the instrument panel were missing.

Reference Number: 40204

O'CONNOR, K. "Carbon Monoxide Poisoning Linked to Piper Crash." *AV Web*, 19 August 2019

Abstract: A 1984 Piper Malibu aircraft carrying soccer star Emiliano Sala and pilot David Ibbitson crashed into the English Channel 22 nautical miles northwest of the island of Guernsey on 21 January 2019. The body of Sala was found, and the post-mortem blood contained 58% COHb. The pilot's body has not been found yet.

> Although the findings aren't yet final, the AAIB [Air Accidents Investigation Branch] says it issued the Special Bulletin, the second related to the accident to raise awareness within the General Aviation community of the dangers of exposure to CO and the measures available to detect its presence in the cabin in order to mitigate this potentially fatal risk. While not mandatory under either EASA or FAA regulations, the AAIB is suggesting pilots carry CO detection devices when flying. It is also asking that pilots familiarize themselves with warning signs of CO leaks and symptoms of CO exposure, along with actions to be taken if the presence of CO is suspected in the cockpit.

Reference Number: 40205

HOWLETT, L., AND R.J. SHEPHARD. "Carbon Monoxide as a Hazard in Aviation." *Journal of Occupational Medicine*, 15: 874–877, 1973 (60 references)

Abstract: The hazards of CO on aviation are discussed. The sources of CO for the air crew include service vehicles, passive smoking, aircraft exhaust (3–8% CO), cabin heaters, and small gasoline-powered generators to start small aircraft in the Arctic. There is also synergism of CO with altitude hypoxia, flight stresses, and task difficulties. Exposure to CO should not exceed 40 ppm for one hour and 15 ppm for eight hours.

> Improvements in vehicle design may reduce the problem of the initial road journey. Smoking should be prohibited in departure lounges, and the air intake should be at a sufficiently high level to avoid local contamination by vehicle exhaust. Runway exposures could possibly be reduced by increased use of electrically operated tractors.

4.03 BOATS/MARINE ENVIRONMENT

> In fact, a vessel can be considered to be just a large floating CO generator.
>
> —J.G. Wigmore, "A Floating Carbon Monoxide Generator" (2023)[1]

Chapter 4

1 J.G. Wigmore, "A Floating Carbon Monoxide Generator," *LinkedIn*, 5 April 2023.

> Carbon monoxide related accidents involving recreational boating constitute an important under-recognized cause of injury and deaths in the United States.
>
> —G. LaSala et al., "The Epidemiology and Characteristics of Carbon Monoxide Poisoning among Recreational Boaters" (2015) (**40302**)

There are many sources of CO on boats—from the engine to the electrical generator, the gas water heater, stove, gas lights, and cabin heater (40301–8). CO tends to accumulate at the stern of the boat and on the swim platform, posing an extra risk for swimming children (40304–6, 40306). There were so many boats idling on Memorial Day weekend on Lake Havasu City, Arizona, that people on the shore obtained elevated CO concentrations (40308).

Reference Number: 40301

ANONYMOUS. "Carbon Monoxide Poisoning on a Motor Launch." *The Hospital*, 17 December 1910, 317–318

Abstract: Five men went fishing on a motor launch powered by a six horsepower (hp) gasoline engine. After a day of fishing, one of the men crawled to the enclosed foredeck past the gasoline engine to sleep. Upon arriving at port, the man could not be awoken, and a medical doctor was called. The doctor thought the dead man was still alive but was unable to be resuscitated. At post-mortem, the vermilion-coloured blood was found to contain 50% COHb.

> With this object, Dr. Barley ventures to suggest the following measures to be observed on motor-boats: (1) Adjust the supply of benzine and air as accurately as possible to secure complete combustion; (2) see that cylinder fittings are not leaking unduly; (3) ensure as much ventilation as possible in the engine-room; (4) whenever possible, sleeping quarters should be entirely and efficiently separated from the engine-room.

Reference Number: 40302

LASALA, G., R. MCKEEVER, J. OKANEKU, D. JACOBS, AND D. VEARRIER. "The Epidemiology and Characteristics of Carbon Monoxide Poisoning among Recreational Boaters." *Clinical Toxicology*, 53 (2): 127–130, 2015 (1 table, 1 figure, 14 references)

Abstract: According to the US Coast Guard Boating Accident Report Database, there are a mean of 14.5 CO-related accidents, 30.9 injuries, and 6.7 deaths each year. There were 29 deaths in a cabin motorboat, 12 in open motorboats, and 9 in houseboats. California had three times more CO boating accidents than the next highest state (Utah), but it also had four times the number of registered CO accidents than Utah as well. The sources of CO were from outboard motors, inboard motors, and cabin heaters or stoves.

Carbon monoxide related accidents involving recreational boating constitute an important under-recognized cause of injury and deaths in the United States.

Reference Number: 40303

BARON, R.L., AND T. RADTKE. "Houseboat Associated Carbon Monoxide Poisonings on Lake Powell-Arizona and Utah, 2000." *Morbidity and Mortality Weekly Report*, 49: 1105–1108, 2000 (1 figure, 3 references)

Abstract: Between 1990 and 1994 and in 2000, there were nine boat-related fatal CO poisonings and 100 non-fatal CO poisonings on Lake Powell. The exhaust of the electrical generator of some houseboats vented to the stern beneath the swim deck into an airspace enclosed by the swim platform. A CO concentration of 6,000–30,000 ppm were observed when the generators were in operation with an oxygen concentration as low as 12%. CO tends to accumulate above the water near the swimming platform and CO concentrations as high as 2,000 ppm were measured at water level 10 feet away from the platform. In one case, three teenage boys were swimming off the stern of the houseboat, sliding off the rear on a mounted slide while the generator was on. After several minutes, one of the boys developed a headache and went back inside the houseboat. The other boy felt dizzy, and his legs became numb, and he fell into the water. One hour later, his body was recovered from the bottom of the lake, and his post-mortem blood COHb was 53.9%.

The outdoor poisonings at Lake Powell and those reported elsewhere probably represent a larger number of deaths not recognized as CO poisoning. Because symptoms of CO poisoning resemble those of other common conditions (e.g., alcohol consumption, motion sickness, heat stress, and nonspecific viral illness), poisonings often go unrecognized. In addition, associating illness with this exposure requires awareness of the problem among EMS [emergency medical services] staff, hospital emergency departments personnel and coroners.

Reference Number: 40304

RICHARDS JR., R J.W., N.N. RICHARDS, K.E. POWELL, R. BARON, J. MCCAMMON, S.C. REDD, J.A. MOTT, AND A.L. STOCK. "Carbon Monoxide Poisoning Resulting from Exposure to Ski-Boat Exhaust: Georgia, June 2002." *Morbidity and Mortality Weekly Report*, 51: 829–830, 2002 (9 references)

Abstract: On 1 June 2002, two parents, three children (ages 2–12 years), and three friends went to a lake in the state of Georgia to water ski. The ski boat was placed in an idling position and the two-year-old girl climbed over the stern of the boat onto the wooden swimming platform and lay in the prone position to kick the water. In less than one minute, she became unconscious and unresponsive. Her father, a family physician, performed emergency breathing. Thirty-five minutes later emergency medical services (EMS) personnel arrived and administered 100% oxygen. Three hours later, her blood COHb was 14.3% (back-calculated to 50–57%). Her youngest brother was also removed from the swimming platform and, several hours later, complained of headache, vomited, and fell asleep. Four hours after exposure, his blood COHb concentration was 10.1% at the hospital (back-calculated 18–21%). Ambient CO concentrations as high as 27,000 ppm have been detected in the stern of boats.

Recreational boaters should be aware that boat exhaust can flow back into the rear of the boat and CO in the exhaust is undetectable because it is odorless and colorless. In addition, they should avoid swimming or body surfing near the exhaust system while the boat or generator is running. Studies of CO concentrations around boats and of COHb levels in recreational boaters are needed to determine the extent of boat-related CO poisonings and public health campaigns to warn of the danger of boat-related CO poisonings require further evaluation. The use of emissions-control devices in recreational boats can reduce CO emissions and the risk for CO poisoning.

Reference Number: 40305

HALL, R.M., G.S. EARNEST, D.R. HAMMOND, K.H. DUNN, AND A. GARCIA. "A Summary of Research and Progress on Carbon Monoxide Exposure Control

Solutions on Houseboats." *Journal of Occupational and Environmental Hygiene*, 11 (7): D92–D100, 2014 (2 tables, 3 figures, 34 references)

Abstract: Between 1990 and 2008, 309 houseboat-related CO poisonings occurred in the United States. Most of the 283 non-fatal CO poisonings were due to the electrical generator exhaust. Various tests were conducted to measure CO build-up in houseboats using detector tubes, evacuated containers, and real-time instruments. Ambient CO concentrations in the exhaust at the rear of the boat have reached 1,200 ppm. Although catalytic converters on the gas-powered generators were able to keep the CO production to less than five ppm, one generator used for 4,565 hours produced 4,000 ppm of CO due to catalyst degeneration.

In order to reduce CO concentrations on a houseboat, it is important that the generator exhaust stack, water separator and associated piping and hoses be designed and installed properly. If the stack exhaust is not designed properly, the performance could be hindered. Rather than hazardous exhaust gas passing through the stack to a height well above the upper deck, high static pressure in the stack could force exhaust gases to pass out the side terminus near the water line. While concentrations on the boat remain low with stack exhaust compared to side exhaust, CO measurements taken directly at the stack outlet indicated a range of 5% to 8.6% CO (50,000 ppm to 86,000 ppm). Because this concentration is 42 to 72 times greater than the immediately dangerous to life and health value to breathing zone concentrations of CO, it is prudent for houseboat manufacturers to clearly identify the exhaust outlet to warn people. A well-and properly designed water/gas exhaust separator and stack is a viable low-cost control that efficiently removes hazardous CO concentrations from any potentially occupied areas on or near the water level of a houseboat.

Reference Number: 40306

WARNER, M. "Carbon Monoxide Density Pattern Mapping from Recreational Boat Testing." *SAE International Journal of Transportation Safety*, 6: 107–131, 2018 (3 tables, 7 figures, 4 references)

Abstract: The CO levels at various locations from actual water testing of three recreational power boats (2001 deck boat, inboard/outboard 395 hp; 1996 wake boat, inboard 395 hp; and 2017 surf boat, 550 hp) were determined. The highest CO levels detected were near the stern of the vessel and can accumulate around the swim platform.

Table: Mean Ambient CO Levels (ppm) of Three Power Boats under Different Test Conditions

Test condition	2001 deck boat	2017 surf boat	1996 wake boat
Idle/neutral crosswise	505	76	457
Engine idle/neutral	124	7	No test
5 miles per hour (mph) (wakeless)	131	27	14
Surf mode (10–12 mph)	43	46	151
Cruise mode (30 mph)	12	4	4

Source: Adapted from Warner 2018.

Exposure to CO gas can occur in any combustion engine-driven vessel and can reach hazardous levels when airflow patterns allow gas to accumulate in or around the vessel. Concentrations of CO gas will vary greatly for a given location depending on the speed of the vessel and airflow conditions. Wind speed (or the absence of wind) and wind direction are key factors in assessing CO exposure risk. When it comes to CO accumulation, concentrations, and exposure relative to the types of vessel tested here, a little air movement goes a long way toward minimizing CO levels.

Reference Number: 40307

EASLEY, R.B. "Open Air Carbon Monoxide Poisoning in a Child Swimming behind a Boat." *Southern Medical Journal*, 93: 430–432, 2000 (23 references)

Abstract: A four-year-old girl wearing a life jacket was swimming near the swim deck of an anchored houseboat, which had a gasoline-powered electrical generator on. The generator exhaust was one-and-a-half inches above the water surface at the rear of the boat. The child became supine in the water and then rigid and unconscious. Paramedics arrived and administered oxygen. At hospital, her blood COHb was 22.2%. She was administered hyperbaric oxygen therapy (HBOT) and showed no adverse sequelae two months later.

The greatest risk factor for accidental CO poisoning is unrecognized exposure. Our case represents a common activity—swimming behind a boat that has a generator running—that unknowingly placed this child at risk for CO poisoning. With the number of families participating in recreational boating steadily increasing every year, physicians should consider

CO poisoning when a child or an adult has unexplained mental status changes, seizures, or lethargy while on a boat or near recreational boating. Carbon monoxide poisoning should also be in the differential diagnosis of unexplained drowning or near-drowning events during recreational boating in both children and adults.

Reference Number: 40308

ROBERTS, P., M. WARD, R.L. BARON, W. HUMBLE, M. HADZIHASANOVIC, R. COX, L. TAPP, J. MCCAMMON, AND R. MCCLEERY. "Carbon Monoxide Poisonings Resulting from Open Air Exposure to Operating Motorboats—Lake Havasu City, Arizona, 2003." *Morbidity and Mortality Weekly Report*, 53: 314–318, 2004 (2 tables, 1 figure, 8 references)

Abstract: CO exposures were determined in 36 Lake Havasu City, Arizona, municipal employees during a busy Memorial Day holiday in May 2003. Substantial outdoor CO exposure occurred mainly in the late afternoon and early evening when the wind speeds were low and the channel was busy with recreational motor boats. The average estimated COHb (from breath) in non-smoking employees increased from 1% in the morning to 6% in the afternoon. The maximum COHb detected was 11%. Breath CO concentrations were also determined in 46 non-smoking vacationers, and the COHb concentration increased from 1% in the morning to 13% in the afternoon.

Table: Percent of Municipal Employees Exceeding CO Limits and Estimated COHb Concentrations during and at the End of the Memorial Day Weekend

CO exposure	24 May 2003 (%)	26 May 2003 (%)
Percent work shifts exceeding maximum 200 ppm CO	100	7
Percent work shifts with air levels exceeding 35 ppm	42	0
Percent wmployees with COHb ≥ 3.5%	100	17

Source: Adapted from Roberts et al. 2004.

Persons in communities with lakes and rivers where boats congregate in large numbers should be aware of the dangers of open air boat-related CO poisoning and the need to evaluate CO exposures during high-traffic periods. Boat manufacturers should improve emission controls to reduce consumer CO exposure. The risk for boat-related CO poisonings should be reduced by considering measures such as limiting the number of boats in certain areas; enforcing a no idle policy when boats are stationary; and warning vacationers of (1) the signs and symptoms of CO poisoning; (2) the hazards related to occupying the back of the boat any time the motor is running; and (3) the risk of CO poisoning in areas of boat congestion, especially during calm weather conditions.

4.04 CHARCOAL BURNING (BARBEQUES)

Revenge of the Barbecue Grill. Carbon Monoxide Poisoning.

—J.D. Gasman, J. Varon, and J.P. Gardner, "Revenge of the Barbecue Grill. Carbon Monoxide Poisoning" (1990) **(40402)**

Charcoal produces large amounts of toxic CO (40401). One major cause of CO poisoning is bringing the charcoal barbeque indoors (40402, 40403, 40408–9). Charcoal has also caused poisoning when used as heating (40405–6). And as indicated in section 3.03, it is commonly used in suicides (40407).

Reference Number: 40401

GAMGEE, A. "On Poisoning by Carbonic Acid Gas, and by Charcoal Fumes." *Journal of Anatomy and Physiology*, 1: 339–346, 1867 (13 references)

Abstract: A study was conducted of the toxicity of charcoal fumes, which have been mainly linked to carbon monoxide. Experiments with CO in blood and hemoglobin and in animals are discussed.

In the case of poisoning by pure carbonic oxide gas, I have, in a course of very many experiments on rabbits, dogs, frogs and mice, found that invariably the remarkable coloration was induced in the blood, and not only observed in the blood, but likewise in specially vascular organs. The remarkable carmine-red injection of the liver in carbonic oxide poisoning is indeed a most striking appearance, and is quite characteristic. The kidneys too, are usually of an exceedingly florid color, and the muscle have a beautiful pink coloration which, if once seen, can never be forgotten. I may incidentally be mentioned that if a portion of muscle be suspended in an atmosphere of carbonic oxide gas, it very readily acquires a most splendid coloration; if placed in a jar which has been very perfectly exhausted and into which, carbonic oxide is subsequently admitted, the occurrence of the decomposition of muscle and other animal tissues is, to a very great extent, postponed and modified.

Reference Number: 40402

GASMAN, J.D., J. VARON, AND J.P. GARDNER. "Revenge of the Barbecue Grill: Carbon Monoxide Poisoning." *Western Journal of Medicine*, 153: 656–657, 1990 (13 references)

Abstract: Four members of a non-English-speaking family attended an emergency room (ER) complaining of headache, lethargy, nausea, and abdominal cramping. They were treated with intravenous fluids and discharged. Several hours later, four other members of the same family attended the same ER with the same symptoms and were treated with intravenous fluids and discharged. Yet again, several hours later, another four members of the same family attended the ER again with the same complaint. This time, through an interpreter, it was admitted that they cooked with charcoal briquettes indoors in the fireplace. Their blood COHb concentration was 6.9%, 15.1%, 17%, and 17.4%, and they were treated with oxygen for two hours. Firefighters were dispatched to check the house.

This case illustrates the importance of always considering CO poisoning when two or more patients are similarly or simultaneously sick. The diagnosis must be excluded by a directed history and physical examination. If suspicion remains, COHb testing should be done and oxygen therapy should be started empirically while results are pending. If CO poisoning is confirmed, the source must be identified and recommendations for correction or avoidance made.

Reference Number: 40403

KIM, H.-J., Y.K. CHUNG, K.M. KWAK, S.-J. AHN, Y.-H. KIM, Y.-S. JU, Y.-J. KWON, AND E.-A. KIM. "Carbon Monoxide Poisoning-Induced Cardiomyopathy from Charcoal at a Barbecue Restaurant: A Case Report." *Annals of Occupational and Environmental Medicine* 27 (1): 13, 2015 (1 table, 1 figure, 24 references)

Abstract: A 47-year-old male restaurant worker who worked as a charcoal recycler at a Korean barbeque grill restaurant collapsed suddenly and lost consciousness. He regained consciousness and complained of a squeezing pain in his left chest. He attended an ER, and his initial blood COHb concentration was 35.8%.

> The patient worked for 14 days at the barbeque grill restaurant from January 5 to 19, 2013, and his primary duty was to handle charcoal that is served to the table for grilling. The patient suffered acute carbon monoxide poisoning while working at the restaurant. The level of blood carboxyhemoglobin was 35%. The extent of myocardial injury is not clear from the findings of the electrocardiography and cardiac angiography. Confirming the diagnosis of the intoxication with elevation of cardiac enzymes and evidence of exposure to significant amounts of carbon monoxide suggest a considerable association and possibility that these factors influenced the development of the cardiovascular complication. Thus, in our patient, the incidence of cardiac injury is clearly and strongly associated with his occupation.

Reference Number: 40404

STERNBACH, G., AND J. VARON. "Carbon Monoxide Poisoning." *Western Journal of Medicine*, 154 (6): 737, 1991 (1 references)

Abstract: Two 31-year-old non-English-speaking women attended hospital due to suspected food poisoning as they had nausea, vomiting, dizziness, and headache. Their other family members had similar symptoms, although they did not eat the same food. It was admitted upon further questioning that they had barbequed their food indoors. Their COHb were 38.3% and 31.9% respectively, and they were placed on normobaric oxygen therapy (NBOT). The other members of the household had COHb of 25.5%, 21.2%, and 18.5%.

> These cases underscore the similarity that carbon monoxide poisoning may display to food poisoning, especially when more than one patient is involved. The difficulty of obtaining an accurate history because of language barriers and social backgrounds may enhance the challenge of arriving at the correct diagnosis.

Reference Number: 40405

ONODERA, M., Y. TSUKADA, T. SUZUKI, K. SORIMACHI, K. EBIHARA, L. SATO, R. ZENDA, S. UENO, K. SUGAYA, AND K. ISEKI. "Development of Delayed Neurological Sequelae in Acute Carbon Monoxide Poisoning Cases

Caused by Briquette-Based Kotatsu." *Medicine*, 100 (16): e25009, 2021 (1 table, 2 figures, 23 references)

Abstract: Briquette-based kotatsu is a traditional Japanese heating system that is still used in rural areas and has caused CO poisoning due to its low-cost, longer burning times and because it does not require electricity. A study was conducted on four men and 13 women (ages 54 to 94 years) treated at hospital for CO poisoning using the kotatsu system for heat between 2017 and 2020. HBOT was conducted up to ten times on patients with abnormal findings. Three patients developed delayed neurological sequelae (DNS), and all had covered their bodies with a kotatsu quilt.

This study demonstrated that DNS occur more frequently in patients who have experienced acute CO poisoning due to briquette-based kotatsu if their entire bodies are covered by the kotatsu quilt at the time of discovery. Hence, we conclude that patients should minimize the extent to which they cover their bodies with the kotatsu quilt.

Reference Number: 40406

TETSUKA, S., T. SUZUKI, T. OGAWA, R. HASHIMOTO, AND H. KATO. "Repeated Unconsciousness Due to Chronic Carbon Monoxide Poisoning in an Older Patient: A Case Report." *Journal of Rural Medicine*, 16: 289–292, 2021 (2 figures, 16 references)

Abstract: A 90-year-old man who lived alone in a rural area of Japan was brought to the ER with a disturbance of consciousness. For several years, he looked dazed in winter and rarely went out of the house. At hospital, his blood COHb was 18.0%. After five hours of high-flow oxygen, his COHb was reduced to 5.5%. The patient had used a kotatsu charcoal briquette heating system that caused chronic CO poisoning.

Because briquettes generate a large amount of smoke and CO for a while after ignition, they should be ignited outdoors and left in a well-ventilated outdoor environment for approximately an hour after ignition. Even if the combustion is stable indoors, users must have plenty of ventilation; for example, windows should be left open three times an hour. In situations such as the current case, as the amount of oxygen in the room decreases, a large amount of CO is generated in reverse proportion to the amount of oxygen. The current patient was probably familiar with the use of briquette kotatsu; however, with age, he may have been unable to manage the briquettes as a result of executive dysfunction caused by dementia associated with Parkinson's disease. In community medicine, this case serves as a reminder that while performing daily medical case, clinicians must consider the patients' living conditions.

Reference Number: 40407

NIELSEN, P.R., A. GHEORGHE, AND N. LYNNERCUP. "Forensic Aspects of Carbon Monoxide Poisoning by Charcoal Burning in Denmark, 2008–2012: An Autopsy Based Study." *Forensic Science Medicine and Pathology*, 10: 390–394, 2014 (2 tables, 24 references)

Abstract: An analysis was conducted of 19 autopsied cases of CO poisoning due to charcoal burning in Denmark between 2008 and 2012. There were 11 cases of suicide, four accidental deaths, and four joint homicide/suicide cases. A distinct cherry-red liver mortis was observed in 14 of the 19 cases (74%). The mean COHB was 67% (range 37–82%) in the postmortem cases with a cherry red colour compared to 61% (range 50–69%) in the four cases without a cherry red colouration. A suicide note was found in only two of the 11 suicides (18%).

Table: Characteristics of Suicide, Accidental, and Homicide/Suicide Deaths Due to Charcoal Burning in Denmark

Characteristic	Suicide	Accidental	Joint homicide/ suicide
Mean age (range) years	46.8 (27–66)	39.5 (38–41)	28.2 (7–48)
Male (%)	64	100	33
Known metal illness (%)	18	0	25
Mean CO (range) %	68.5 (58–82)	49.3 (37–57)	73 (69–78)
% positive blood alcohol content (BAC) (%)	64	25	25

Source: Adapted from Nielsen, Gheorghe, and Lynnercup 2014.

In conclusion, our study found 19 autopsied cases all deceased from CO poisoning by charcoal burning in Denmark, 2008–2012. In the view of the rising trend of charcoal-burning suicides in Asia and a possible increasing incidence of this suicide method in Denmark, it is highly relevant to give this subject attention, and prevent this method of suicide from mobilizing a hidden group of potential victims.

Reference Number: 40408

LYNESS, J.R., AND J. CRANE. "Carbon Monoxide Poisoning from Disposable Charcoal Barbeques." *American Journal of Forensic Medicine and Pathology*, 32: 251–254, 2011 (6 figures, 13 references)

Abstract: Four cases of fatal CO poisoning involving portable disposable charcoal barbeques were described. A 54-year-old woman and 34-year-old woman intentionally inhaled fumes from a portable charcoal barbeque indoors. The post-mortem COHbs were 74% and 74% respectively. The other two cases were accidental and involved a 41-year-old woman and 49-year-old man who used the charcoal barbeque for heat and to cook food inside a car and a tent respectively. Their post-mortem blood COHbs were 51% and 65%. The post-mortem COHbs were measured with a hemoximetre.

Interestingly, both of these deaths occurred during the summer months which is contrary to the recognized increase of nonintentional deaths from CO poisoning during the colder winter months from CO poisoning during the older winter months. This is possibly due to the availability of disposable BBQs during the summer when they are commonly used for outdoor cooking.

Reference Number: 40409

WINDER, C. "Carbon Monoxide Induced Death and Toxicity from Charcoal Briquettes." *Medical Journal of Australia*, 197: 349–350, 2012 (9 references)

Abstract: A 42-year-old man and his friend cooked a barbeque lunch using charcoal briquettes on the veranda of a small apartment in Sydney, Australia. After lunch, they brought the still burning briquettes inside the apartment for warmth. The doors and windows of the apartment were closed due to airport traffic noise. The man was found dead. At autopsy, his post-mortem blood COHb was 61%, and he was found to have underdiagnosed coronary artery disease. The friend survived but had his right arm amputated below the elbow as he had collapsed and was unconscious on his right arm for many hours. The concentration of carbon in briquettes is 85%. The lethal ambient CO concentration is 10,000 ppm (1%). Therefore, for a 100-cubic-metre (m^3) closed apartment, the burring of 629 grams of charcoal briquettes could produce this toxic CO concentration.

The burning of BBQ charcoal can give off carbon monoxide, which has no odour and can be lethal. NEVER BURN CHARCOAL INDOORS OR IN VEHICLES OR TENTS. When burning BBQ charcoal ensure BBQ area is well VENTILATED. Keep out of reach of children. Please take care when burning.

4.05 CIGARETTES/OTHER TOBACCO PRODUCTS

A cigarette is a pinch of tobacco in paper with a fire on one end and a fool on the other.

—George Bernard Shaw, Irish playwright (1856–1950)

Cigarettes and especially cigars produce high CO concentrations. Non-smokers generally have a blood COHb of less than 2%. Cigarette smokers, especially those in bingo halls, obtain toxic COHbs of up to 22.3% (40501–2, 40507–8). Cigar smokers have been found to have even higher COHb levels of up to 38.6% (40504–5). Alternative cigarettes such as bidis produce high concentrations of CO (40503). Even vapours can obtain elevated blood COHb concentrations for supposedly "safer" e-cigarettes (40510, 40511). Firefighters who smoke have a higher exhaled breath CO concentration than non-smoking firefighters who have fought a fire (40509).

Reference Number: 40501

WATSON, W.C. "Bingo Brain." *Canadian Medical Association Journal*, 126 (11): 1266, 1982.

Abstract: A 69-year-old two-pack-a-day female smoker was admitted to hospital with a history of chest pains and confusion. She also played bingo three nights a week.

It so happened that 2 weeks previously I had paid my first visit to a bingo hall, not because I was in dire need of funds or camaraderie but because I was helping to raise money for a charitable organization. I had never been in such a smoke-filled, polluted, acrid atmosphere in my life. Of the registered 310 bingo players, in the hall, 304 were smoking; the other 6 were sitting in a tiny no-smoking area in a distant corner of the hall. Within 20 minutes my eyes were burning, by the end of the evening I was fatigued and had a headache, and when I went home every item of clothing had to be washed to rid it of the smoke. I have no doubt that this patient, who is a heavy smoker and spends hours in a bingo hall 3 nights a week, has recurrent carbon monoxide poisoning. Chest pain and confusion are two of the commonest symptoms of carbon monoxide poisoning.

Reference Number: 40502

SEN., S., C. PELTZ, J. BEARD, AND B. ZENO. "Recurrent Carbon Monoxide Poisoning from Cigarette Smoking." *American Journal of Medical Science*, 340: 427–428, 2010 (4 references)

Abstract: A 48-year-old female chronic cigarette smoker presented with a one-day history of headache and dizziness. She also showed significant digital clubbing and diffuse wheezing. Her COHb was 21.8%, and she was treated with 100% oxygen until her COHb was 1.9%. She returned one week later with similar complaints, and her COHb was 24.2%. She admitted to smoking up to two packs per day. She was again treated with 100% oxygen until her COHb was 3.2% and was recommended for tobacco cessation therapy.

Significant tobacco use can increase COHb levels and lead to signs and symptoms consistent with CO poisoning.

Reference Number: 40503

KUMAR, R., S. PRAKASH, A.S. KUSHWAH, AND V.K. VIJAYAN. "Breath Carbon Monoxide Concentration in Cigarette and Bidi Smokers in India." *Indian Journal of Chest Disorders and Allied Sciences*, 52: 19–24, 2010 (4 tables, 23 references)

Abstract: Breath CO concentrations were measured in 241 cigarette, 148 bidi smokers, and 40 non-smokers using a Smokerlyzer between 2001 and 2004. The mean breath CO concentrations were 4.07, 13.6, and 18.9 ppm for the non-smokers, cigarette smokers, and bidi smokers respectively.

The general perception among the people especially smokers is that bidi is less harmful compared to cigarettes because the tobacco content in a bidi is less compared to a cigarette. However, results of the present study prove that in spite of having lower tobacco content compared to cigarette, bidi is more harmful. Our conclusion is based on a single parameter of CO level in breath; therefore, it has its limitations in comparing the hazards of cigarette and bidi but our study provides a platform for further research involving multiple parameters for comparing the hazards of cigarette and bidi. As of now, we can to a certain extent, negate the popular assumption that bidi is less harmful and a safe alternative to cigarette and one bidi is equal to one cigarette for calculating pack-year smoking.

Reference Number: 40504

HAMILL, W., AND R.P. O'NEILL. "Carbon Monoxide in Cigar Smokers." *Journal of Medical Science, Seventh Series*, 2: 273–277, 1969 (36 references)

Abstract: Two cigar smokers attended the Ohio State University hospital as a result of headaches and dizziness of long-term duration. One patient smoked up to 20 cigars a day. Their blood COHb concentrations at hospital were between 12% and 15 %. All clinical and laboratory findings cleared as a result of the cessation of smoking.

This report combines the known effects of carbon monoxide with a heretofore unreported source of exposure. The role of carbon monoxide intoxication from smoking has been little emphasized in producing decreased arterial oxygen saturation particularly as determined by the present laboratory methods and may well be more common than recognized in polycythemia. Carbon monoxide also appears to have a direct effect on intracellular function. This effect in addition to the presence of the hypoxia may tend to aggravate symptoms in patients with chronic cardiac, pulmonary, and vascular disease.

Reference Number: 40505

DOREY, A., P. SCHEERLINCK, H. NGUYEN, AND T. ALBERTSON. "Acute and Chronic Carbon Monoxide Toxicity from Tobacco Smoking." *Military Medicine*, 185: e61–e67, 2020 (2 tables, 54 references)

Abstract: A 65-year-old male veteran with chronic obstructive lung disease (COPD) and a long history of cigarette smoking switched from cigarettes to smoking small unfiltered cigars. His hemoglobin increased from 15.1 to 19.2 grams (g) per decilitres (dL), and his COHb was 19.4% and one minute later was 18.5%. He stopped smoking cigars, and his COHb decreased to 7.6%. Other cigar smokers with headaches, polycythemia, and tremors had COHb concentrations that ranged from 13% to 38.6%.

Many active duty military veterans are tobacco smokers and these patients and their health care providers should be aware of the adverse effects of CO toxicity from tobacco smoking. Patients may have symptoms such as headache, syncope, and ataxia in the setting of acute CO toxicity; however, the chronic effects of CO may not be completely understood. Future work could explore chronic CO toxicity and its effects on strength and exercise tolerance in military personnel and provide education to service members, veterans, and health care providers.

Reference Number: 40506

WHINCUP, P., O. PAPACOSTA, L. LENNON, AND C. HAINES. "Carboxyhemoglobin Levels and Their Determinants in Older British Men." *BMC Public Health*, 6: 189–197, 2006 (4 tables, 1 figure, 27 references)

Abstract: The blood COHb concentrations were determined in 4,252 older British men (ages 60–79 years) from 24 British towns between 1998 and 2000. The COHb distribution was positively skewed (mean 0.46%, median 0.50%). A blood COHb concentration of 2.5% or greater was detected in 9.2% of the men, and a blood COHb of 7.5% or greater was found in 0.1% of the men.

Table: Mean Blood COHb Concentrations in Elderly British Men According to Smoking Status

Current Smoking Status	Adjusted Mean Blood COHb (%)
Non-smoker	0.32
Ex-smoker	0.35
Pipe or cigar (light)	0.61
Pipe or cigar (heavy)	1.53
Cigarettes (< 5 per day)	1.39
Cigarettes (5–10 per day)	2.18
Cigarettes (10–20 per day)	3.07
Cigarettes (> 20 per day)	3.18
No passive smoking	0.45
Passive smoking	0.47

Source: Adapted from Whincup et al. 2006.

Factors which were independently related to mean COHb level included season (highest in autumn and winter), region (highest in Northern England), gas cooking (slight increase) and central heating (slight decrease) and active smoking, the strongest determinant. Mean COHb levels were more than ten times greater in men smoking more than 20 cigarettes a day (3.29%) compared with non-smokers (0.32%); almost all subjects with COHb levels of 2.5% and above were smokers (93%). Pipe and cigar smoking was associated with more modest increases in COHb level. Passive cigarette smoking exposure had no independent association with COHb after adjustments of other factors. Active smoking accounted for 41% of the variance in COHb level and all factors together for 47%.

Reference Number: 40507

NEILSEN, B.K., J. ALOI, AND A. SHARMA. "Acute Carbon Monoxide Poisoning Secondary to Cigarette Smoking in a 40-Year-Old Man: A Case Report." *American Journal on Addictions,* 28: 413–415, 2019 (17 references)

Abstract: A 40-year-old man with a medical history of depression, anxiety, panic attacks, and substance use disorder smoked three packs of cigarettes within three hours and developed chest pains, palpitations, dizziness, headache, weakness, shortness of breath, and vomiting. The CO-oximeter reading by first responders showed a COHb of greater than 20%. At hospital, his COHb was 22.3%. He was treated with HBOT and benzodiazepines.

While cigarette smoking is a common, well established source of chronic carbon monoxide exposure and significantly increases the levels of carboxyhemoglobin in the blood, it is not commonly associated with acute or chronic carbon monoxide toxicity. However, it is important to recognize cigarette smoke as a possible source of carbon monoxide exposure particularly in heavy smokers or individuals that smoke in a confined area. While the negative effects of cigarette smoking are often perceived as being chronic and only coming to fruition after numerous years of exposure, it is important for both physicians and patients to recognize the possibility for potentially life-threatening acute toxicity secondary to carbon monoxide exposure.

Reference Number: 40508

HAMASHOJI, M., K. KUNITOMO, T. SHIMIZU, AND T. TSUJI. "Carbon Monoxide Poisoning in a Psychiatric Patient Mistaken for Indefinite Complaints." *European Journal of Case Reports in Internal Medicine,* 9 (3): 003242, 2022 (1 figure, 5 references)

Abstract: A 42-year-old man with psychiatric disorders, including schizophrenia and panic disorders, attended the ER four times in one month due to dizziness, abdominal pain, and nausea. Since his laboratory tests were normal, it was felt that these symptoms were due to his pre-existing psychiatric problems. On his fifth visit, his blood COHb was determined to be 11%. He was a current smoker who smoked cigarettes in a poorly ventilated room. It was recommended that he stopped smoking and stay in well-ventilated areas. Since then, there has been no further ER attendance.

It is apparent that this patient was misdiagnosed because of his medical history and standard analysis was overlooked. When patients with psychiatric disorders have nonspecific symptoms. It is important to check for urgent underlying conditions during diagnosis.

Reference Number: 40509

VINNIKOV, D., Z. TULEKOV, Z. ROMANOVA, I. KRUGOVYKH, AND P.D. BLANC. "Smoking Practices in Relation to Exhaled Carbon Monoxide in an Occupational Cohort." *BMC Public Health*, 20 (1): 1894, 2020 (3 tables, 1 figure, 19 reference)

Abstract: Exhaled CO concentrations were measured in 787 male firefighters (329 active) at 18 fire stations using a Smokerlyzer. The subjects were asked to hold their breath for 15 seconds and exhale smoothly into the mouthpiece. The mean exhaled breath CO concentration was three ppm for never smokers, three ppm for ex-smokers, and 10 ppm in current smokers. In bivariate analysis, breath CO concentrations were statistically greater among those who were older, smoked a cigarette or used a waterpipe in the last 12 hours, were exposed to second-hand smoke (SHS), lived in a rural area, and/or used coal or biomass for cooking.

Cigarette and recent waterpipe smoking are associated with increased exhaled CO in firefighters. Firefighting itself was a less potent contributor to exhaled CO when measured at an annual screening but an age interaction was manifested.

Reference Number: 40510

GRECH, A.K., D.T. KEATING, D.J. GARNER, AND M.T. NAUGHTON. "A Case of Extreme Carboxyhemoglominemia Due to Vaping." *Respirology Case Reports*, 10 (5): e0942, 2022 (7 references)

Abstract: A 52-year-old woman went to the hospital on 12 occasions over eight years. Although the last 11 times were in the last two years of her life. All presentations were similar with weakness, atypical chest pains, headache, and drowsiness. She vaped two bottles of e-liquids daily for the last year and smoked up to three packs a day since the age of 16 years. Her COHb was elevated and ranged from 7.1% to 32.9% (mean 23.3%). Twelve other smokers admitted to hospital had a range of COHb from 3.1% to 11.1% (mean 6.3%). It required on average 47 hours with oxygen therapy for her blood COHb to return to normal. On her final admission, she had a systolic collapse at home and could not be resuscitated.

This case highlights two important observations. First, that the combination of vaping and cigarette smoking or vaping alone was associated with nearly four times greater COHb levels than are seen with cigarette smoking. Second, such abnormal COHb levels causes acute chronic high output right heart failure.

Reference Number: 40511

SON, Y., C. BHATTARAI, V. SAMBUROVA, AND A. KHYLSTOV. "Carbonyls and Carbon Monoxide Emissions from Electronic Cigarettes Affected by Device Type and Use Patterns." *International Journal of Environmental Research and Public Health,* 17 (8): 2767, 2020 (3 tables, 5 figures, 62 references)

Abstract: Carbon monoxide can be formed by electronic cigarettes (ECs) through thermal decomposition of propylene glycol (PG) and vegetable glycerin (VG) and some flavouring compounds found in e-liquids. CO and carbonyl emissions were determined in four types of ECs, (1) cig-a-like; (2) top coil; (3) sub ohm mod; (4) pod (that is, JUUL). The highest CO emission occurred with fruit flavour e-liquids in a top coil, followed by fruit flavour in a mod, then tobacco flavours and the cig-a-like, followed by JUUL.

Emissions increased nearly linearly with puff duration, while puff flow had a relatively small effect. Flavored e-liquids generated more carbonyls and CO than unflavored liquids. Carbonyl concentrations and CO in e-cigarette aerosols were found to be well correlated. While e-cigarettes emitted generally less CO and carbonyls than conventional cigarettes, daily carbonyl exposures from e-cigarette use could still exceed acute exposure limits with the top-coil device potentially posing more harm than conventional cigarettes.

4.06 COFFEE ROASTING

Coffee, the favorite drink of the civilized world.

—Thomas Jefferson (1824)[2]

An unusual source of CO is due to the coffee roasting of green beans, but the CO is produced not during roasting but, rather, during the storage ("degassing") of freshly roasted coffee beans (40601, 40605). Ethiopian coffee ceremonies may expose young pregnant women to unsafe ambient CO concentrations (40603). Workers at large coffee extract plants have obtained high COHb concentrations of 21.7% COHb from the coffee grounds and extract-holding plants, and some workers have even died of CO exposure there (40602, 40604).

Reference Number: 40601

MCCARRICK, A., AND B. LETTER. "Carbon Monoxide Release from Whole Bean Roasted Coffee in Storage." Paper presented at the 49th International Conference on Environmental Systems, Boston, 7–11 July 2019 (3 tables, 11 figures, 10 references)

Abstract: Coffee roasting causes the green beans to expand as gases such as CO_2 and CO are generated within the beans and held in pockets within the beans. Once the beans cool, the surface hardens, and the gases become trapped in the beans. These gases (typically 80–90% CO_2) begin to diffuse out of the bean over time into the environment. This process is also known as de-gassing. The current best practice method for packaging roasted beans use an impermeable flexible bag with a one-way pressure relief/check valve. The release of trapped gases can continue for 60 to 100 days. The production of CO and CO_2 were determined in bags of coffee beans stored at cool, warm, and room temperatures by electrochemical CO cell and infrared CO_2 sensors in a test chamber. Freshly roasted beans off-gassed more CO than beans roasted earlier. An ambient CO concentration of 1,200 ppm and greater was found in the storage at 70 degrees and 100 degrees Fahrenheit.

The Navy has an interest in determining the CO off-gassing rate and total capacity as it applies to shipboard storage involving large quantities of

---

2 Thomas Jefferson to Edmund Rogers, 14 February 1824, Tucker-Coleman Papers: Series 2 Thomas Jefferson Correspondence, Special Collections Research Center, Swem Library, College of William and Mary.

bagged coffee in order to ensure the safety and wellbeing of Navy sailors. An environmental chamber was configured to mimic the volume/mass conditions of coffee aboard US Navy ships. Increases in CO and CO_2 concentrations were measured over time at three different temperatures, 40°F, 70°F, and 100°F. Off gassing rates were capable of rapidly generating hazardous levels of CO and found to be significantly affected by variations in temperature. Diffusion occurred more rapidly with increasing temperature. A second test at room temperature released CO at much higher rates that the first test. Further investigations indicated that the second set of coffee had been shipped significantly closer to the roasting date. It is known that the rate of gas release is substantially higher immediately after roasting.

Reference Number: 40602

HAWLEY, B., J.M. COX-GANSER, AND K.J. CUMMINGS. "Carbon Monoxide Exposure in Workplaces, Including Coffee Processing Facilities." *American Journal of Respiratory and Critical Care Medicine*, 196: 1080–1081, 2017 (7 references)

Abstract: At one coffee-processing facility, elevated personal CO exposure levels were found not in the roasting facility but in the area where the roasted beans were stored or ground. Workers had personal exposure levels of 7–30 ppm when averaged over a shift, but the worker who stood next to an open hopper of ground coffee had a peak CO level of 584 ppm. At processing facilities that make coffee extracts, several workers have died of CO poisoning in the storage tank.

Recognizing an occupational source of CO has diagnostic and therapeutic implications for a patient with CO poisoning and provides an opportunity for prevention in coworkers. We encourage clinicians to consider workplace CO exposure when evaluating patients with symptoms and signs compatible with CO poisoning. Our data indicate that coffee processors should be added to the list of occupations at risk.

Reference Number: 40603

KEIL, C., H. KASSA, A. BROWN, A. KUMIE, AND W. TEFERA. "Inhalation Exposures to Particulate Matter and Carbon Monoxide during Ethiopian Coffee Ceremonies in Addis Ababa: A Pilot Study." *Journal of Environmental and Public Health*, 2010 (1): 213960, 2010 (6 tables, 13 references)

Abstract: The Ethiopian coffee ceremony generally involves a young woman of the household who washes the coffee beans then roasts the beans in a flat pan over a small charcoal fire or three-stone fire. The coffee beans are roasted until the beans turn black and shiny. The roaster then takes the freshly roasted beans in the pan around the room to each guest until the room is filled with a coffee aroma. During the ceremony, incense (ital and sendel) is usually burned. The beans are then placed in a large heavy wooden bowl, cracked, ground, and placed into a small pot to which water is added. The pot is put over the heat source until the mixture boils. The boiled coffee is allowed to settled and then is served in small cups. Water is added to the coffee grounds two more times, boiled, and served. After the third time, an elderly person says a prayer or blessing. It takes approximately 70 minutes to conduct the ceremony. Ambient CO and particulate matter were measured in 10 homes in Addis Ababa where the coffee ceremony was conducted. The highest ambient 30-minute average ranged from 25 to 88 ppm (mean 45 ppm).

Coffee preparers are often young women who may at some point be pregnant. Considering that one health concern from CO exposure is fetal hypoxic effects these exposures may be cause for concern. The potential outcomes from fetal hypoxia are manifold. Depending on the developmental stage of the fetus when hypoxia occurs, the impacts on individuals and in the community as a whole may be significant. The prevalence of coronary artery disease in Ethiopia is unknown. Given the typical Ethiopian diet the prevalence of coronary artery disease among the young is expected to be low. The prevalence may be higher in the elderly who are often frequent participants in the coffee ceremony. The results show that the one-hour CO guideline is often exceeded and the shorter term guidelines are also exceeded, though less often. This represents an increased risk of heart attacks in the susceptible population.

Reference Number: 40604

NEWTON, J. "Carbon Monoxide Exposure from Coffee Roasting." *Applied Occupational and Environmental Hygiene*, 17: 600–602, 2002 (1 table, 1 figure)

Abstract: A worker at a company that produces coffee extract after roasting and grinding coffee beans collapsed while cleaning the inside of a tank that had contained liquid coffee extract. He was rescued by the fire department and administered oxygen immediately. At hospital, he was found to have a blood COHb of 27.1%. It is estimated that the ambient CO concentration was 2,500 ppm, and the fire department detected greater than 999 ppm on their CO detector after rescue. The employer installed better ventilation, CO detectors, and installed an automatic system to clean the inside of the tank.

Table: Ambient CO Concentrations at Various Locations in a Coffee Roasting and Extraction Plant

Sample location	Ambient CO detected (ppm)
Green coffee bean feed hopper	0 ppm at start of day, 25 ppm average over day
Coffee roaster and bean cooling tray	7–11 ppm
Coffee grinder and feed hopper	> 999 ppm
Coffee grounds and storage bin	3,000 ppm inside bin
Open extracting tank	600–1,000 ppm
Filter	> 100 ppm
Coffee extract holding tank	7,500 ppm
Process tank	590 ppm when valve opened

Source: Adapted from Newton 2002.

Carbon monoxide is generated and held in the coffee bean during roasting and is released after roasting and grinding. One theory is that during the 400°F roasting process, in a roaster atmosphere with a limited supply of oxygen present, green coffee undergoes a color change from green to brown due to the oxidation of organic substances present in the bean. These browning reactions lead to the formation of carbon monoxide and carbon dioxide, some of which is given off directly by the roasted bean, the rest is held in the very small spaces of the internal bean structure, ready to be released during grinding.

Reference Number: 40605

LEBOUF, R.F., AND M. ALDRIDGE. "Carbon Monoxide Emission Rates from Roasted Whole Bean and Ground Coffee." *Journal of Air and Waste Management Association*, 69: 89–96, 2019 (1 table, 5 figures, 13 references)

Abstract: The CO emission rate was determined in whole bean; freshly ground and packaged ground coffee was determined in a microchamber with an electrochemical CO monitoring for 1.5 hours. Ground coffee emitted more CO than whole been coffee. The roast level did not affect CO emissions from whole bean coffee but was increased in medium dark to dark roasted ground coffee. CO emissions decreased with age from the time of roasting and after the package was opened. Fresh ground coffee emissions were more than two times greater than ground coffee in a sealed package. Higher CO emissions of ground coffee are a result of increased surface area available for diffusion in the release of trapped gases in the pore structure of whole bean coffee.

> Worst-case emission factors from commercially available whole bean and ground coffee measured in unpackaged, loose form showed that roasted coffee can rapidly emit CO. Emission factors were higher for ground than whole coffee. Darker roasts emitted more CO than lighter roasts when the coffee was ground. CO concentrations should be monitored in storage spaces in service and coffee manufacturing facilities as well as transport vessels to ensure exposures do not exceed occupational exposure limits. Storage spaces may need to be ventilated to control CO concentrations to safe levels.

4.07 ENVIRONMENTAL (URBAN)

> Suicide by carbon monoxide used to be done in the garage. Now all you have to do is go to Mexico City and inhale.
>
> —Richard Bayan, "azquotes.com"

Although there are natural sources of CO (40706), carbon monoxide is the most abundant air pollutant produced by humans in the atmosphere except for CO_2 (40701). CO levels in most cities peak around 7:00 to 9:00 a.m. and in the winter months (40701). City dwellers have higher blood COHb than rural folk (40705). High ambient CO concentrations have been associated with poor brain growth and cardiac deaths (40702, 40704). The main sources of indoor CO are gas-fired appliances, attached garages, faulty furnaces, and tobacco smoking (40703).

Reference Number: 40701

JAFFE, L.S. "Ambient Carbon Monoxide and Its Fate in the Atmosphere." *Journal of the Air Pollution Control Association*, 18: 534–540, 1968 (2 tables, 1 figure, 39 references)

Abstract: Carbon monoxide is the most abundant air pollutant found in the atmosphere except for CO_2. Most of the CO produced is due to human activity, and 90% of the CO emissions are due to the burning of fossil fuels in motor vehicles. CO levels in most cities peak between 7:00–9:00 a.m., corresponding to morning rush hour and also peak during the winter months. In the upper atmosphere, CO is converted into CO_2 by ultraviolet light. CO is also converted to CO_2 by microorganisms in the soil. The oceans may also act as a sink for CO as the solubility of CO is 0.032 to 0.017 (volume of gas standard temperature and pressure dry that is absorbed by a unit volume of water when the pressure of the gas equals one atm). The duration of CO in the lower atmosphere has been estimated to vary from 0.3 to 5 years.

More than 87 million tons of CO are emitted annually in the United States alone while the annual world-wide estimate of CO emission is approximately 200 million tons per year from technological sources. Sustained significant exposures of 50 ppm CO are encountered in the central business areas of large cities due primarily to motor vehicles. Additional quantities of CO while comparatively small and difficult to estimate, are produced by vegetation (algae and higher plants) and by millions of marine hydrozoan jellyfish found in the oceans of the world. This large tonnage of CO in the absence of scavenging process is sufficient to raise the atmospheric concentration by 0.03 ppm or more, yet the background levels of CO do not appear to be increasing significantly.

Reference Number: 40702

LEVY, R.J. "Carbon Monoxide Pollution and Neurodevelopment: A Public Health Concern." *Neurotoxicology and Teratology*, 49: 31–40, 2015 (3 figures, 112 references)

Abstract: Vehicle exhaust contributes to 75% of all CO emissions in the United States and up to 90% in urban settings. The outdoor urban CO air level in the United States ranges from 2 to 5 ppm, although, depending on the circumstances, spikes can occur. The global background CO level in the troposphere ranges between 50 and 120 parts per billion. Indoor CO concentrations can range up to 100 ppm due to gas stoves and 5–35 ppm in smoking rooms.

Exposure to indoor and outdoor pollution including tobacco smoke may negatively impact brain development. It needs to be reiterated, however, that air pollution is a complex mixture of heterogeneous gaseous and particulate neurotoxins. Therefore, attributing pathologic effects to an individual pollutant is fraught with limitations. Recognizing this caveat, there is a suggestion that CO, a major component of air pollution, is associated with impaired neurodevelopmental outcome and experimental perinatal exposure has been shown to disrupt a number of critical processes in the developing brain. Although much is known about the toxic effects of CO at symptomatic levels, the potential public health impact of subclinical CO exposure has, by-and-large, been underappreciated.

Reference Number: 40703

MORANDI, M., AND M. EISENBUD. "Carbon Monoxide Exposure in New York City: A Historical Overview." *Bulletin of the New York Academy of Medicine*, 56: 817–828, 1980 (3 figures, 3 tables, 43 references)

Abstract: Ambient CO concentrations in New York City traffic sites have decreased from an hourly maximum of 60 ppm in the late 1960s to 35 ppm in 1978. A similar decrease has been observed in other cities and is due to the elimination of coal as a fuel to heat buildings, the banning of apartment dwelling incinerators, and a decrease in CO emissions in cars from 7% in the late 1930s, to 3.8% in the 1960s, to 0.5% in the late 1980s. The main sources of indoor CO emissions are gas-fired cooking appliances, attached garages, faulty furnaces, and tobacco smoking.

Table: The Consumption of Principal Fossil Fuels in New York City for 1934, 1965, and 1975

Fuel	1934	1965	1975
Anthracite coal (times 10^6 tons)	10.6	0.72	0.3
Bituminous coal (times 10^6 tons)	9.1	5.4	0
Fuel oil (times 10^6 gallons)	611	4,086	3,387
Natural gas (times 10^9 cubic feet)	0	167	98
Manufactured gas (times 10^9 cubic feet)	54	0	0
Gasoline (times 10^6 gallons)	500	1,250	1,453

Source: Adapted from Morandi and Eisenbud 1980.

CO concentrations in the downtown areas of large cities have remained stable or decreased since the beginning of the century in spite of increases in total gasoline use. These decreases can be explained by such factors as the improvement of the efficiency of the internal combustion engines, the change in the type of fuel used for space heating and electrical generation and replacement of manufactured by natural gas. Indoor exposures have decreased also as indicated by the large reduction in accidental CO poisoning deaths.

Reference Number: 40704

BURNETT, R.T., S. CAKMAK, M.E. RAIZENNE, D. STIEB, R. VINCENT, D. KREWSKI, O. PHILLIPS, AND H. OZKAYNAK. "The Association between Ambient Carbon Monoxide Levels and Daily Mortality in Toronto, Canada." *Journal of the Air and Waste Management Association,* 48: 689–700, 1998 (4 tables, 6 figures, 46 references)

Abstract: The effects of ambient levels of CO and total suspended particles on cardiac and non-cardiac non-accidental mortality in Toronto was determined between 1980 and 1994. Eighty-six percent of CO emissions were from motor vehicles. The mean daily ambient CO concentration was 1.18 ppm and ranged up to 6.1 ppm from the four urban testing sites. Of the 40 mean daily non-accidental deaths in Toronto, 4.7% could be attributable to CO and 1% to total suspended particles. The relative risk of death for CO exposure was 1.055 times for non-cardiac and 1.086 times for cardiac deaths.

Although the association between CO and death for cardiac diseases was stronger than the association with deaths from noncardiac conditions, CO was a statistically significant predictor of noncardiac mortality. CO is known to affect people with cardiac problems through a well understood pathophysiological mechanism. Thus, it is not surprising that an association was detected with death due to heart diseases. The immediate cause of death for many people dying of noncardiac causes may also have often been cardiac in nature. Therefore, it is also plausible that CO may play a role in exacerbating these latter conditions. Since only the underlying cause of death was available in a computerized format, we could not directly examine the association between CO and the immediate cause of death.

Reference Number: 40705

GETTLER, A.O., AND M.R. MATTICE. "The Normal Carbon Monoxide Content of the Blood." *Journal of the American Medical Association*, 100: 92–97, 1933 (10 tables, 1 figure, 20 references)

Abstract: The blood COHb concentration was determined in the blood of various individuals in and around New York City. The persons with the highest blood COHb were smokers.

Table: Mean Blood COHb Concentrations in Various Individuals Living in and around New York City

Subjects	Mean blood COHb (%)
18 persons living in New York City	1.0–1.5
12 inmates of rural state institutions	< 1.0
12 New York City street cleaners	3
2 taxi cab drivers	8.0–19.0

Source: Adapted from Gettler and Mattice 1933.

Emil Bogen collected and analyzed all the smoke and fumes from burning tobacco. Figures are given for mainstream smoke (that drawn through the mouth) and sidestream smoke (that escaping from the end). The sidestream smoke is rich in carbon dioxide and is alkaline because of its ammonia content. The mainstream smoke is acid and rich in carbon monoxide. The more rapid the smoking, the greater the concentration of carbon monoxide. From the literature it would appear that tobacco smoke might be a prominent factor in determining the normal level of carbon monoxide in the blood.

Reference Number: 40706

JOHN, K.S., AND K. FEYISAYO. "Air Pollution by Carbon Monoxide (CO) Poisonous Gas in Lagos Area Southwestern Nigeria." *Atmospheric and Climate Science*, 3: 510–514, 2013 (5 figures, 15 references)

Abstract: The largest natural source of CO is due to photochemical reactions in the troposphere. Other natural sources are forest fires, volcanoes, and other combustions. CO has a molecular weight of 28, which makes it slighter lighter than air (average MW of 28). CO was first identified by Joseph Priestley in the eighteenth century, but it was Claude Bernard in 1870 who discovered the binding of CO to hemoglobin. Ambient CO concentrations were measured using a handheld CO detector at various locations in Lagos, Nigeria. The largest emitters of CO were trucks, cars, and motorcycles followed by diesel and gasoline generators. Firewood cooking locations, eateries, and households were much lower sources of CO. The CO emitted by diesel trucks ranged from 107 to 855 ppm.

> For as long as fossil fuels are being used to run automobiles; domestic generators, cooking and other related activities, carbon monoxide will be emitted into the environment thereby making it unsafe. From the results, automobiles dominate other sources of carbon monoxide within Lagos environs. The CO emissions from the domestic generators might not be unconnected to the rate at which families are wiped out due to fumes from generators kept within flats or at an enclosure in the residential buildings. With this report Government needs to take decisive steps toward adequate enlightenment program to educate the Nigerian people on the need to take ample preventive measures and curb the destruction to lives via this menace.

4.08 FIREFIGHTERS

> At the present time, the only reasonable method for attaining this goal is by the insistence that breathing apparatus be used by firefighters whenever work is being done in a burning or recently burned structure.
>
> —T.R. Griggs, "The Role of Exertion as a Determinant of Carboxyhemoglobin Accumulation in Firefighters" (1977) **(40801)**

Non-smoking firefighters who wear a mask while fighting fires obtain only low levels of COHb of less than 0.5%, which is much less than the blood COHb obtained in smoking firefighters who wear a mask (40801–3).

Reference Number: 40801

GRIGGS, T.R. "The Role of Exertion as a Determinant of Carboxyhemoglobin Accumulation in Firefighters." *Journal of Occupational Medicine*, 19: 759–761, 1977 (2 figures, 12 references)

Abstract: Twenty male firefighter trainees (ages 19 to 48 years) participated in a search and rescue drill in a smokehouse for approximately five minutes with and without a breathing apparatus. The smoke was created by burning damp straw in a barrel and blowing it through the building with a fan. The mean ambient CO concentration was 300 ppm. Blood COHb and heart rate were determined before and after exposure. The mean heart rate using a breathing apparatus increased from approximately 105 to 175 beats per minute (bpm), compared to an increase of 95 to 150 bpm.

The breathing apparatus weighed 37 pounds, and the increased exertion in carrying this extra rate may have caused a higher heart rate post-exposure. No cardiac arrhythmias were detected.

Table: Blood COHb Concentrations in Smoking and Non-Smoking Firefighters' Exposure to Smoke with and without a Breathing Apparatus

	Smoker (mask) (%)	Smoker (no mask) (%)	Non-smoker (mask) (%)	Non-smoker (no mask) (%)
Pre-smoke exposure	4.5	4.5	< 0.5	< 0.5
Post-smoke exposure	3	6.2	< 0.5	4.0

Source: Adapted from Griggs 1977.

Specifically, levels as low as 4–6% may indeed accelerate myocardial ischemia in a working firefighter who has subclinical coronary heart disease. It seems likely that this mechanism accounts in part for the extremely high rate of death from myocardial infarction among on-duty firefighters. Limitation of CO exposure during firefighting to the lowest possible level is therefore required if the risks of firefighting are to be reduced. At the present time, the only reasonable method for attaining this goal is by the insistence that breathing apparatus be used by firefighters whenever work is being done in a burning or recently burned structure.

Reference Number: 40802

STEWART, R.D., R.S. STEWART, W. STAMM, AND R.P. SELLEN. "Rapid Estimation of Carboxyhemoglobin Level in Fire Fighters." *Journal of the American Medical Association*, 235: 390–392, 1976 (1 table, 1 figure, 19 references)

Abstract: Breath CO and blood COHb were measured in 177 firefighters. The breath samples were collected after a breath hold of 20 seconds, and the last portion of the exhaled breath was collected in a Saran bag and then analyzed by an electrochemical device (Teflon-bonded diffusion electrode), the Ecolyzer model 2100. There was a good correlation between breath results and blood COHb (+/− 0.5% COHb). After a 24-hour tour of duty, the mean COHb was 5.0% in 55 non-smoking firefighters and 7% in smoking firefighters.

The use of the described rapid electrochemical method of breath analysis remedies this defect in the surveillance program. The quantity of

CO absorbed by any workman can be determined by breath analysis performed with a frequency necessary to define the magnitude of the CO exposure.

Reference Number: 40803

LOKE, J., W.C. FARMER, R.A. MATTHAY, J.A. VIRGULTO, AND A. BOUHUYS. "Carboxyhemoglobin Levels in Fire Fighters." *Lung*, 154: 35–39, 1976 (2 tables, 11 references)

Abstract: The blood COHb concentration and 12 lead electrocardiograms (ECG) were measured in 23 non-smoking and 28 smoking West Haven, Connecticut, firefighters. The blood COHb was also determined in 16 firefighters who used and did not use self-contained breathing apparatus after three building fires. The mean baseline COHb was 2.1% in non-smoking and 5.3% in smoking firefighters. The mean blood COHb was 3.4% in 16 firefighters before the fire and 6.3% after the fire. The duration of firefighting ranged from 15 to 150 minutes. Only one firefighter with a post-fire COHb of 13.2% had symptoms of CO poisoning (that is, headache, malaise, and dyspnea).

Half of 16 fire fighters involved in the building fires did not use the breathing apparatus. The difference in mean COHb (pre and post exposure to CO) between fire fighters who used the breathing apparatus and those who did not was minimally significant. There would have been a greater difference had the three smoking fire fighters who used the breathing apparatus not smoked after the fire. The risk of a dangerously high blood COHb level is relatively small in fire fighters who used the self-contained air breathing apparatus during a fire. The increase in resting heart rate following acute exposure to CO noted in this study has been reported by others. However, we did not observe any other electrocardiographic abnormalities.

4.09 GAS APPLIANCES/HEATERS

The suppliers of gas, whether companies or municipalities, are permitted to tear up our streets at strange and inconvenient times by Act of Parliament, but no Act of Parliament has ever given them leave to foul the subsoil with coal gas or to allow so deleterious a substance to escape into our dwelling-homes. Such an escape is undoubtedly a nuisance and injurious to health. But to prove its origin is not easy and this is the barrier behind which public bodies shielded themselves from attack.

—Anonymous, *British Medical Journal*, 2 November 1895, 1117

Now you're cooking with gas!

—an advertising slogan used to promote the use of gas stoves, Deke Houlgate, American Gas Association (1930s)

Just like tobacco companies, the coal gas industry has downplayed health risks for years. Even natural gas still poses risks of CO poisoning, especially in cooking stoves (40902–3, 40908–9, 40912). Other gas appliances have caused CO poisoning as well (40904, 40906–7, 40910–11). Using a flange-bottom kettle produces more CO than a flat-bottom frying pan (40905).

Reference Number: 40901

FIDDES, F.S. "Accidental Carbon Monoxide Poisoning: Dangers of Inadequate Ventilation." *British Medical* Journal, 22 September 1956, 697–698

Abstract: The victims of accidental CO poisoning have been described as the four Ds: the decrepit, the diseased, the drugged, and the drunks. Three fatal cases of CO poisoning are described. The three cases were all due to a lack of ventilation (closed doors and windows) in a room with a combustible fire.

> These fatal cases of accidental carbon monoxide poisoning provide a timely reminder of (1) the danger of using stoves in confined areas, even though these are in good order and fueled with good quality coal, unless adequate positive ventilation is ensured at all time, whether the doors or windows be closed or not; and (2) the danger which may arise from gas-fires, even if these and their fittings are in perfect order and they appear to be burning satisfactorily, if there is not also adequate ventilation and extraction of the combustion products.

Reference Number: 40902

PULLINGER, R. "Something in the Air: Survival after Dramatic, Unsuspected Case of Accidental Carbon Monoxide Poisoning." *British Medical Journal*, 312: 897–898, 1996 (16 references)

Abstract: An unconscious woman in her seventies was brought by ambulance to an ER. She was found on the bathroom floor; her husband was dead in bed, and their two pets were found dead. Her husband had advanced atherosclerotic coronary artery disease and bronchopneumonia, and his post-mortem COHb was 56%. Her hospital admission COHb was 40%, and she had hypothermia (rectal temperature 29 degrees Celsius. She recovered after medical treatment, including 100% oxygen and rewarming.

> An examination by British Gas of the cooker in the patient's kitchen showed deposits of lint around the air inlet ports of the grill burner. When lit, the flame was yellow, indicating incomplete combustion, and the carbon monoxide concentration in the vicinity of the burner was in excess of 2000 parts per million. The carbon monoxide concentration on the first floor after the grill had been alight for one hour was 500 parts per million and increased to about 8000 parts per million after two hours.

Reference Number: 40903

LEHR, E.L. "Carbon Monoxide Poisoning: A Preventable Environmental Hazard." *American Journal of Public Health*, 60: 289–293, 1970 (8 references)

Abstract: Three residents of a small house in Savannah, Georgia, awoke on a cold morning in 1968 with feelings of extreme stress and, one by one, became incapacitated. They were rushed to hospital by neighbours. The ER physician suspected botulism and administered botulinum antitoxin, which resulted in an extreme reaction in the patients. But a high blood COHb was detected in one of the patients, which lead to the source of the CO (and their symptoms) in the home as due to two unvented gas heaters.

> Carbon monoxide poisoning, both fatal and nonfatal is more prevalent than is generally recognized. There is increasing evidence that its chronic low-level form may be a cause of much ill health. Accidental carbon monoxide poisoning in the residential environment could be almost totally controlled if existing standards were improved, if effective codes were enforced and if the public had a better understanding of how to avoid this hazard. Some communities had made special efforts to control the hazard and they have succeeded. Unfortunately, there is too often a slackening of attention, after an immediate situation has been controlled. The Public Health Service is now undertaking a comprehensive national program to assist communities in sustained prevention programs.

Reference Number: 40904

TAM, C.W., R.J. BEVAN, P.T.C. HARRISON, L.C. YOUNGS, AND D. CRUMP. "Public Health Impacts of Exposure to Carbon Monoxide from Gas Appliances in UK Homes: Are We Missing Something?" *Indoor Built Environment*, 21: 229–240, 2012 (3 tables, 54 references)

Abstract: Over 22 million homes in the United Kingdom use gas appliances with the potential of CO exposure as people generally spend about 90% of their time indoors. It is estimated that, in 2008, a CO exposure occurred in 0.2% of the homes with gas (that is, 40,000 homes).

> It is clear that the most effective form of protection from CO poisoning is the prevention of inappropriate exposure. Although CO intoxication can be treated effectively with normobaric (or hyperbaric) oxygen, in some cases treatment does not reverse all the symptoms and in severe cases long-term effects are apparent. The literature also highlights the possibility of an underlying epidemic of tiredness, reduced cognitive function and dizziness resultant of low-level chronic exposures to CO. Occult CO poisoning may greatly affect the happiness and mental well being of individuals but remain unnoticed throughout the population.

Reference Number: 40905

ANONYMOUS. "Flange-Bottomed Kettle Danger." *British Medical Journal*, 24 March 1956, 694.

Abstract: A 37-year-old woman was found dead in her bath due to CO poisoning. Her post-mortem blood COHb was 40–45%. The source of the CO was not the gas geyser but the base of the flange-bottomed kettle used to boil water being the same diameter as the gas burner. An ambient CO of 0.1% was produced boiling water with the flange-bottomed kettle compared to 0.002% with the gas burner alone and 0.003% with a flat-bottom frying pan.

> Mrs. Stone was therefore exposed to a high concentration of carbon monoxide at the sink while washing up, was kept topped up while she was in the living room, and then returned to the bathroom which contained a concentration still dangerous although not so high as the during the washing up. There the exertion of bathing after her heavy meal caused her to succumb.

Reference Number: 40906

CROXFORD, B., E. HUTCHINSON, G.S. LEONARDI, L. MCKENNA, L. NICHOLSON, G. VOLANS, AND P. WILKINSON. "Real Time Carbon Monoxide Measurements from 270 UK Homes." *Air and Waste Management Association and US EPA's Office of Research and Development*, 1: 13, 2006 (2 tables, 5 figures, 38 references)

Abstract: Ambient CO concentrations were monitored in 270 mainly low-income homes in East London for one week during the winter of 2005–6. About 5% of the homes were found to have faulty gas appliances. The overall mean ambient CO concentration was 1.7 ppm but increased to 6.3 ppm in the 10 hones that exceeded the World Health Organization limit of 52 ppm. No significant difference was found in the mean CO in smoker's homes (2.3 ppm) compared to non-smoker's homes (1.8 ppm).

Table: World Health Organization Guidelines for Ambient CO Exposure, 1999

Ambient CO (mg/m^3)	Ambient CO (ppm)	Duration
100	87	15 minutes
60	52	30 minutes
30	26	1 hour
10	8.6	8 hours

Source: Adapted from Croxford et al. 2006.

Nearly 5% of the total sample was found to have at least one element of at least one gas appliance that could be considered dangerous by the visiting gas engineer. The exposure to the occupants from these appliances depends strongly on how they are used. However, if an appliance is found to emit high levels of CO, then the risk is present to the occupant. In this study, the occupants of these homes were often elderly and vulnerable people. The WarmZone program aims to find and help vulnerable homes that are in fuel poverty. The measures that can be installed under this program include new central heating systems, and draft proofing to improve ventilation heat lose. However, neither WarmZone nor the larger scale Warm Front Scheme includes provisions to replace cookers or remove old room heaters. In theory CO concentrations could increase after fuel poverty reduction measures have been installed due to reduced air infiltration while keeping old inefficient gas appliances. Currently in the UK there is no program of gas appliance replacement or grants to help families with little spare money that are at risk from CO exposure. One occupant when asked how old her gas cooker was, replied that it had been in her home for 25 years and it was second hand when she got it.

Reference Number: 40907

RIEZEBROS, R.K., E. DE MELKER, AND W.G. DE VOGT. "Pitfalls in Cardiology: Myocardial Ischemia in Carbon Monoxide Poisoning." *Netherlands Heart Journal*, 14: 183–186, 2006 (2 figures, 24 references)

Abstract: A 26-year-old man presented at hospital with syncope after having fainted several times. He also complained of nausea, was incontinent, and vomited. His ECG showed sinus tachycardia. His blood COHb was 25.5%, and he was admitted to a coronary care unit and administered 100% oxygen. Twelve hours later, his blood COHb was zero. He suffered no further syncope or cardiac arrhythmias and quickly recovered. The fire department discovered a poorly functioning gas heating system in his flat, which was dismantled and replaced.

As the mortality and morbidity in CO intoxication are considerable, early diagnosis and treatment is crucial. While pulse oximetry cannot distinguish between carboxy- and oxy-hemoglobin, peripheral saturation is most likely to be normal in CO poisoning. Co-oximetry analysis of arterial blood should be the method of choice in determining the presence of potential life-threatening levels of carboxyhemoglobin. Prompt treatment with 100% oxygen by a high-flow device should be initiated when the diagnosis of CO poisoning is suspected. In severe cases, treatment with hyperbaric oxygen can be considered. High levels of oxygen can decrease the CO half-life and thereby prevent further damage.

Reference Number: 40908

HAYHURST, E.R. "Domestic Carbon Monoxide Poisoning from Gas Stoves." *American Journal of Public Health*, 13: 462–465, 1923

Abstract: Since fire was first discovered, carbon monoxide has caused the most widespread poisoning in humans. The dangers and risks of gas stoves/water heaters are described as well as 50 deaths that occurred in the winter of 1922 in Ohio, including the death of a family of six overnight while sleeping and a bridal couple in their new house on their wedding night. This situation recalls the time of another CO epidemic, when

Ben Franklin invented the Franklin stove to counter CO poisoning from the use of fireplaces and charcoal braziers.

> One of the public health aspects of the case is the wayside hotel. In Steubenville, Ohio, according to the press, four fatalities occurred over a period of eleven months in the same hotel, two of which occurred in the same room and, indeed from the same stove! Thus, will people refuse to accept the evidence even where death may be the answer! A similar situation of doubt exists in the case of the gas-fired water heater of the instantaneous type, which is too commonly set up in bath rooms, kitchens, etc., without a chimney attachment. Here, Prof. Schoch of the University of Texas found a family refused to accept the carbon monoxide explanation of the cause of death of its members despite chemical air analysis and continued to use the unvented water heater! Water heaters are especially dangerous because carbon monoxide is yielded whenever a carbonaceous flame touches a cool surface.

Reference Number: 40909

BARON, R.C., R.C. BACKER, AND I.M. SOPHER. "Fatal Unintended Carbon Monoxide Poisoning in West Virginia from Nonvehicular Sources." *American Journal of Public Health*, 79: 1656–1658, 1989 (3 tables, 1 figure, 14 references)

Abstract: Between 1978 and 1984, there were 62 unintended deaths in West Virginia due to CO from non-vehicular sources. Post-mortem COHb was measured and ranged from 50% to 88%. Fatal cases increased sharply in the fall and peaked in the wintertime. Furnaces or space heaters were the source of CO in 89% of the cases, cooking ranges (mainly natural gas) in 8%, and generators in 2%. Natural gas, methane, butane, and propane in the presence of sufficient oxygen will produce relatively non-toxic carbon dioxide and water vapour. But a decrease in the oxygen supply in a closed space and a lack of proper maintenance or cleaning of the burners can result in CO formation.

> Prevention, therefore, depends largely on assuring proper function and appropriate use of gas-burning appliances and can be achieved by professional installation, annual servicing, and education to avoid burning fuels in confined or tightly sealed enclosures. A coordinated educational and inspection program involving health departments, gas companies, and

appliance distributors—conducted in the late summer and early fall—would be desirable. Because there is a tendency to conserve heat by closing or sealing windows, terminating the production of unvented units has also been suggested.

Reference Number: 40910

CRAWFORD, R., D.G.D. CAMPBELL, AND J. ROSS. "Carbon Monoxide Poisoning in the Home: Recognition and Treatment." *British Medical Journal*, 301: 977–979, 1990 (1 tables, 1 figure)

Abstract: A 84-year-old grandmother lived in a flat with her daughter, son-in-law, and two grandchildren and was admitted repeatedly to hospital as a result of chronic CO poisoning, which was misdiagnosed. The source of the CO was a poorly maintained, inefficient gas-powered, wall-mounted heater, which had an offset flue. During a gale force wind of 123 kilometres per hour, 13 inhabitants of the flat were poisoned by CO fumes and had to be treated in hospital. Their COHb ranged from 9.6% to 29.6%. Four patients were treated with hyperbaric oxygen (HBO_2).

Chronic carbon monoxide poisoning can be difficult to diagnose, but failure to recognize it may cause substantial disruption of medical services. Treatment consists of administering 100% oxygen using a leak tight non-rebreathing circuit with supportive therapy. Hyperbaric oxygen treatment should be used when appropriate. Failure to use existing facilities has resulted in successful litigation by patients in the United States.

Reference Number: 40911

WILSON, E., C. TOMASALLO, AND J. MEIMAN. "Occupational Carbon Monoxide Exposure in an Industrial Kitchen Facility, Wisconsin, 2017." *Morbidity and Mortality Weekly Report*, 67: n.p., 2018 (5 references)

Abstract: On 5 September 2017, a manufacturer of frozen appetizers suspected a CO leak in the area of the plant with gas-burning fryers. First responders detected a CO concentration of 313 ppm, and the plant was evacuated. Forty employees had symptoms of CO exposure and 37 were taken to the ER. The mean blood COHb concentration within six hours of exposure was 11.7% (range 4.1–21.4%). The most common symptoms were headache, dizziness, and nausea. No employee required HBOT, and none stayed overnight.

An Occupational Safety and Health Administration (OSHA) investigation identified a CO course associated with gas burners on the fryer appliances that had been replaced 4 days earlier. OSHA found that ventilation was inadequate to clear combustion products from new burners, which resulted in a buildup of CO in the facility. Gas-burning appliances in industrial kitchen facilities are not common occupational causes of CO-related morbidity and mortality. However, improperly maintained and ventilated appliances can be a source of CO exposure. Because symptoms of CO poisoning are nonspecific, CO poisoning might be underreported.

Reference Number: 40912

ROSENTHAL, D.J., D.W. JONES, R. DUNCAN, R. CROSBY, D. ENGELBERG, J.M. BODIE, R.W. EMMONS, T. MIDURA, AND R. WOOD. "Epidemiologic Notes and Reports, Carbon Monoxide Poisoning—California." *Morbidity and Mortality*, 21: 247–252, 1972.

Abstract: A wife and her 62-year-old husband with heart disease spent the night of 26 April 1972 in a mountain cabin. During the night, the wife felt nauseous, vomited, and fainted. Her husband tried to help, but she fainted again. When she woke, her husband was dead. He had a post-mortem COHb of 65%. The doors and windows of the cabin were shut due to the cold weather. The sources of CO in the cabin were a lantern, an old oil heater, a well-vented gas water heater, and an unvented gas refrigerator. The refrigerator caused an ambient CO of 1,000 ppm after 30 minutes of operation and 1,300 ppm after 45 minutes.

Epidemiologic investigation revealed that the couple had previously used the cabin only in the summer and had either slept outside or with the doors and windows open for ventilation. On this occasion however, the doors and windows were kept closed because of the cold weather.

4.10 GEYSERS (HOT WATER HEATERS)

Because carbon monoxide is an odorless, colorless gas, it is easy to forget that when a victim is discovered s/he must be removed from the source environment lest the rescuer also become a victim of carbon monoxide poisoning.

—J. Brandenburg, "Inhalation Injury" (1980)[3]

Almost everyone needs hot water, and, unfortunately, most hot water is obtained from a flame of some sort, which, if not maintained properly, will start producing CO, which can result in a fatal CO concentration in the small unventilated bathroom or kitchen (41001–9). Even a change in weather or wind can cause a gas hot water heater to start producing CO (41002). In one case in a motel in Portugal, a young English couple was found dead in their motel, but their two children were still alive. The cause of death was initially thought to be clam poisoning. Instead, it was due to CO poisoning caused by the unvented gas water heater in their motel room (41006). Unfortunately, in countries such as India, the electrical supply is erratic and expensive, and so most water heaters are geysers (41003–4, 41009).

Reference Number: 41001

JAISWAL, S., A. BATRA, A. VERMA, M. HALDAR, AND W.R. SHEIKH. "Carbon Monoxide Poisoning Secondary to Gas Geyser." *Indian Journal of Case Reports*, 5: 423–425, 2019 (2 figures, 15 references)

Abstract: A 34-year-old woman with no pre-existing medical conditions was found unconscious and unresponsive in her bathroom by her family. At the ER, she was drowsy, confused, and slow, but the rest of her systematic examination was unremarkable. She was administered high-flow oxygen via a face mask. A magnetic resonance imaging (MRI) brain scan showed changes suggestive of CO poisoning, but no COHb was measured. The family used a gas geyser in the house to heat water. The patient was continued on high-flow oxygen and discharged in a week. Two weeks later, the patient had no neurological sequelae, and her MRI brain scan was normal.

> Gas geysers have emerged as a cost-effective; easy to use and an efficient method for heating water at homes especially in India. They run on liquid petroleum gas, the combustion of which leads to the generation of CO, hydrocarbons, and nitrogen dioxides.

---

3 J. Brandeburg, "Inhalation Injury: Carbon Monoxide Poisoning," *American Journal of Nursing*, 80 (1): 98–100, 1980.

Reference Number: 41002

SEDDA, A.F., AND G. ROPSSI. "Death Scene Evaluation in a Case of Fatal Accidental Carbon Monoxide Toxicity." *Forensic Science International*, 164: 164–167, 2006 (2 figures, 22 references)

Abstract: Low concentrations of CO (0.05–0.12 ppm) are considered a normal constituent of the atmosphere, but 60% of the CO found in the non-urban troposphere is attributed to human activities. A 21-year-old woman was found dead, crouching in a shower cabin of a small building located beside the country house where the victim lived with her family. A jet of warm water was still running over the pale cherry pink-coloured naked body. As it was cold, she had turned the unvented propane/butane water heater 10–15 minutes earlier and had died 40–50 minutes before the body was discovered. Her blood COHb was 60%. The shower had been used for 10 years before without any incident. The air-exhaust pipe of the heater was one metre long and 10 centimetres (cm) in diameter and had an elbow joint to the outside. Reconstruction showed that, when the wind was not blowing, the room CO concentration reached 1,000 ppm after 35 minutes of the water heater working. When the wind was blowing in a certain direction, the room CO concentration reached 6,300 ppm after 35 minutes.

> The wind that was present at the date and time of the accident, directed in the same direction of the air-exhaust horizontal pipe, not only hindered the elimination of CO from the room but contributed to raise the CO_2 concentration. This fact decreased the O_2 concentration and increased the CO production rate in a catastrophic circle which rapidly raised the respirable CO concentration of the room to lethal values, leading to confusion, drowsiness, lethargy and finally to the death of the victim. After the report of the above mentioned measurement, the court attributed the death to accidental poisoning from carbon monoxide, due to a sum of unfortunate circumstances, caused from the widespread underestimation of an insidious life threatening hazard.

Reference Number: 41003

CHOU, C.-H., C.-H. LAI, S.-H. LIOU, AND C.-H. LOH. "Carbon Monoxide: An Old Poison with a New Way of Poisoning." *Journal of the Formosan Medical Association*, 111: 452–455, 2012 (1 figure, 7 references)

Abstract: Two tragic incidents of CO poisoning occurring in high-rises in Taiwan as a result of obstructed vents of gas heaters are reported. In the first case, CO was produced from a gas water heater on the sixth floor that spread to the third and ninth floors as well. The fire department detected 1,800 ppm CO, and the building was evacuated, but three people died, and eight suffered CO poisoning. In the second case, the vent outlets of the gas water heater on the thirteenth floor were blocked by steel plates on the roof. The CO spread to the fourteenth, fifteenth, and eighteenth floors as well. Three people died, and three suffered CO poisoning.

In conclusion, there are at least two important lessons we should learn from these two tragic events: (1) Incomplete combustion exhaust generated from a gas-powered water heater and obstructed ventilation pipes can result in CO poisoning of neighboring populations. This unique method of spreading CO poisoning has not been reported previously and might have been underestimated because the vast majority of residences in densely metropolitan worldwide are apartments; (2) clinicians ought to suspect chemical-related casualties when they encounter clusters of patients from a single locale. Furthermore, public health authorities, governmental authorities, and the residents of buildings must be made aware of this problem.

Reference Number: 41004

CHONG, C.K., P. SENAN, AND G.V. KUMAR, "Carbon Monoxide Poisoning from Gas Water Heater Installed and Operated in the Bathroom." *Medical Journal of Malaysia*, 52 (2): 169–171, 1997 (3 references)

Abstract: Between 1978 and 1988, two-thirds of the CO poisonings at home were due to liquid petroleum gas (LPG) water heaters. LPG is domestic cooking gas in pressurized tanks and is mainly propane and butane but may contain small amounts of CO and methane. In the presence of adequate ventilation, LPG undergoes complete combustion to form water and CO, but, when the oxygen supply is inadequate, CO can be produced. Gas water heaters are used extensively in the Cameron Highlands of Malaysia, as they heat water faster and are cheaper to operate than electric heaters. Two fatal cases due to gas water heaters occurred in August and September 1995. The fatal blood COHb concentrations were 59% and 70% respectively. The victims were found dead in the bathroom where they were taking a bath. In one case, a minimum ambient CO of 1,200 ppm occurred in 16 minutes of operation of the gas water heater.

> In summary, CO produced by gas water heaters during mechanical malfunction is a significant hazard.

Reference Number: 41005

MOHANKUMAR, T.S., T. KANCHAN, K.S. PINAKINI, R.G. MENEZES, M. SINGH, P. SIROHI, AND N. ANWAR. "Gas Geyser: A Cause of Fatal Domestic Carbon Monoxide Poisoning." *Journal of Forensic and Legal Medicine*, 19: 490–493, 2012 (4 figures, 23 references)

Abstract: A 21-year-old woman was found dead in the bathroom of her home in which a LPG-fueled water heater was located. The bathroom measured 2.7 by 3.9 metres and was poorly ventilated. The LPG gas geyser was recently installed, and she complained of headaches one month before. She was treated for anemia and weakness by her family physician. At autopsy, cherry pink/red colour of hypostasis, tissues, and organs were observed. The cherry red colour may be due to CO-induced vasodilation and tissue ischemia. Her post-mortem blood COHb was 41.6% as determined by spectrophotometry.

> Domestic carbon monoxide poisoning is rarely reported in India and remains an under recognized problem. Effective preventive measures can help in bringing down the mortality and morbidity associated with carbon monoxide poisoning. Development of carbon monoxide detector is potentially the most important advance in the prevention of carbon monoxide poisoning. This however should be considered a secondary prevention method and more cost-effective measures need to be taken up in developing countries like India. It is suggested that cross ventilation be ensured in order to prevent such mishaps in future when any combustible device (like gas geyser) is used in a closed environment along with emphasis on the need for a regular maintenance of any fuel burning appliances. Immediate medical help should be sought if poisoning symptoms are suspected and a high index of suspicion on clinician's part remains vital. With increasing use of gas water heater, public education about the dangers of carbon monoxide with emphasis on safety measures at home remains the key to effective prevention.

Reference Number: 41006

SPECIAL CORRESPONDENT. "Carbon Monoxide Deaths in Portugal: A Mistake Rectified." *British Medical Journal*, 5127: 979–980, 1959.

Abstract: A dental surgeon (age 36 years), his wife (age 34 years), and their two children (ages two and four) stayed at a motel near Lisbon, Portugal. The parents ate clams and veal, but not the children. At the motel, both children were put to bed in another room, and both parents used the bathroom. At 10 a.m. the next morning, a maid knocked on the door with no response. She knocked again at 12:00 p.m. and 4:00 p.m. before the manager was called, and the room was forced open. Both the parents were found dead on the floor with traces of vomit around their mouths. But the children were alive. The Portuguese Tribunal of Almada determined the deaths were due to clam poisoning. A British pathologist, Dr. Keith Simpson, was dispatched to investigate. The post-mortem blood COHb of the parents was between 50% and 54%. The butane gas water heater in the bathroom was not vented, and an ambient CO concentration of 2,800 ppm was detected after one hour of water heater use.

> The case illustrates what is well known in England as a result of the concern felt in recent years over the rising toll of accidental deaths from domestic gas. Both the B.M.A. committee and the Gas Board's figures show that inadequate ventilation often unrecognized, accounts for 7–8% of these accidental CO deaths. It is not uncommon, when vomiting occurs in the last stages of CO poisoning, to have suspicion of food-poisoning aroused. Many people do not notice the stale smell of gas fumes and the distinctive cherry-pink colour is commonly overlooked—especially when the possibility is not borne in mind.

Reference Number: 41007

SIMPSON, D., M.W. CALNAN, AND C. DE FONSEKA. "Factors Contributing to Bathroom Accidents from Carbon Monoxide Poisoning." *Community Health*, 5: 271–278, 1974 (2 figures, 4 references)

Abstract: A 21-year-old woman was taking a bath and was found by her husband in the bathtub unconscious but with her head above water. She was admitted to hospital for 12 days and was originally diagnosed as having viral meningitis. But an examination of her bathroom showed a room ambient CO concentration of 2,800 ppm after nine minutes of operation of the gas water heater. The windows had been painted shut. In another case, another 21-year-old woman developed CO poisoning due to the gas water heater in the bathroom and was diagnosed as hysteria or pregnancy. A representative sample of 66 gas water heaters in the same area showed that 12% produced a toxic room CO concentration of 500 ppm or greater.

> The incidence of reported cases of people obtaining medical attention due to carbon monoxide poisoning from bathroom gas water heaters in the home accident study area is 2 cases per year for every 100,000 people. The experience of the research team would suggest that the true incidence of poisoning from this cause is appreciably greater in that: (1) Patients who were visited after an accident often refer to others who have also been unwell in the bathroom but have not sought medical care. (2) Incorrect diagnosis by doctors of some patients who collapse in a bathroom. (3) In our random sample of 500 electors it was found that one person had fainted during the previous year in a bathroom which under the survey test conditions had a highly toxic atmosphere. She had not reported the incident to the doctor. It was interesting to discover that a previous occupant of the flat also used the bathroom and collapsed remaining unconscious for 2 hours. It is believed that she too had not obtained medical care.

Reference Number: 41008

HEATH, K., AND R.W. BYARD. "Lethal Carbon Monoxide Toxicity in a Concrete Shower Unit." *Forensic Science Medicine and Pathology*, 15: 133–135, 2019 (3 figures, 15 references)

Abstract: A 47-year-old woman was found dead on the floor of an outside concrete shower cubicle in a rural area of South Australia. At autopsy, the victim had significant atherosclerotic narrowing of the coronary arteries (90%), and her viscera showed a cherry pink discolouration. Her postmortem COHb was 55%. The CO source was a poorly maintained gas water heater, showing soot stains, on the inside of the cubicle.

> The reported case demonstrates a rare event in Australia of accidental death in a shower from a faulty gas water heater system. Although the shower was separate to the main farm dwelling, the use of an impermeable concrete tank had created the potential for gas accumulation when

the water heater malfunctioned. The unit had the appearance of some age and seemed not to have been regularly serviced. Had the cherry pink discoloration of tissues not been noted at the time of autopsy, it is unlikely that carbon monoxide toxicity would have been suspected as the police history was of collapse in the shower, and significant stenosing coronary artery arteriosclerosis was identified. If the cause of death had been incorrectly attributed to coronary artery disease with myocardial ischemia it is probable that others on the farm would have been at risk if they used the same shower unit.

Reference Number: 41009

METHA, A., R. MAHALE, A.A. JOHN, M.M. ABBAS, M. JAVALI, P. ACHARYA, AND S. RANGASETTY. "Odorless Inhalant Toxic Encephalopathy in Developing Countries Household: Gas Geyser Syndrome." *Journal of Neurosciences in Rural Practice*, 7: 228–231, 2016 (1 table, 1 figure, 11 references)

Abstract: A study was conducted of seven male and 17 female patients who lost consciousness in the bathroom while bathing due to a gas geyser hot water heater in India. The electrical supply is typically erratic in developing countries, and, as a result, cheaper gas geysers for heating water while bathing are often utilized. Twenty-one patients had recovered spontaneously, one died within one week and two were found dead in their baths. No blood COHb concentrations were reported.

Awareness regarding CO intoxication due to usage of ill-fitted, ill-ventilated gas geyser is necessary as they are entirely preventable conditions. The clinical features range from mild altered level of consciousness with spontaneous recovery to death. It is also important from medicolegal point of view as most of the patients were female and were found unconscious in the bathroom.

4.11 HOTELS, MOTELS, RESORTS, AND RESTAURANTS

1. A young Irish tourist travelling in Argentina died of CO poisoning from a lit stove in his room.
2. An American professional football player died from CO from a faulty water heater in Prague.
3. Two Chinese tourists died in Katmandu caused by a malfunctioning hotel room heater.

4. An elderly American tourist died in a Rome guest house due to a malfunctioning radiator.
5. A Canadian tourist died of CO from a natural gas water boiler in an illegal hostel in Taipei.

—James Wigmore, "CO: The Silent Killer Does Not Take Vacation" (2023)[4]

People tend to feel when they are on vacation, especially in a southern climate with no heating systems, that there is no danger of CO poisoning. However, CO can be produced by many sources such as a gas-powered electrical generator (41101), portable gas heater (41103), swimming pool heater (41105), and restaurants (41109–9). Thus, I recommend that one should bring a small, portable CO detector, although it would be best if motels, hotels, and so on were required by law to install CO detectors in every room (41102–4). CO is such a will o' the wisp that any CO produced by restaurants easily passes through gaps in the walls and floors to adjacent properties (41106–8).

Reference Number: 41101

ALBAUGH, R.P. "Gasoline Engine Exhaust Gas Poisoning." *Monthly Review of the US Bureau of Labour Statistics*, 4: 272–275, 1917

Abstract: A small hotel in western Ohio had its own private gasoline-powered 14 hp electrical generator in its basement. A three-inch exhaust pipe from the generator led underground and exited 20 feet from the building. Over a period of two weeks, the night clerk who stopped the generator at midnight and added oil experienced extreme dizziness and shortness of breath. It was determined that the exhaust pipe had collapsed and was blocked with water and dirt.

> There were 19 other persons in the hotel, and as they came into the corridor—all on the second floor—they also became so weak and dizzy that they were forced to either sit or lie down. One man was able to reach a telephone and summon a physician, who decided that they had eaten poisoned food, and accordingly administered warm milk and other emetics and later a large dose of Epsom salts. All complained of practically the same symptoms, viz, dizziness, general muscular weakness, throbbing

4 James Wigmore, "CO: The Silent Killer Does Not Take Vacation," *LinkedIn*, 19 December 2023.

temporal headache, palpitation, and shortness of breath. Two were in a comatose condition, four complained of nausea, while in two instances there were evidence of relaxation of the sphincters. Although some were rather indisposed for a day or two, all recovered in a short time.

Reference Number: 41102

WEAVER, L.K., AND K. DERU. "Carbon Monoxide Poisoning at Motels, Hotels, and Resorts." *American Journal of Preventative Medicine*, 33: 23–27, 2007 (1 table, 1 figure, 32 references)

Abstract: Each year, more than 200 people die of CO poisoning in the United States. Between 1989 and 2004, there were 68 incidents of CO poisoning at hotels, motels, and resorts, resulting in 711 guests, 41 employees/owners, and 21 first responders being affected. There were 27 deaths (3.5%), and 66 people had confirmed sequelae. Six had sequelae resulting in a jury verdict. Public verdicts involving CO poisoning resulted in a range of between $1 and $17.5 million in verdicts (mean $4.8 million).

It may be more practical to adopt measures that place CO alarms where CO is produced. CO can be produced by any fuel if it is burned incompletely or ventilated inadequately. Alarms should be placed in every room where there is a fuel-burning appliance, including those not accessible to guests so that employees are protected. In addition, CO can spread to other rooms and be distributed throughout a building via a forced air system, or it can be vented outside and entrained. Alarms should therefore be placed in rooms adjacent to those with fuel-burning appliances as well as exhaust vents and air intakes. Due to the number of cases related to faulty pool/spa boilers, the pool/spa areas and pool/spa boiler rooms would be appropriate sites for CO alarms.

Reference Number: 41103

TAPEANTONG, T., AND N. POUNGVARIN. "Delayed Encephalopathy and Cognitive Sequelae after Acute Carbon Monoxide Poisoning: Report of a Case and Review of the Literature." *Journal of the Medical Association of Thailand*, 92: 1374–1379, 2009 (1 table, 4 figures, 18 references)

Abstract: A 50-year-old Thai women and her husband and daughter developed CO poisoning while on vacation in the United States at a hotel in which a small portable gas heater was added to the room to provide warmth. All were found comatose, and the 50-year-old woman had vomited. They were administered oxygen and HBOT. She was admitted to intensive care and was discharged three days later without apparent problems. She returned to Thailand three weeks later and, after another two weeks, was admitted to hospital due to encephalopathy. The MRI revealed abnormal white matter changes in the brain. She was administered HBOT again and had some improvement in memory and other cognitive functions.

> Most patients with CO poisoning, whether incidental or accidental, initially visit the Emergency Department. Prompt recognition and proper management of this condition is essential. Most of them recover without any complication, with hyperbaric or high oxygen therapy. Delayed encephalopathy occurring from 14 to 45 days after recovery from the acute state is well recognize. Clinical manifestations included cognitive impairment, akinetic mutism, sphincter incontinence, gait ataxia, and extrapyramidal syndromes such as chorea, dystonia, and parkinsonism. It is estimated that as high as 50% of individuals with CO poisoning will develop neurological, neurobehavioral, or cognitive sequelae. CO related cognitive impairments included impaired memory, attention, executive function, motor, visual, spatial, and slow mental processing speed. The delayed neuropsychiatric impairment showed that 2.75% of patients suffered from this complication with a mean lucid period of about 3 weeks.

Reference Number: 41104

HAMPSON, N.B., K.L. HAUSCHILD, K. DERU, AND L.K. WEAVER. "Carbon Monoxide Poisonings in Hotels and Motels: The Problem Silently Continues." *Preventive Medicine Reports* 16: 100975, 2019 (2 tables, 13 references)

Abstract: Between 2005 and 2018, 115 incidents of unintentional CO poisonings in US hotels, motels, or resorts were identified, involving 905 victims. Sixty-five percent of CO poisoning incidents occurred at national chains.

Table: Source of CO Poisoning, Usual Fuel, and Number of Incidents in Hotel and Motel CO Poisonings between 2005 and 2018

CO source	Usual fuel	Number of incidents
Boiler	Natural gas	24 (1 fatality)
Clothes dryer	Natural gas	1
Electrical generator	Gasoline	2
Furnace	Natural gas	5
"Heater"	Natural gas	3
In-wall heater (guest room)	Natural gas	1 (1 fatality)
Motor vehicle (hotel garage)	Gasoline	1 (5 fatalities)
Oven (hotel kitchen)	Natural gas	1
Portable heater	Kerosene or propane	2
Swimming pool heater	Natural gas	28 (12 fatalities)
Water heater	Natural gas	22 (3 fatalities)

Source: Adapted from Hampson et al. 2019.

The lodging industry has opposed installation of CO alarms in guestrooms in the past on the basis of cost. It is now possible to purchase a sealed residential CO alarm with a ten-year battery for under forty US dollars. Amortizing that price over ten years yields a cost of only about one cent per guestroom per day. Optimally in the case of public lodgings, hardwired and centrally monitored combination smoke and CO alarms should be required for maximum safety.

Reference Number: 41105

GEORGE, D.J. "Chapter 8: Carbon Monoxide." *Poisons: An Introduction for Forensic Investigators* (Boca Raton, FL: CRC Press, 2017), 75–82 (4 tables, 7 references)

Abstract: Carbon monoxide is a leading cause of poisoning deaths in the United States. Hundreds die each year, and many thousands are treated in hospital and suffer neurological damage. In 2013, an elderly couple in a 72-room hotel in Boone, North Carolina, died of CO poisoning in Room 225. Three weeks later, 10 girls attending a birthday party in Room 325 (one floor up) became violently ill. A few weeks later an 11-year-old boy died, and his mother developed permanent brain injury of CO poisoning while in Room 225. The source of the CO was an indoor swimming pool heater that was not installed or maintained properly. The exhaust pipe from the water heater was corroded and had several leaks and ran under Room 225. The medical examiner determined that the elderly couple had died of CO poisoning but did not notify the hotel or police and resigned. The first responders did not check out the elderly couple's room for CO, but the same responders did after the mother/son CO poisoning and found high CO concentrations. The president of the company that owned the hotel was indicted for involuntary manslaughter and assault. Wrongful death suits were filed for damages against the president and owner of the hotel, the hotel franchise, and other companies and individuals who worked on the pool heating system. The hotel was closed for renovations and Room 225 was permanently closed.

There is reason to believe that the pool heater was not designed, installed, maintained, or inspected properly. The heater was originally fueled by propane but this was converted later to natural gas. The elderly couple's initial death investigation was apparently considered to have been simultaneous heart attacks. Relatives and people within the community suspected carbon monoxide in the deaths but officials did not investigate. Two or more people in the same environment having simultaneous fatal heart attack are extremely unlikely. It is not clear why further investigation was not carried out. It is noteworthy that there have been at least 200 deaths in hotel rooms throughout the country during recent years from carbon monoxide exposures.

Reference Number: 41106

CHUA, I.S.Y., K.B.K. TAN, AND R. PONAMPALAM. "Carbon Monoxide Poisoning in a Group of Restaurant Workers: Lessons Learnt and How to Prevent Future Occurrences." *Singapore Medical Journal*, 65 (5): 288–290, 2021 (2 figures, 13 references)

Abstract: The majority of CO poisonings (73%) occur in homes and 13% are work related. The exhaust fan in a restaurant malfunctioned. The chief and three other restaurant workers initially attended the hospital. The chef's COHb was 18.9%, and his lactate concentration was 10.5 millimoles (mmol) per L. All of the restaurant staff—a total of 30 in all—were assessed at hospital. Most were treated with NBOT. The paramedics were not equipped with portable CO detectors or pulse CO-oximeters.

Carbon monoxide is dubbed the silent killer as it is colourless, odourless and tasteless. Clinical symptoms are non-specific and many patients are unaware that they have been exposed until a CO detector alarms or when the patient presents to hospital and is found to have an elevated carboxyhemoglobin (COHb) level. CO is produced from the incomplete combustion of hydrocarbons and malfunctioning heating systems, improperly ventilated vehicles, generators, grills, stoves and residential fires are common sources of CO poisoning. Anybody is at risk of CO poisoning and vulnerable groups like infants, elderly and those with chronic medical issues are especially at risk. More than 400 Americans die from unintentional CO poisoning not linked to fire annually, more than 20,000 end up in the emergency department and more than 4,000 are hospitalized.

Reference Number: 41107

KESHISHIAN, C., H. SANDLE, M. MELTZER, Y. YOUNG, S. WARD, AND S. BALASEGARAM. "Carbon Monoxide from Neighbouring Restaurants: The Need for an Integrated Multi-Agency Response." *Journal of Public Health*, 34: 477–482, 2012 (2 figures, 15 references)

Abstract: CO can travel through any cavities or gaps in walls and floors separating properties, and so residents of neighbouring properties may be at risk of CO poisoning from combustion appliances there. Three incidents are reported of residents being poisoned with CO above restaurants with a charcoal-burning tandoor oven left smouldering overnight with the ventilation turned off. The most serious incident involved the CO poisoning of seven people from three flats above a restaurant who were taken to hospital. Their COHb concentrations ranged from 7.4% to 21.0%. In all three incidents, the replacement of the charcoal-burning oven with a propane oven and CO detectors being placed in the apartments resolved the problem.

Residential flats are frequently built above restaurants in urban areas. It is therefore possible, and indeed likely, that similar CO poisoning events are occurring that enforcement bodies and the HPA (Health Protection Agency) are unaware of. Traditional style ovens are becoming fashionable in UK restaurants as people seek an authentic gastronomic experience. In addition to tandoor ovens, other ethnic ovens have been implicated in CO safety issues as many do not meet CE (European Directive) standards. The current emphasis on making buildings more energy efficient makes it easier for indoor gases to build up and travel through any gaps or cavities within walls and floorboards. Legal requirements for the installation of CO alarms in restaurants and mandatory ventilation of charcoal-burning ovens could be considered to prevent similar incidents occurring.

Reference Number: 41108

OJIMA, J. "Generation Rate of Carbon Monoxide from Burning Charcoal." *Industrial Health*, 49: 393–395, 2011 (1 table, 2 figures, 10 references)

Abstract: Charcoal is used as a cooking fuel at some restaurants in Japan and has caused CO poisonings in cooks and serving staff, due to the large amounts of CO it produces. The generation of CO from burning charcoal—from *Quercus phillyraeoides* and *Quercus glauca* raw timber—was determined as well as the ventilation rate required. The generation of CO ranged from 137–85 millilitres (mL) per minute per kilowatt (kW), and the ventilation rate of between 41.2 and 55.6 m3/h/kW was required.

Considering the airtightness of modern Japanese houses or buildings, the natural ventilation in some restaurants can't be expected to adequately remove generated CO. Therefore, workers in such restaurants should be informed that opened windows or doors will not provide sufficient ventilation and some additional mechanical ventilation should be installed. Without some practical and effective precautionary measures, the number of CO poisoning cases in the Japanese food industry will probably increase in the near future because the annual consumption of charcoal in Japan has recently been increasing.

4.12 INDOOR SKATING RINKS/ARENAS

Control of carbon monoxide exposure can be achieved by the elimination of its emissions, the use of mechanical ventilation, good administrative practices and routine air monitoring.

—P.W. Kwok, "Evaluation and Control of Carbon Monoxide Exposure in Indoor Skating Arenas" (1983) **(41207)**

The vast majority of CO poisonings in arenas have been due to ice-resurfacing machines powered by propane, gas, or diesel (41201–4, 41207, 41210). The natural gas compressor for the refrigeration of the ice in an arena was not vented properly and caused CO poisoning in 171 players and spectators (41205). However, indoor tractor pulls and monster truck races can overwhelm the ventilation system and have been banned in some arenas (41208–9). Ironically, one CO poisoning event in an area was due to an ambulance idling just outside the arena (41206).

Reference Number: 41201

JOHNSON, C.J., J.C. MORAN, S.C. PAINE, H.W. ANDERSON, AND P.A. BREYSSE. "Abatement of Toxic Levels of Carbon Monoxide in Seattle Ice-Skating Rinks." *American Journal of Public Health*, 65: 1087–1090, 1975 (7 tables, 3 references)

Abstract: In November 1971, 15 children who had been skating in an indoor ice rink in Seattle became ill. All had headaches, four became nauseous, and two vomited. At first, it was thought it was some sort of food-borne illness, but it was soon realized that it was due to a large ice resurfacing machine fueled by propane. The ice rink was also inadequately ventilated. Testing showed that CO levels could be as high as 304 ppm. In addition, children have a higher rate of metabolism than adults, resulting in a higher uptake of CO. Physical activity such as skating can also cause a 10-fold increase in respiratory volume compared to being sedentary. The Environmental Protection Agency recommends a maximum nine ppm CO exposure for eight hours. Catalytic mufflers were found to greatly reduce CO and CO_2 emissions of gasoline-powered ice resurfacing machines.

There are an estimated 600 to 700 ice-skating arenas in the United States as well as large numbers in Canada and in Europe. Most of these use ice-resurfacing machines that produce considerable amounts of CO. Arenas using these machines should be checked to evaluate the hazard to the public from carbon monoxide poisoning. Nitrogen dioxide levels should also be determined. Adequate ventilation is probably a better solution than the catalytic mufflers, especially where NO_2 [nitrogen dioxide] emissions are also a problem. The best solutions to this problem of CO and NO_2 emissions is probably the use of the electrically powered ice-resurfacing machines.

Reference Number: 41202

PELHAM, T.W., L.E. HOLT, AND M.A. MOSS. "Exposure to Carbon Monoxide and Nitrogen Dioxide in Enclosed Ice Arenas." *Occupational and Environmental Medicine*, 59: 224–233, 2001 (8 tables, 72 references)

Abstract: A general review and summary was conducted of the adverse effects of poor air quality in enclosed indoor ice rinks on patrons and employees. A detailed literature search was conducted involving 72 studies. The gas most often reported for poisoning was CO, although CO poisonings are probably under-reported. Nitrogen dioxide (NO_2) from combustion engines can also cause poisonings.

In most cases of poisonings, the source of the contamination has been shown to be the ice resurfacer. As mentioned earlier, these machines are driven by an internal combustion engine fueled by propane, gas or diesel. The combustion of fuel necessarily results in the formation of contaminants. However, the concentration of the contaminants is determined by the efficiency of combustion of motor fuel. Well kept resurfacers fueled by propane usually generate less CO than those fueled by gasoline. As for NO_2, the differences are minimal. Finally, resurfacers that use diesel have numerous secondary pollutants (poor odor). A simple solution would be the purchase of an electric resurfacer. A North American company has introduced a series of these machines for commercial use. However, the cost of one such machine is 60% higher than that of the internal combustion engine machines.

Reference Number: 41203

LUCKHURST, D.G., AND W. FRENCH. "Carbon Monoxide in Indoor Skating Arenas." *Canadian Medical Association Journal*, 121: 1053–1054, 1979 (7 references)

Abstract: CO exposure incidents have occurred widely across Canada in indoor skating arenas. It is recommended that the ambient CO concentration should not exceed 25 ppm for one hour or 12 ppm for eight hours for skaters, spectators, or employees. This would result in a blood COHb concentration of 3%, which should not cause any symptoms.

Corrective measures investigated by our group are aimed at improving ventilation and reducing emission levels at the source. Since ventilation equipment is expensive to modify in existing structures, the corrective approach has been aimed primarily at raising exhaust pipes above the Plexiglas barrier for better dispersal of the exhaust, opening the end doors of the arena during and after flooding, and shunting the exhaust outside during warm-up of the machine.

Reference Number: 41204

SALONEN, R.O., A.S. PENNANEN, M. VAHERISTO, P. KORKEILA, S. ALM, AND J.T. RANDELL. "Health Risk Assessment of Indoor Air Pollution in Finnish Ice Arenas." *Environment International* 34: 51–57, 2008 (5 tables, 33 references)

Abstract: The vast majority of the 6,000 indoor ice arenas worldwide use resurfacers powered by propane or gasoline that produce exhaust fumes and emissions that include CO and NO_2. Four cases of epidemic CO poisoning in Finnish indoor arenas that occurred between 1988 and 1997 involving 373 people are discussed. The exposure time was between one and nine hours. The COHb levels measured were between 8% and 24%.

Table: Four Major Cases of CO Poisoning Involving Ice Resurfacing in Indoor Arenas in Finland

Case number	Ice resurfacer	Reason for excess emission	Number of people affected	Headache	Dizziness
1	Propane	False tuning	111	86%	39%
2	Propane / gasoline	False tuning, gasoline use	76	96%	18%
3	Propane	False tuning	28	97%	78%
4	Propane	Lambda our of use	110	>90%	-

Source: Adapted from Salonen et al. 2008.

1. An electric resurfacer is the best solution to abate indoor air pollution in ice arenas. 2. Retrofitting of emission control technology in propane-fueled resurfacers is an efficient temporary option to reduce engine emissions. However, this technology requires regular (twice per season) service and checks of the equipment as well as the measurement of exhaust emission once a year according to regulations given for low emitting person cars. In addition, a CO monitoring and alarm system should be installed in the arena. 3. All other types of combustion engine powered resurfacers are not acceptable and should no longer be used in ice arenas.

Reference Number: 41205

SASLOW, A.R., AND P.S. CLARK. "Carbon Monoxide Poisoning: Report of an Outbreak Involving 171 Persons." *Journal of Occupational Medicine*, 15: 490–492, 1973 (1 figure, 2 tables, 8 references)

Abstract: At 3:00 p.m. on 20 March 1969, 247 girls from 17 grammar schools attended a broomball tournament in an indoor ice rink in Anchorage, Alaska. Illness was first noticed at 4:00 p.m. but, by 5:30 p.m., over half the girls complained of severe headache and nausea. The ice surface was refrigerated by a natural gas compressor, but the exhaust pipe to the outside had split in half and was leaking exhaust into the skating rink. It was estimated that the ambient CO concentration was 900–1,000 ppm. A total of 131 broomball players, 30 hockey players, and 10 spectators had CO-poisoning symptoms. Even when the exhaust pipe was sealed with asbestos tape, the CO concentration was still 40 ppm.

Table: CO Symptoms Experienced in Arena Attendees

Symptom	Broomball players (%)	Hockey players (%)	Adults (%)
Headache	91	57	100
Dizziness	77	43	13
Nausea	49	43	44
Tinnitus	43	14	0
Disorientation	31	14	0
Numbness of feet	26	29	11

Source: Adapted from Saslow and Clark 1973.

Rapid recognition of the presence of CO is imperative to avert tragedy. In this episode, the quick implication of the ruptured exhaust pipe and the subsequent attempts at airing out the area avoided a major catastrophe. The fact that additional children were allowed to enter the arena and engage in strenuous activity after the outbreak began is inexcusable and was obviously due to the unawareness of the school supervisors, and especially the employees of the arena, of the dangerous conditions, which existed and its implications for future illness and possibly death. Since CO is ubiquitous in our society, vigilance from unsuspected sources of contamination is imperative and must be constantly maintained. Natural gas engines or heating plants are uniquely capable of exposing large groups of people to subtle toxic levels of CO and major disaster is a recognizable possibility whenever large numbers of people are confined in a closed, poorly ventilated area.

Reference Number: 41206

PAULOZZI, L.J., F. SATINK, AND R.F. SPENGLER. "A Carbon Monoxide Mass Poisoning in an Ice Arena in Vermont." *American Journal of Public Health*, 81 (2): 222, 1991 (4 references)

Abstract: On 6 December 1989, 25 spectators and/or participants in a local indoor ice arena in which a high school hockey game was playing attended a hospital due to CO exposure. The mean blood COHb was 9% (range 5.5–13.2%). During the same time, the arena's ventilating system had been switched off. The Zamboni was found to be emitting only 47 ppm of CO. The main CO source was probably an ambulance that was idling in an alcove during the game in which there was a significant influx of air around the garage door into the arena.

Ice arenas should use mechanical ventilation capable of keeping the carbon monoxide level at a minimum. Twenty ppm has recently been suggested as an acceptable level. Monitoring of the air quality in arenas should be considered. Ambulances on standby outside buildings should not be left running or should be parked far enough away to ensure that their exhaust fumes do not enter the building.

Reference Number: 41207

KWOK, P.W. "Evaluation and Control of Carbon Monoxide Exposure in Indoor Skating Arenas." *Canadian Journal of Public Health*, 74: 261–265, 1983 (26 references, 2 tables)

Abstract: The ambient CO concentration was measured with a Drager Multi-Gas Detector Model 21/31 at 48 inches above the ice surface after the operation of an ice-resurfacing machine in four indoor skating arenas in Etobicoke, Ontario.

Table: Ambient CO Concentrations before and after Use of a Zamboni and Natural Ventilation Rates in Four Indoor Skating Arenas

Arena #	Ambient CO before Zamboni use	Ambient CO during Zamboni use	Ambient CO after Zamboni use	Natural ventilation rate (cubic feet per minute)
1	50	110	90	826
2	3	5	5	0
3	8	35	25	1,930
4	8	15	15	1,600

Source: Adapted from Kwok 1983.

The ice resurfacing machine is a significant source of carbon monoxide. Poor natural ventilation and non-use of mechanical ventilation contribute to maintain high concentrations of carbon monoxide in the arena. Exercising young skaters are incurring physiologically undesirable levels of carboxyhemoglobin at relatively low average 22.5 ppm carbon monoxide concentrations, and have experienced such symptoms as headache, nausea, chest tightness and fatigue without recognizing the cause to be carbon monoxide because of the similarity of symptoms to other minor diseases. There are other unrecognized or neglected health effects of low level carbon monoxide exposure. These include decreased ability to sustain strenuous exercise, more risk of early development of arteriosclerotic heart disease, decreases in coordination, judgment, cognitive and psychomotor tasks and visual acuity, and increases in choice discrimination errors and reaction time. Control of carbon monoxide exposure can be achieved by the elimination of its emissions, the use of mechanical ventilation, good administrative practices and routine air monitoring.

Reference Number: 41208

BOUDREAU, D.R., L.R. WOLF, AND E. SIEGEL, "Carbon Monoxide Levels during Indoor Sporting Events, Cincinnati, 1992–1993." *Morbidity and Mortality Weekly Report*, 43: 21–23, 1994 (8 references)

Abstract: Indoor ambient CO concentrations were determined in an arena in Cincinnati during a tractor pull, monster truck jump, and mud race sporting events. None of the vehicles in the events had emission control devices. The arena was, on average, 40% full of spectators. The arena's ventilation system was on high, and the ground-level commercial truck entrances were completely opened during the events. The peak CO concentrations were 140 ppm, 250 ppm, and 283 ppm respectively for these events. Measured CO concentrations varied inversely with seating level in the arena.

Because the existing ventilation system could not supply a quantity of fresh air sufficient to eliminate excess levels of CO produced by the internal combustion engines operated during events, at the midpoint of each event the time interval was increased between each run to lower CO levels. However, this measure did not lower concentrations of CO.

Reference Number: 41209

LUCKHURST, D.G., AND G.R. SOLKOSKI. "Carbon Monoxide Levels in Indoor Tractor-Pull Events, Manitoba." *Canadian Medical Association Journal*, 143: 647–648, 1990 (5 references)

Abstract: A tractor-pull event involves a modified tractor powered by an aircraft turbine or supercharged car engines with pollution or noise controls. The event involved 25 pulls and lasted about 2.5 hours. At the start of the event, the ambient CO was 68 ppm and increased to 262 ppm after 2.5 hours of the event. At the second event, the initial ambient CO was 77.5 ppm, which increased to 435.7 ppm. A third event was cancelled due to health concerns.

> Our studies of tractor-pull events in a large indoor arena showed that the CO levels in the seating area were markedly higher than those recommended. Reports of CO-related illness may be unreliable in determining the toxic effects of the air at such events. As tractor-pulls proceed on a circuit in the United States and Canada, some being held in much smaller arenas than the one we studied, health units are advised to closely examine whether they should be permitted. The effects of similar programs such as moto-cross events, should also be evaluated.

Reference Number: 41210

CRESWELL, P.D., C.M. VOGT, R.J. WOZNIAK, J. CAMPONESCHI, M.A. WERNER, AND J.G. MEIMAN. "Improving Outreach and Surveillance Efforts Following a Large-Scale Carbon Monoxide Poisoning in Wisconsin." *Journal of Public Health Management and Practice*, 23: S67–S71, 2017 (1 figure, 16 references)

Abstract: On the second Saturday of December 2014, a hockey player lost consciousness during a tournament at the Lake Delton, Wisconsin, arena. Elevated CO levels were caused by a poorly maintained propane-fueled resurfacing machine. In total, 74 individuals had elevated COHb concentrations, and two required hyperbaric oxygen. As a result, the Wisconsin Environmental Public Health Tracking program increased its educational material and surveillance about the risks of CO poisoning.

> WIEPHT [Wisconsin Environmental Public Health Tracking] found that 42% of stakeholders reviewed the outreach materials and 1 ice arena had installed a CO detector as a result. CO alerts were developed using WPC data and are now routinely used in statewide public health surveillance.

4.13 MILITARY/WAR

> I remember reading that after one of our successful offensives on the Western Front, pillboxes were found full of dead Germans untouched by shell fire. It is likely that these were cases of carbonic oxide poisoning and were not due to shell concussions as was said at the time.
>
> —J.F. Mayne, "The Danger of the Concrete Pill-Box" (1920) **(41302)**

> British, French and German observers have proved this by finding CO in the blood of men killed or wounded during the shelling and by observing symptoms of poisoning- headache, giddiness, sickness, loss of consciousness—in a number of those exposed. These symptoms were generally regarded by the men as the result of the noise and the shock of shells bursting in their vicinity."
>
> —D.D. Logan, "Carbon Monoxide Poisoning from the Use of Petrol Engines: Some Experiences during the War" (1939)

It is unexpected that a large number of military personnel did not die as a direct result of bullets, shells, or shrapnel but, rather, as a result of CO poisoning from the CO produced in explosions, firing of bullets, military vehicles, tanks, airplanes, and even ambulances (41301–7, 41310, 41317). Attempts were made to use CO as a poison gas in the First World War, but its will o' the wisp nature made it a poor poison to be used in open air (41301, 41320). Although toxic CO was produced by napalm and flame throwers, which may have contributed to its lethality when used in bunkers or confined areas (41319, 41224). Military personnel have been poisoned by CO in their barracks (41309, 41311), submarines (41313–4), C-130 transports (41318), and air raid shelters (41312, 41325). Between 2009 and 2019, 6,042 US military personnel suffered from CO poisoning, mainly from vehicle exhaust (41322). Military environments such as high altitude (41321) and the Arctic (41324) pose special risks for CO poisoning. Chronic CO poisoning (known to cause paranoia, fear, and delusions may have caused a lack of "moral fibre" (41307) and contributed to the Gulf War Syndrome (41315, 41317).

Reference Number: 41301

ANONYMOUS. "Carbon Monoxid Poisoning." *Science*, 66: xii–xiii, 1717, 1927

Abstract: A short article was published regarding CO poisoning due to automobile exhausts and modern warfare. To avoid CO poisoning by a motor vehicle, good ventilation is required to make the place perfectly safe as CO is slightly lighter than air and vanishes almost immediately through an open door or window.

> Carbon monoxid would be a prime factor in modern warfare if it were not so easily dispelled in the open air. The high explosives with which shells are loaded generate great quantities of it; the gas from a TNT explosion contains as much as 60 per cent of carbon monoxid. If a shell happens to explode in a dugout or other confined space, the soldiers who escape injury from the fragments and the concussion are very likely to fall victims to the gas. It is the more deadly since even the most modern types of gas mask afford far less protection against it than they do against the heavier varieties of military poison gas.

Reference Number: 41302

MAYNE, J.F. "The Danger of the Concrete Pill-Box." *BMJ Military Health*, 34 (3): 250–254, 1920 (1 figure)

Abstract: Concrete machine gun emplacements or pillboxes in the First World War tended to be airtight to prevent poison gas from entering. But CO produced by the firing of machine guns could escape along the breach of the gun and also originate from disused cartridges. Some of the British gunners complained about giddiness and faintness after firing machine guns in a pillbox, and so a simulation was conducted. The test was nearly too successful after the firing of the machine guns in the test pillbox as one officer was badly gassed, and three others were unconscious. The officer with more severe CO poisoning probably obtained a higher COHb due to his exertions of removing three unconscious men from the pillbox. The air in the pillbox was found to contain 0.3% CO (3,000 ppm).

> This is an interesting description of an experiment which undoubtedly saved many lives. Had an attack been made by the enemy in the Struma Valley, and had we been using these pillboxes, there is no doubt that within twenty minutes, a whole line of machine guns would have been out of action, with perhaps serious consequences.

Reference Number: 41303

RUTHERFURD, W.J. "Carbon Monoxide Poisoning in Warfare." *BMJ Military Health* 36 (2): 115–124, 1921

Abstract: Various explosive compounds such as gun powder, blasting powder, and gun cotton, when exploded, produce a large amount of gas, including CO. Gun cotton is also known as nitrocellulose and is the replacement for gunpowder. This is because there is insufficient oxygen in the explosives to oxidize completely all the carbon atoms in the charge, which causes CO formation. One kilogram of gun cotton can produce 859 L of gas, of which about 25% is CO, and, when the gun cotton burns rather than explodes, it can produce up to 40% CO. CO poisoning also occurred when charcoal braziers were used in the ill-ventilated dugout in the winter for warmth. CO was also produced by military mining or tunneling operations due to the use of explosives that can force CO gas through cracks or breaks in the rock into other areas, even the surface, which can cause an accumulation of CO and poisonings.

> Note: These papers were originally written in 1916, but permission to publish was withheld by the War Office as active warfare was then in progress.

Reference Number: 41304

SUNDELL, C.E. "The Clinical Manifestations of Gas in Military Mines." *The Lancet*, 187 (4836): 957, 1916

Abstract: The number of explosives used in military mines during the First World War were much larger and the air space much smaller, and the ventilation was poorer than in a typical industrial mine. More casualties tend to occur for a few days after a hostile mine is exploded than due to the explosion itself. This is because of the large amount of CO produced and the fact that the chalky soil is shattered over a wide zone due to the blast, which allows the CO to collect in numerous crevices throughout the battle area.

> The following characteristics of carbon monoxide should be borne in mind: the gas is odourless, it does not affect a candle's flame, its presence in the atmosphere at a strength of 0.1 per cent, may be fatal, and its action is cumulative. Men may thus gradually come under its influence without any warning of its presence.

Reference Number: 41305

LOGAN, D.D. "Carbon Monoxide Poisoning from the Use of Petrol Engines: Some Experiences during the War." *Journal of State Medicine*, 28: 306–319, 1920.

Abstract: The CO poisoning that occurred during the First World War in connection with submarines, tanks, airplanes, and other petrol engines is discussed. Gasoline-powered air compressors for ventilation in mine galleries caused CO poisoning. In one case, the overheating of a cylinder in a compressor caused a fire and the CO-laden air was delivered to soldiers underground. Six men were affected, and two died. There was also CO poisoning caused by gasoline-powered electric generators. Several deaths occurred in ambulances due to the exhaust gas discharged through pipes on both sides of the ambulance out the back. These accidents generally occurred when the ambulance was held up and the engine was left running for some time.

> Although not of frequent occurrence poisoning from the exhaust has taken place in closed motor vehicles such as cars, lorries, and ambulances. In long runs in motor lorries, men have been rendered unconscious by the exhaust gas being swirled into the lorry when in motion or finding its way through the floor. Definite proof that CO is the poisonous agent in such cases was afforded by the poisoning of a number of white mice belonging to a Mine Rescue School, which were on the floor of a 3-ton lorry. All were dead and examination of the bodies showed that the blood was saturated to a high degree with CO which apparently had found its way through the floor.

Reference Number: 41306

BROWN, D. "Never Mind the Heat, Never Mind the Noise: Understanding the Working Conditions of Tank Crews during the First World War." *Journal of the Society of Historical Research*, 98: 175–194, 2020 (3 tables, 4 figures, 68 references)

Abstract: Tanks were first used in the Battle of Flers-Courcelette in September 1916. The early tanks were just metal boxes that contained the engine, weaponry, and crew together with no compartments, no sound proofing, and minimal ventilation. The operating conditions of the tank were difficult and harsh, and the crew members developed CO poisoning, not only from the engine but also from firing the machine guns. No compressed oxygen tanks were available to treat the CO poisoning, and so recovery from CO toxicity was a slow process for the tank crew.

Table: CO Concentrations Measured on a Mark V Tanks, 14 October 1918

Location	% CO detected
6 feet from back of tank, 4 feet from ground	nil
6 feet from back of tank, 6 feet from ground	0.16
12 feet from back of tank, 6 feet above ground	nil
Behind engine in tank	0.16
t driver's seat	0.12

Source: Adapted from Brown 2020.

It has been found by experience that tanks with a full crew on board and closed down for action show, after running a certain time, both the presence of CO in the air and a high temperature and that the crew develop headache, giddiness, breathlessness, rise of body temperature, vomiting and nausea, sometimes mental confusion, and even collapse and unconsciousness. Some of these arise partly from a sort of sea-sickness caused by the motion of the tanks, and the symptoms are often aggravated in action by the fumes from the firing of the Hotchkiss and 6 pounders.

Reference Number: 41307

LOGAN, D.D. "War Wounds and Air Raid Casualties: Detonation of High Explosive in Shell and Bomb, and Its Effects." *British Medical Journal*, 28 October 1939, 864–866

Abstract: CO has a specific gravity of 0.96 (slightly lighter than air) and when shells and bombs explode in the open, it is quickly dissipated, which makes it useless as a poison gas. But, in deep trenches, the explosion may cause a CO build-up that can poison the soldiers sheltering there. CO has been found in the blood of soldiers killed by shelling, although not close enough to be killed by the explosion itself. CO-poisoning symptoms experienced by soldiers such as headache, giddiness, sickness, and loss of consciousness have generally been regarded as being due to the noise and shock of shelling.

Gas from the detonation of shells, bombs, mobile charges, minenwerfers etc., has been known to gain an entrance into cellars and dugouts at some distance from the burst through fissures in the strata or through cracks in the walls of buildings. In an intensive bombardment by heavy shells or by bombing, the ground in the affected area may become saturated with gas.

This is more likely to occur if there has been imperfect detonation of the explosive, or if heavy shells and bombs with time-fuses have penetrated so deeply that when detonation occurred no craters formed. In all these cases gas may find its way into cellars, basements, or dugouts by percolating through the porous ground or through crevices formed in the ground.

Reference Number: 41308

VALMAN, B. "Alexander Russell." *British Medical Journal*, 326 (7399): 1149, 2003.

Abstract: Dr. Alexander Russell was born in 1914 and died in 2003.

During the Second World War, when he served in the Royal Air Force, he defined the syndrome of carbon monoxide poisoning in the gun cockpits at the rear of Whitley bombers, where space was confined. He showed that the poisoning accounted for air sickness that had previously been ascribed to weakness of moral fibre in veteran gunners. For his discovery and his field research in malaria and hepatitis, he was mentioned twice in dispatches and awarded the OBE [Officer of the Most Excellent Order of the British Empire].

Reference Number: 41309

FRANKLIN, R.V., AND F. COURTNEY. "Carbon Monoxide Poisoning in Barracks." *BMJ Military Health*, 57: 299–301, 1931.

Abstract: During the months of February and March 1931, about seven soldiers suffered CO poisoning in the barracks at an isolated Hill Station in northern India, due to the burning of charcoal inside by sigrees. All of the men had similar symptoms of headache, giddiness, and vomiting. One soldier was unconscious and stopped breathing but was revived, administered oxygen, and survived.

Coming to the conclusion from the history that they were probably being poisoned by carbon monoxide, the following Station Order was issued—The use of sigrees and the burning of charcoal in barrack rooms is strictly prohibited. Sigrees may be used in officers only with the special permission of the Station Commander.

Reference Number: 41310

WEYANDT, T.B., AND C.D. RIDGELEY JR. "Chapter 11 Carbon Monoxide." *Occupational Health: The Soldier and the Industrial Base, Part 3*, 2: 397–427, 1993 (3 tables, 11 figures, 57 references)

Abstract: An extensive chapter about CO and its effect on the military is presented. Some of the worst workplace exposures occurred during the nineteenth and early twentieth centuries when industrial miners were frequently overcome by white damp (CO and other products of partial combustion) after mine fires, explosions, or blasting operations. During the First and Second World Wars, CO hazards occurred from gunfire from closed pillboxes, tanks, or other military vehicles, mine explosions, explosions of enemy projectiles, portable generators, heaters, or gasoline-powered motors. Army field exposure commonly occurs in tanks, the troop compartment of armored vehicles, enclosed communication vans, portable gasoline-powered generators, indoor firing ranges, and indoor operation of motorized vehicles. In 1984, the test firings of the M1E1 tank caused the four-man crew to suffer symptoms of CO poisoning such as dizziness and even unconsciousness. At hospital, the loader had a COHb of 33% and the commander 27.8%. High CO concentrations were also found in the M109 howitzer and other armored vehicles. In 1988, the firing of the M134 mini-gun mounted in the UH-60A Black Hawk helicopter resulted in a CO concentration of 79.4 ppm. The effects of CO are primarily due to anoxia (lack of oxygen) as a result of the reduced oxygen-carrying capacity and impaired oxyhemoglobin dissociation and causes greater anoxia than the equivalent reduction in ambient oxygen alone. After dissociating from CO, the regenerated hemoglobin shows no impairment.

> Exposure to carbon monoxide can cause acute clinical illness. Military exposures to the chemical occur in homes, vehicle, and weapons system. The profile of exposure during the performance of military duties is typically episodic, with high peak exposure followed by prolonged periods of limited, minimal exposure. Although the primary effect of carbon monoxide poisoning appears to be the consequence of hypoxia associated with the production of carboxyl, cellular enzyme systems are also adversely affected by carbon monoxide poisoning. Individuals with coronary heart disease appear to be at excess risk of adverse effect, such as myocardial infarction following acute exposure.

Reference Number: 41311

HENZ, S., AND M. MAEDER. "Prospective Study of Accidental Carbon Monoxide Poisoning in 38 Swiss Soldiers." *Swiss Medical Weekly*, 135: 398–406, 2005 (3 tables, 3 figures, 55 references)

Reference Number: 41311

Abstract: In October 2000, 38 male soldiers (mean age 20.7 years) were accidentally exposed to CO while sleeping in a long stable, adjacent to a small room in which a gasoline-powered 5 kW generator had been recently installed. At 1:30 a.m., an alarm was sounded by a guard who could not wake a colleague to a change in the guard. The soldiers were immediately evacuated from the stable, and all those comatose regained consciousness spontaneously within 15 minutes of exposure to fresh air. All the soldiers were transported to the local hospital for assessment. Soldiers who had lost consciousness were administered oxygen during transportation, and all soldiers were administered 100% oxygen at hospital. The mean COHb was 14.29%, and the mean extrapolated COHb (using half-life of five hours for ambient and one hour for oxygen treatment) was 38.0%. The highest COHb concentration was not found in the soldiers closest to the generator but increased 0.5% for each metre further away, up to five metres, probably due to the convection of the hot exhaust gases in the room. Nine soldiers reported residual symptoms after one year.

Table: Characteristics of Soldiers Reporting More Than One Symptom One Year after CO Poisoning

Number	Initial COHb (%)	Coma 0 (none) to 3 (severe)	Smoker	Residual symptoms
1	44.4	3	yes	Headache, dizziness, weakness, clumsiness, fatigue, nausea
2	31.7	2	yes	Headache, visual changes, dizziness, nausea
3	–	1	yes	Headache, dizziness, nausea
4	26.6	0	yes	Headache, auditory disturbances, lack of energy
5	26.5	1	no	Headache, dizziness, confusion
6	28.2	0	yes	Visual changes, auditory disturbances
7	29.6	1	yes	Confusion, reduced strength
8	21.7	0	yes	Headache, confusion
9	27	3	no	Headache, confusion

Source: Adapted from Henz and Maeder 2005.

Six soldiers (16%) were initially unconscious, five others (13%) could be woken up but were unable to get out of their sleeping bags without help and eight soldiers (21%) could leave the stable but were unable to help others. Within the first day the most prevalent symptoms were headache and dizziness, followed by nausea and a sensation of weakness. The prevalence of most symptoms peaked at two to three hours. The latency from the alarm to the onset of each symptom differed considerably. Clumsiness and visual disturbances were already present when the patient woke up from sleep or coma. Confusion, dizziness, weakness, headache, palpitations and shortness of breath were noted after a mean latency of 10 to 20 minutes; nausea and hearing impairment started within an hour, whereas vomiting and chest pains were late symptoms.

Reference Number: 41312

CUI, L.L., J.W. ZHOU, X.Y. GENG, Z. LUI, G.A. CONWAY, AND J. ZHANG. "Occupational Carbon Monoxide Poisoning Associated with Diesel Motor Operation in an Air-Raid Shelter in June 2015, Jinan City, China: An Epidemiological Investigation." *Biomedical and Environmental Sciences* 30: 777–781, 2017 (2 tables, 2 figures, 10 references)

Abstract: On 2 June 2015, three male water channel clean-up workers set up a diesel motor at the left branch of an air raid shelter (ARS) and began to pump out the water. After two hours of pumping, two workers developed headaches, dizziness, and fatigue and evacuated the ARS after turning off the diesel engine. The third worker was unconscious and was rescued by 33 firefighters using 10 handovers. Fourteen of the firefighters were exposed to CO and suffered similar symptoms, but all refused a blood test for COHb. All were treated with 100% oxygen. The three workers had blood COHb of 25.4%, 28.4%, and 33.1% respectively.

This study is the first reported large scale investigation of occupational CO poisoning related to a diesel motor in an air raid shelter in Jinan City, China. A total of 2,770 CO poisoning accidents were documented in Jinan from 2006 to 2014; most were non-occupational poisoning accidents caused by heating and warming of homes. Rarely were the occupational poisoning accidents and none was related to diesel motor. A total of 17 cases were identified in this accident, three confirmed cases among workers and 14 possible cases for firemen, the first known CO poisoning of firemen in Jinan. This epidemiological investigation indicated that exposure to a high level of CO, combined with a lack of effective individual respiratory protection, might be the main cause of this CO poisoning accident. Blood carboxyhemoglobin percentage is the most frequently used biomarker of CO exposure. In this accident, the blood carboxyhemoglobin level of the three worker cases exceeded 25%, which is defined as severe CO poisoning.

Reference Number: 41313

LIGHTFOOT, N.F. "Chronic Carbon Monoxide Exposure." *Proceedings of the Royal Society of Medicine*, 65: 798–799, 1972 (1 figure, 7 references)

Abstract: Normally, body tissues extract about 25% of the available oxygen from the blood, but the cardiac muscles extract about 75% and so are more susceptible to CO poisoning. A simple equation to determine a person's COHb percentage at equilibrium is to multiply the ambient CO (ppm) by 0.16. The ambient CO concentration in early Royal Naval submarines could be detected as high as 100 ppm and, hence, a maximum permissible level was set at 50 ppm. Improvements in the air-filtering system brought the levels down to 25 ppm, and, recently, levels of 8–10 ppm have been detected throughout the entire Polaris submarine patrols. CO on the Royal Naval submarines are mainly due to smoking. A survey on a recent Polaris nuclear submarine showed that the COHb in five non-smoking submariners was 1.1%, which increased to 8% in smokers who smoked 20 cigarettes a day and 18% in two heavy smokers.

> As the smoker does not normally expose the non-smoker to the same risks as himself perhaps, he should be allowed to continue to smoke to his satisfaction. All submariners, however, live in an unnatural environment in which many other factors may affect health, and the question arises whether one can permit any abnormal constituent which may be simply removed from the atmosphere, in this case by ceasing to allow unlimited smoking, to be present.

Reference Number: 41314

TINGLE, C. "Submarine Accidents: A 60 Year Statistical Assessment." *Professional Safety*, 54: 31–39, 2009 (3 tables, 8 figures, 19 references)

Abstract: Since the eighteenth century, more than 1,800 submarines have been sunk, and 60,000 crew killed worldwide in peace and war. An analysis of fatal submarine accidents between 1946 to 2005 were analyzed. It was determined that 21% of the accidents were due to flood, 18% due to collision, 11% due to explosions, 10% due to fires, and 4% due to toxic substances exposure such as CO.

The reasons for K-129's demise point to one of the following: an explosion (ammunition or battery); a faulty snort (diesel powered submarines when dived and charging, induct combustion air through a mast with a float vale designed to prevent ingress of water in the event that the vessel plunges due to heavy weather or poor trimming) with subsequent carbon monoxide poisoning (similar to what is believed to have occurred with Chinese submarine 361 in 2003).

Reference Number: 41315

EMMEROVA, M., AND F. JIRAVA. "Is Gulf War Syndrome Really a Mystery?" *Medicine, Conflict and Survival*, 20: 209–217, 2004 (32 references)

Abstract: Since the end of the 1991 Gulf War, about 20,000 US veterans and other allied troops have been affected by a variety of symptoms known as the Gulf War Syndrome or Desert Fever. During the war, 1.5 million barrels of oil were consumed every day by burning oil wells, producing black clouds of soot and highly elevated ambient concentration of sulphur, nitrogen, and carbon oxides (about 12 times that found in London). The symptoms of chronic CO poisoning resemble the symptomatology of many of the veterans.

It is proposed that the common sense is the toxic smoke from incomplete combustion of oil from burning wells, and this hypothesis is related to the known toxicology of two likely combustion products, nitric oxide, and carbon monoxide.

Reference Number: 41316

ANONYMOUS. "Gas in Military Mines: Symptoms of Carbon Monoxide Poisoning." *The Hospital*, 20 May 1916, 169

Abstract: Military mines in the First World War used much larger charges in a smaller air space with less ventilation than civilian mines and so created more CO. When exploded in chalky and friable soil, CO can linger and cause casualties in different parts of the trench line days later. CO is known as fire damp.

When all the perils of the military mine are taken into consideration, it is not to be wondered at that a high proportion of military crosses, distinguish conduct medals, and other greatly prized decorations in recent lists have been awarded to officers and men of the tunneling companies to which these operations are confided. Civil mining is no child's play, but military mining must be a nerve-wracking business indeed.

Reference Number: 41317

TEICHMAN, R. "Health Hazards of Exposures during Deployment to War." *Journal of Occupational and Environmental Medicine*, 54: 655–658, 2012 (1 table, 1 figure, 42 references)

Abstract: A brief history and review of the poisons used in military campaigns for over 3,000 years was conducted. Around 1000 BCE, Chinese writings described numerous poisons or irritating smoke for military use. Solon of Athens used hellebore root to poison the water in the aqueduct of an enemy city. Alexander the Great encountered poisonous arrows in India. Arsenic, sulfuric, tallow, rosin, turpentine, and antimony were used in incendiary shells. The top 10 environmental exposures of concern in Gulf War veterans are listed.

Table: Top 10 Environmental Exposures of Concern to US Gulf War Veterans

Concern	%
Protective gear/alarms	82.5
Diesel, kerosene, other petrochemicals	80.6
Oil well, fire smoke	66.9
Local food	64.5
Insect bites	63.7
Harsh weather	62.5
Smoke from burning trash or feces	61.4
Within one mile of missile warfare	59.9
Repellants and pesticides	47.5
Paint, solvents	36.5

Source: Adapted form Teichman 2012.

Military medical professionals are first known to be involved with systematically correcting exposures to environmental toxins during World War I. A historical review noted that medical professionals in the French Army identified and attempted to correct carbon monoxide poisoning from machine guns firing in enclosed spaces, such as a tank.

Reference Number: 41318

MARTINI, E.A. "World on Fire: The Politics of Napalm in the Global Cold War." *Cold War History*, 16: 463–481, 2016 (54 references)

Abstract: Napalm was originally designed in 1942 by a team of Harvard chemists and was named for the combination of aluminum naphthenate and palmitate used to bake the jellied gasoline mixture engineered to stick to and burn whatever it lands on. It is ignited by phosphorous and burns at 1,000 degrees Celsius. Burning napalm rapidly deoxygenates the area and produces large amounts of CO and CO_2. Napalm was use in Germany and Japan as firebombs during the Second World War and in many conflicts since then. Napalm has become a global weapon of terror.

> The smell of napalm in the morning: seemingly everyone knows the line, but relatively few would actually recognize the smell. Colonel Kilgore (Robert Duvall) did not find any bodies in Apocalypse Now, but more often than not, his real-life contemporaries would have found traversing a hill attacked with napalm to smell not simply of gasoline but of seared flesh and rotting corpses burned at temperatures reaching 1,000 degrees Celsius. Many of the victims would have died as a result of asphyxiation or carbon monoxide poisoning caused by napalm's consumption of oxygen in the area attacked. Others may have died quickly from shock. They were the lucky ones.

Reference Number: 41319

DOIR, A., R. POKROV, L. GOLDSTEIN, E. BABENBOIM, AND M. ZILBERBERG. "Heat Stress and Carbon Monoxide Exposure during C-130 Vehicle Transportation." *Aviation, Space and Environmental Medicine*, 76: 399–402, 2005 (2 tables, 3 figures, 13 references)

Abstract: Loading the C-130 by driving gasoline-powered military vehicles onto the platform may expose the cabin crew to heat and CO. Data was collected on two night and two day C-130 flights in Israel transporting either two or three vehicles in the summer of 2003. CO dry and wet bulb temperature (WBGT) were measured. The three vehicles hot engines and bodies added 5.9 degrees Celsius to the cabin WBGT and the cabin WBGT peaked at 37.5 degrees Celsius after 140 minutes into the flight. The ambient CO concentration detected was not hazardous for the longest 160-minute flight.

Table: Cabin Mean (Range) CO Levels (ppm) of the C-130

Day or night flight	3 vehicles transported	2 vehicles transported
Day	12.4 ppm (1–266)	4.2 (0.8–27.3)
Night	25.5 ppm (1–401)	7.7 (0.2–52)

Source: Adapted from Dor et al. 2005.

In hot climates, C-130 vehicle transportation may exacerbate heat stress. The in-flight heat stress can be predicted by the ambient temperature, duration of vehicle transportation, and number of transported vehicles. The cabin CO level is related to the number of transported vehicles. We recommend the use of effective environmental control systems during the C-130 vehicle transportation in hot climates.

Reference Number: 41320

COPE, R.B. "Carbon Monoxide: Can't See, Can't Smell, Body Looks Red But They Are Dead." *Handbook of Toxicology of Chemical Warfare Agents* (Cambridge, MA: Academic Press, 2020), 353–371 (5 tables, 2 figures, 125 references)

Abstract: A detailed review of CO and its role in the military over the years was conducted. The ambient concentration of CO varies from 0.1 ppm in a clean atmosphere to 5,000 ppm in the proximity of a domestic wood fire chimney. The atmospheric lifetime of CO is one to two months. CO has been added to meat products because it reacts with myoglobin to form red-colour COHb and increases the visual appeal of red meat. Because combustion is an inevitable consequence of warfare, CO will always be present to some degree in battlefield conditions. CO was used in the initial Holocaust-related gassing apparatus using mobile gas vans and diesel exhaust fumes. CO may account for some of the fatalities associated with napalm and the explosion of jellified gasoline produces large amounts of CO_2 and CO. CO is not listed on the Chemical Schedules of the 2005 Convention on the Prohibition of the Development, Production, Stockpiling and Use of Chemical Weapons and on Their Destruction. The limiting factor for the absorption of inhaled CO is its ability to diffuse across alveolar membranes and is diffusion limited. Elimination occurs mainly by exhalation and is dependent on the amounts of oxygen and CO present, the magnitude of ventilation and quality of the alveolar diffusion barrier. Alveolar diffusion barrier thickens with age resulting in slower rates of elimination. The CO body elimination curve is a two-compartment one with an initial sharp decline in blood levels over 20 minutes after exposure due to a distribution phase and a slower whole body elimination phase.

Being an inevitable consequence of the combustion of organic materials, battlefield CO exposures, whether deliberately applied or incidental, are likely an ancient phenomenon of warfare dating back to the very beginnings of recorded human history and beyond. Given that combustion is an integral part of the operation of modern military vehicles and weapons systems, incidental CO exposure will continue to be part of the battlefield experience for the foreseeable future. Given the simplicity, ease and low cost of its production and other factors, the deliberate use of CO as a weapon in particular military circumstances will continue. While there is now a very substantial medical and toxicological understanding of simple CO exposures there remain important gaps in the understanding of the implications and role of CO in the effects of complex gas mixtures that are likely to occur in association with modern battlefields and modern weapons systems.

Reference Number: 41321

JINDAL, A.K. "The Highest Battlefield of the World: Medical Problems and Solutions." *Medical Journal Armed Forces India*, 65: 170–172, 2009 (2 figures, 8 references)

Abstract: The medical issues encountered at the highest battlefield in the world on the Siachen Glacier by the Indian Armed Forces Medical Services is discussed. They include high altitude pulmonary edema, acute mountain sickness, hypothermia, and CO poisoning.

Carbon monoxide is a colorless, odorless gas. Cases of poisoning can take place when one lights the bukhari at night in a closed space like a fiberglass hut and goes to sleep. This is particularly dangerous because the symptoms are not immediately apparent. The men inside the tent or the hut will be found dead in the morning in case of severe carbon monoxide poisoning. Orders on the glacier are that regardless of the cold, all heating devices will be put off prior to sleeping in any closed space. Lives are lost when these instructions are violated.

Reference Number: 41322

WILLIAMS, V.F., A. FEDGO, AND S. STAHLMAN. "Carbon Monoxide Poisoning, Active, and Reserve Component Service Members and Non-Service

Member Beneficiaries of the Military Health System US Armed Forces, July 2009–June 2019." *Medical Surveillance Monthly Report*, 27 (1): 2–16, January 2020 (11 tables, 6 figures, 41 references)

Abstract: Between July 2009 and June 2019, there were 1,288 confirmed/probable CO-poisoning cases among active component service members, 366 among the reserve component members, and 4,754 among non-service beneficiaries of the US navy, army, air force, and marines corps. Of the CO poisonings in the active component members, 1.9% returned to service with limitations, and 9.2% did not return to duty. Deaths occurred in 24 active component service members within one week of CO-poisoning hospitalization. The highest monthly number of CO poisonings occurred during January 2017 (61 cases). Most of the CO poisonings occurred in the army (41.4%), followed by the air force (29.2%), navy (21.0%), and marine corps (8.4%). Work in repair/engineering was the most frequent occupation associated with CO poisoning. Motor vehicle exhausts were the most common source of CO (17%), followed by utility gas (5.8%).

Table: Number and Percent of Confirmed CO Cases According to the Military in Active Component Members of US Armed Forces, 2009–19

Military occupation	Number	Percent
Combat specific	158	12.5%
Motor transport	40	3.2%
Pilot/Air crew	79	6.2%
Repair/engineering	383	30.2%
Communications/intelligence	254	20%
Health care	111	8.7%
Other/unknown	244	19.2%

Source: Adapted from Williams, Fedgo, and Stahlman 2020.

This report also documents that confirmed/probable cases among active component service members and non-service member beneficiaries exhibited the expected cold weather seasonal trends commonly associated with CO poisoning. This seasonal pattern generally corresponds to times of increased use of indoor heating. Suspected active component cases demonstrated a broadly similar trend.

Reference Number: 41323

VAN WYCK, D.W. "Beyond the Burn: Studies on the Physiological Effects of Flamethrowers during World War II." *Military Medical Research* 7 (1): 8, 2020 (5 figures, 31 references)

Abstract: The first soldiers' portable mobile flamethrowers were developed by the Germans during the First World War. In Second World War, the United States developed the most advanced flamethrower of the war, the M2-2, which was used extensively in the Pacific during the last two years of the war. Napalm was added to the flamethrowers, which increased its lethality. Flamethrowers were initially touted as providing a quick and painless death and as mercy killing. This view has been shown to be incorrect in numerous animal experiments. Fire and shock caused the majority of fatalities rather than CO poisoning or asphyxiation. The ambient CO concentration in simulated bunkers caused by flamethrowers was 0.8%, and the ambient oxygen concentration was 14.6%. In one animal test, a blood COHb of 30% was detected. The US military voluntarily withdrew flamethrowers from active service in 1978.

> All weapons are by their nature, ghastly in purpose but there is something uniquely awful and inhumane about the flamethrower. Fire is popularly and quite logically held as one of the worst ways to die. It is not quick, the victim taking long seconds or even minutes to succumb as the flesh, nerve, muscle and eventually organs are charred to destruction. That intelligent human beings have turned their minds to developing instruments specifically to achieve this goal represents another tragic failure in society. These evolving views were ultimately factors considered in the US military's decision to voluntarily withdraw flamethrowers from active service in 1978. Nevertheless, a number of other nations still stock flamethrowers in their military inventories and there is no international law specifically banning their use.

Reference Number: 41324

YOUNG, E.J. "Military Hygiene Problems Encountered in the Arctic and Subarctic." *Canadian Journal of Public Health*, 41: 123–127, 1950

Abstract: Approximately, 83% of Canada's landmass is within the Arctic and subarctic. The cold and permafrost causes problems of disposing of waste, frostbite, hypothermia, and keeping accommodations warm. In the summer, there are hordes of insect pests.

> Occupational hazard peculiar to cold climates are frostbites, snow, and fog blindness. Carbon monoxide is another hazard which must be considered because of the necessity of burning fuel in confined places. Carbon monoxide poisoning caused a few temporary casualties on the Muskox expedition because of operating in vehicles completely closed down.

Reference Number: 41325

LUCAS, K.A., J.M. ORIENT, A. ROBINSON, H. MACCABEE, P. MORRIS, G. LOONEY, AND M. KLINGHOFFER. "Efficacy of Bomb Shelters: With Lessons from the Hamburg Firestorm." *Southern Medical Journal*, 983: 812–820, 1990.

Abstract: On the night of 27–8 July 1943, a terrible firestorm was ignited caused by Allied incendiary bombing. Approximately 40,000 people died. The death toll was low in bomb shelters, and no one there died of oxygen deficiency due to the fire. Most of the casualties occurred in basement shelters due to CO poisoning. Much greater fatalities occurred in the firebombing of Dresden as the authorities did not consider the city to be a target, and so no concrete and steel bomb shelters were constructed. Dresden was also full of refugees.

> Of those who did die in basements in Hamburg, the majority probably perished from carbon monoxide poisoning. The victims who were fully clothed, appeared to have fallen asleep peacefully. There was no sign that they had been struggling to escape from unbearable heat. Elevated blood levels of carbon monoxide in firestorm victims were documented by chemical analysis as reported by the Police President of Kassel. In many victims, this cause was suspected but could not be proved because of the unreliability of the tests in persons dead for more than a few hours. Other causes of death in basement shelters included heat exhaustion and mechanical causes, such as burial under rubble from the collapse of the burned-out houses above.

4.14 MINING/BLASTING/EXPLOSIONS

> On one occasion, however, when a rat was being used, as a mouse could not be obtained, the exploring party evidently got into a somewhat

> dangerous atmosphere. A rat, on account of its much greater size, will take fully twice as long as a mouse to show the effects of carbonic oxide [CO]; and possibly this fact was not sufficiently allowed for.
>
> — Anonymous, "Poisoning by Carbonic Oxide" (1898)[5]

CO is the major cause of death in mining explosions (41401–5). In an operating open pit coal mine, blasting caused CO to seep into the homes of nearby residences (41406). Nearby blasting in construction and other projects also cause the will o' the wisp to leak into nearby homes causing CO poisoning (41407–11, 40413–14). Two men were found in a grain silo after a grain explosion, unconscious but with no sign of trauma The ambient air was found to contain 4,000 ppm of CO. The silo was ventilated, and the men recovered and were administered 100% oxygen and recovered (41412)

Reference Number: 41401

ANONYMOUS. "Poisoning by Carbonic Oxide. The Snaefell Mining Disaster." *British Medical Journal*, 2 July 1898, 32–34 (1 table)

Abstract: The death of 20 miners occurred on the slopes of Snaefell on the Isle of Man in 1897 due to CO poisoning. The shaft of the mine was 1,000 feet deep and was accessible by ladders. A fire had occurred in the timber supports, which caused the roof of the mine to fall, blocking ventilation and producing large quantities of CO that poisoned the miners. A bright red colour was observed on the face and hands, characteristic of CO poisoning, of all the miners post-mortem, and all the miners had a calm and peaceful expression on their faces. An analysis of gas in the mine by John Scott Haldane showed that a miner would be unconscious within seven to eight minutes of breathing it in, and death would occur shortly afterwards.

> It was of great importance to obtain a sample of the gas for analysis, and this Mr. Williams succeeded in doing. He descended with a rope round him to a point just below which a candle would not burn and filled two bottles with air by emptying water from them. He had just corked the second bottle and handed it to a miner who stood above him, when he suddenly became unconscious. He was then dragged up the ladders 80

5 Anonymous, "Poisoning by Carbonic Oxide: The Snaefell Minning Disaster," *British Medical Journal*, 2 July 1898, 33.

feet by Mr. Jones, the other assistant inspector, and several miners, who themselves almost overcome. By this time, it was found that his respiration had entirely ceased and he appeared to be dead. Mr. Jones at once applied artificial respiration, at the same time holding his head to a hole which had been punched in a pipe conveying compressed air from the surface. After a time natural respiration returned, and he was sent up to the surface where he soon recovered.

Reference Number: 41402

POLLOCK, C. "The Canary in the Coal Mine." *Journal of Avian Medicine and Surgery*, 30: 386–391, 2016 (1 table, 8 figures, 31 references)

Abstract: Coal is the most plentiful fossil fuel, and coal miners faced death from the formation of toxic gases (methane, CO, CO_2, and hydrogen sulfide), rockfalls, and explosions. One historic detection method was completed before a shift began when one miner wearing a wet blanket carried a long lit wick in front of him and walked through the mine shaft. Another detection method was the miner's lamp, in which the flame would change colour or go higher or lower depending on the type of gas. The third method was carrying a canary in a cage to detect mainly CO. A canary would faint about 20 minutes prior to humans to give a warning to the miners. Female canaries were often used as they were cheaper than male canaries due to their inferior singing ability. Some mining companies even built aviaries to breed canaries for their mines. In 1987, there were more than 200 canaries used in British mines before they were phased out over the next year.

The ideal sentinel animal serves as a sensitive indicator of an environmental hazard. Coal mines are particularly hazardous working environments. Thanks to their small size and highly efficient respiratory tracts as well as the ease of access, canaries were once regularly used in coal mines as an early warning system for human health. The signs of carbon monoxide poisoning in a canary were dramatic and quite recognizable to the miners since an affected bird would appear visibly ill and even fall off its perch. Canaries were used in coal mines well into the 20th century, but fortunately modern mines have sensors far more sensitive than these hard-working birds.

Reference Number: 41403

HATTON, W.A. "The Hulton Colliery Explosions." *British Medical Journal*, 1 (2629): 1164, 1911 (1 table, 2 figures)

Abstract: The dull explosion lasting several seconds could be heard up to four miles from the Hulton Colliery (coal mine) on 21 December 1910 at about 7:50 a.m. due to gas accumulation from a roof collapse the previous day. Three hundred and forty-three men died in the second worst mine disaster in England, of which CO poisoning accounted for 287 deaths (81%).

Table: Causes of Death in Hulton Colliery Disaster, 1910

Cause of death	Number killed (%)
Carbon Monoxide	224 (65%)
Explosion	53 (15%)
Explosion and CO	63 (18%)
Asphyxia	2 (0.5%)
Unstated (probably explosion and CO)	3 (0.8%)
Died in infirmary (shock and burns)	1 (0.3%)

Source: Adapted from Hatton 1911.

> Poisoning by CO is a very merciful death; there is no pain or distress; the person simply feels tired and sits down. Dr. Turner, the medical hero of the Darran disaster who was decorated by the late King for his noble rescue work had no symptoms until he exerted himself trying to rouse his companions, he then felt tired, wanted to lie down, falling down several times and finally losing consciousness.

Reference Number: 41404

DAVIES, I.J. "Carbon Monoxide Poisoning in the Senghenydd Explosion." *Proceedings of the Royal Society of Medicine*, 21 May 1914, 49–68 (3 figures, 1 reference)

Abstract: The worst mining disaster in Britain occurred at the Senghenydd mine at 8:10 a.m. on 14 October 1913, which killed 440 men. Most of the miners working on the eastern side of the mine were rescued, but all of the miners on the western side where the explosion and fire occurred died mainly due to CO poisoning, although some were killed by burns, blunt force trauma, or a falling roof.

> This refers to external appearances for the most part, inasmuch as permission could not be obtained to do any dissections. Although some of the bodies were not recovered for several months and were consequently in an advanced state of decomposition, yet typical signs of carbon monoxide poisoning were present in nearly every case. The colour in most instances was bright pink and was evident even in portions of flesh exposed by injury, such as the torn ends of muscles around joints from which the limbs had been blown off. This indicates the firm stability of the compound formed by the union of carbon monoxide with hemoglobin. In one instance were false teeth found to be loose and this points to the absence of convulsions in most instances. In one only was the tongue found to have been bitten.

Reference Number: 41405

EMERYS-ROBERTS, E. "Smoke as a Factor in Complicating Carbonic Oxide Poisoning in Colliery Disasters." *British Medical Journal*, 20 March 1915, 500

Abstract: In addition to CO and CO_2 produced in mine explosions/fires, numerous irritant gases such as acetic acid, ammonia, sulfurous, and nitrous fumes are produced as well. Irritant gases give rise to a stinging of the eyes, coughing, and choking can increase the danger of CO poisoning.

> Since the irritant gases present in smoke are freely soluble in water, it might be advantageous to emphasize the value during exposure of tying a wet cloth over the mouth and inhaling through it, and at the same time to point out the fallacy of employing a dry cloth for the same purpose.

Reference Number: 41406

ELTSCHLAGER, K.K., W. SHUSS, AND T.E. KOVALCHUK. "Carbon Monoxide Poisoning at a Surface Coal Mine: A Case Study." *Proceedings of the Annual Conference on Explosives and Blasting Technique*, 2: 121–132, 2001 (3 tables, 5 figures, 8 references)

Abstract: In April 2000, two adults and their newborn infant were poisoned by CO in their home near an open pit coal mine in Pennsylvania.

Twenty blasts were detonated between 7 March 2000 and 20 April 2000. Their blood COHb concentration at hospital were 31% for the infant, 28% for the father, and 17% for the mother. They were treated by 100% oxygen for the adults and HBO_2 for the infant. The CO gas from the explosions had migrated through the geologic structure that contained fractures and acted as a conduit for CO to the basement of their house. CO gas detectors had warned them of the danger.

> While the blaster was concentrating on controlling fly rock to protect a structure less than 50 feet away, he inadvertently created a situation where the gas could be contained in the ground under pressure. Unique site geology and an unfortunate alignment of blast holes with the sandstone fracture system allowed for the gas to accumulate in the fractures of the aquifer and migrate into the residence. Thus, combining all the factors which includes geologic conditions, blasting practices, mining practices, well type and foundation construction, blasting at the Milliren mine generated the carbon monoxide that entered the home and caused the family to become ill.

Reference Number: 41407

HARRIS, M.L., J.H. ROWLAND III, AND R.J. MAINIERO. "Blasting-Related Carbon Monoxide Incident in Bristow, Virginia." *International Society of Explosive Engineers*, 2: 319–327, 2004 (2 figures, 8 references)

Abstract: On 8 November 2002, a trench blast of 97.5 metres in length was conducted using 1,170 kg of bulk emulsion and 52.6 kg of primer. Later that evening, a CO monitor in a home approximately 27.4 metres away from the blast sent out an alarm. The gas company and the fire department measured a CO concentration greater than 800 ppm near the water sump in the basement of the home, 100 ppm on the first floor, and 200 ppm in the basement. The National Institute for Occupational Safety and Health's recommended exposure CO limit is 35 ppm for eight hours. The house was ventilated by opening the doors and windows, and the homeowners were able to return home later that evening. The fire department returned at 7:00 a.m. the next morning and the CO reading was 300 ppm in the basement and 680 ppm at the water sump. The CO readings in three neighbouring houses ranged from 23 ppm to 370 ppm. When the fire department covered the water sumps with plastic and sealed then with duct tape, at the four houses, no CO monitor alerted after another blast. All explosions generate CO and nitric oxide (NO), but if the explosive is oxygen balanced by the manufacturer and used according to instructions, the CO generated should be minimal.

During pre-blasts surveys, the blaster should determine where the likely or open pathways exist for potential gas entry into nearby homes. Possible entries may include, but are not limited to, sumps, open plumbing, an open wall (dirt floor or unsealed stone wall), an uncased well or cracks in the basement floor or wall. The CO monitor should be located at the most likely point of entry. Each CO migration occurrence is unique and depends on the route of entry, distance of site from CO generation source, and geology. Therefore, possible monitoring of nearby underground enclosed spaces may continue for an extended period of time from several hours to several days. It is imperative for the construction personnel to coordinate their efforts closely with the blaster. Shots should be excavated as soon as possible following detonation. Venting is important for relief of potentially harmful toxic gases. CO poisonings were most likely prevented by early warning provided by a homeowner's installed CO detector. Because of the early warning, the source of CO was determined and affected homes were evacuated and closely monitored. Fortunately, no one had to be treated for CO poisoning.

Reference Number: 41408

DEITCHMAN, S., J. DECKER, AND L. SANTIS. "A Novel Source of Carbon Monoxide Poisoning: Explosives Used in Construction." *Annals of Emergency Medicine*, 32: 381–384, 1998 (15 references)

Abstract: During a municipal sewer project, a new manhole was built and finished at 2:00 p.m. At 3:40 p.m., 265 pounds of explosives were detonated about 50 feet south of the manhole. At 4:30 p.m., a 45-year-old worker entered the manhole and collapsed. A 32-year-old worker and a 25-year-old worker retrieved their co-worker. The 32-year-old worker made it to the surface, but the 25-year-old worker collapsed into the bottom of the manhole. The 25-year-old worker died and had a COHb concentration of 33%. The other two workers had a COHb of 33% and 23% and were administered HBO_2 and released several days later. A gas probe six feet below the surface showed a CO of 600 ppm and hydrogen sulfide (H_2S) of 12 ppm, which was probably false as the detector also recorded high levels of CO as H_2S. Later analysis showed a CO gradient with the greatest CO concentration (1.901 ppm and no H_2S) near the bottom of the sewer. A concentration of 19.5% oxygen and 3% CO_2 were also detected. A test detonation of the same explosives used found that 27–9 litres of CO were produced from each kilogram detonated.

> The danger of CO exposure from use of explosives in confined areas such as underground mines is well known. We believe this to be the first report of a fatal CO poisoning in the construction industry from occupational exposure to CO migrating through soil and accumulating in a confined space after surface detonations. Although the physiology of CO intoxication is understood, poisonings and deaths still occur. These incidents transpire in part because new technologies and applications present new or unrecognized sources of CO exposure.

Reference Number: 41409

AUGER, P.L., H. PRUD'HOMME, D. BELLEMARE, C. BARBEAU, M. RHAINDS, L. BENOIT, G. PHILLIPE, AND R. LEROUX. "A Public Health Case Study: Carbon Monoxide Sever Intoxication in a Residential Home Following Trench Blasting." Presentation at the Major Accident Council of Canada Conference: Prevention, Preparedness, and Response, Toronto, October 1995, 95.

Abstract: On 27–9 April 2004, blasting at a new residential development in Quebec occurred at about six metres from the residential home of the victims of CO poisoning. Approximately 228, 318, and 227 kgs of explosives were detonated each day respectively. The next day at the house, the father, mother, and infant were taken to hospital and had COHb concentrations of 29%, 24%, and 0.1% respectively. Fortunately, the child had been playing outside most of the day. The parents were treated with HBOT and recovered without incident. The CO was found to be concentrated at the east side of the house at a 250 ppm concentration. The CO dissipated slowly even with ventilation, and so three air shafts were dug to vent a possible CO pocket in the soil. After two weeks, the occupants were allowed back into the house.

Table: Combustion Yield Data of the Explosives Used

Combustion gas	APEC Ultra 40 yield (L/kg)	Power FRAC yield (L/kg)
Water	577.0	425.0
Carbon monoxide	15.9	24.0
Carbon dioxide	60.0	160.0
Nitrous oxide	0	3.36
Nitrogen dioxide	0	0
Nitrogen	213.0	189.0

Source: Adapted from Auger et al. 1995.

The Quebec region public health agency emergency response was essential during this event for two main reasons: (1) Possible causes of these unusual cases of carbon monoxide poisonings were rapidly searched for. If it was not from our intervention, carbon monoxide would have been present for a longer period increasing the stress to the occupants. (2) Firefighters could have become new victims adding up to causalities. The fire department did not have the instruments and the expertise to deal with potential exposure to carbon monoxide.

Reference Number: 41410

AUGER, P.L., B. LEVESQUE, R. MARTEL, H. PRUD'HOMME, D. BELLEMARE, C. BARBEAU, P. LACHANCE, AND M. RHAINDS. "An Unusual Case of Carbon Monoxide Poisoning." *Environmental Health Perspectives*, 107: 603–605, 1999 (2 figures, 13 references)

Abstract: On 30 April 1995, at 8:00 p.m., a 33-year-old man and a 29-year-old woman were brought to the ER of a Quebec City hospital, suffering from extreme fatigue, severe headache, nausea, and palpitations. The symptoms had lasted from the previous day, and the couple thought they were suffering from food poisoning. Blood samples were collected two hours after they had left their house, and the COHb concentrations were 29.6% and 24.7% respectively. The patients were administered HBOT. On 27–9 April, explosives had been used nearby (12.3 metres distance) for blasting a trench. A pocket of CO was found under the foundation of the house, and an ambient CO concentration of 500 ppm was found in the basement. The windows of the house were opened, and three ventilation pipes were installed under the foundation of the house. The ambient

CO concentration decreased until, on 6 May 1995, it could no longer be detected.

> We therefore believe that other factors also interacted to increase the quantities of CO produced. For example, the contractor was behind schedule and set the explosives under a thick layer of overburden. These conditions are clearly very different from those of the laboratory tests. The presence of overburden very likely produced an oxygen deficiency and increased concentrations of CO generated by increasing negative oxygen balance. Another variable to be considered is the type of rock where the blasting occurred; this rock was a limestone high in carbonates which, with intense heat, may generate CO. In this case, the heat produced by the explosives may have been sufficient to produce this phenomenon.

Reference Number: 41411

JONES, C.T.A., AND H.A.F. MACKAY. "Carbon Monoxide Poisoning in a Former Mining Community." *British Medical Journal*, 286: 603–604, 1983 (1 reference)

Abstract: Four women and two men (ages 60–72 years) were exposed to CO in their bungalow homes from an abandoned coal mine that was in operation from 1780 to 1933. The mine shaft was filled with rubble, but an industrial waste-burning fire outside a factory ignited the material in the shaft. The CO spread through the fissures and faults in the underlying sandstone formation into the bungalows. Two of the exposures were fatal and involved a husband and wife who died at home with a blood COHb of 75%. The other victims experienced typical signs of occult CO poisoning. The mine shaft was reopened, the gas extracted, and the shaft was refilled. The affected bungalows were demolished.

> In 1981, in a circular to health authorities, the medical officers of British Gas reminded doctors of the vague symptoms of carbon monoxide poisoning. Old people with pre-existing cerebral, cardiac, or pulmonary disease are at a particular risk from incomplete combustion of household gas especially if confined to badly ventilated houses in adverse weather. An alternative explanation emerged in the cases reported here, but this must be regarded as a rare cause of carbon monoxide intoxication even in former mining communities.

Reference Number: 41412

KLIS, J. "Case Study: Carbon Monoxide Poisoning after Grain Explosion." *Flight Rounds*, Winter 2010, 1–8 (12 references)

Abstract: Two men with unknown injuries were trapped in a grain silo following a grain dust explosion. Both men were unconscious but showed no signs of trauma. The ambient air of the silo was found to contain over 4,000 ppm. The confined space in the silo was ventilated, and the men administered 100% oxygen. Both were transported to a hyperbaric oxygen facility where they both recovered the next day.

Carbon Monoxide (CO) is a colorless, odorless gas produced by burning material containing carbon. Carbon monoxide poisoning can cause brain damage and death. Commonly overlooked or misdiagnosed, CO intoxication often presents a significant challenge as treatment protocols, especially for hyperbaric oxygen therapy remains somewhat controversial.

Reference Number: 41413

DECKER, J., S. DEITCHMAN, AND L. SANTIS. "Carbon Monoxide Poisoning and Death Following the Use of Explosives." *Applied Occupational and Environmental Hygiene*, 14: 7–14, 1999 (3 tables, 18 references)

Abstract: A municipal project to upgrade the sewer system involved setting-off explosive blasts using Goma, to loosen the soil in order to excavate for a sewer pipe. Several of the crew worked in a 12-foot manhole. The manhole was pumped clear of water using an electric sump pump that was powered by a diesel electric generator on the surface, with its exhaust about 10 feet from the opening of the manhole. Three workers who descended into the manhole were overcome by gas but were rescued by firefighters, one of whom did not use a self-contained breathing apparatus (SCUBA) but climbed back out of the manhole in time. The other firefighter used SCUBA and a harness to rescue the workers. There were administered 100% oxygen and taken to hospital, but one worker was already dead. The two workers who survived had blood COHbs of 33% and 23% respectively and were treated with HBOT. The worker who died had a post-mortem blood COHb of 32.6%. The firefighter without SCUBA had a blood COHb of 14%, was administered 100% oxygen, and released. The firefighter with SCUBA had a blood COHb of 1.4%. Tests showed the source of the CO was not the diesel generator on the surface as no CO was detected around the vicinity of the manhole when it was operating, and CO is somewhat lighter than air. The CO source was from the multiple explosions that also fractured the soil, allowing the CO to leak into the manhole.

Table: CO Levels Measure in the Manhole after the Blast on 4 August 1997.

Location	6 August 1997	8 August 1997	11 August 1997
Surface	0	77 ppm	0
45 inches depth	15 ppm	72 ppm	0
85 inches depth	42 ppm	130 ppm	2 ppm
118 inches depth	1905 ppm	472 ppm	6 ppm

Source: Adapted from Decker et al. 1999.

We hypothesize that the August 4 explosive blast at 3:30 pm resulted in elevated levels of CO in the surrounding earth; this CO migrated into the manhole and resulted in the exposures to the construction workers and rescuer. Although some products of detonation would have been released to the atmosphere, some of the gas would have been trapped in the voids in the sandy clay soil and fractures in the rock created by the explosion. Furthermore, the rock and soil between the manhole and the trench blast had been previously fractured and loosened by the blasts 1–2 weeks earlier. Given that the explosives were detonated as close as 40 feet from the manhole, it is probable that the high pressure of the explosion forced the detonation gases through the loosened soil and fractures in the underlying rock. The manhole openings had not been sealed, and water and gases could readily enter the manhole from the surrounding earth.

Reference Number: 41414

HARRIS, M.L., AND R.J. MAINIERO. "Monitoring and Removal of CO in Blasting Operations." *Safety Science*, 46: 1393–1405, 2008 (2 tables, 12 figures, 10 references)

Abstract: CO and other toxic gases can be produced by detonating explosions in construction and mining operations. One study in 2001 found nearby blasting operations caused 39 cases of confirmed CO poisoning and one fatality. Five blasts were 6.1–15.2 metres away, three were 30.5–45.7 metres, and one was 122 metres away from the affected residences.

In open pit operations usually, the wind dilutes the formed CO from the explosion. But trench blasting near occupied buildings is done with heavy explosive confinement to prevent fly rock, and so there is no surface venting of the explosions. Various test explosions were conducted to determine the behaviour of CO produced by blasting operations.

> After a blast, CO can migrate through the ground and into underground enclosed spaces. CO may not appear for several hours, but once it does appear, it can linger for several days. The results reported here suggest that application of negative pressure to a borehole is the most effective way to remove CO from the ground. Negative pressure applied away from occupied buildings can minimize CO migration. An available source of negative pressure is the dust collector on the drill rig. By placing the drill boom over an existing open hole near the blast, the dust collected would be large enough to create a negative pressure at a high flow rate. Immediate evacuation also removes the CO before it has time to migrate, but this is not as effective as application of negative pressure. The most promising cost-effective and feasible methods appear to be application of negative pressure near the blast site to mitigate CO and immediate excavation to prevent CO accumulation and migration.

4.15 MOTOR VEHICLES

> Catalytic converters, however, do not remove all CO from exhaust. Motor vehicles emit toxic levels of CO when their emission controls malfunction or after cold starts, and the engine can be fooled into putting out more CO when the vehicle is running in a tightly controlled space. Emission controls therefore have not eliminated the MVCO poisoning problem.
>
> —B.J. Strife and L. Paulozzi, "To Make Further Progress against Carbon Monoxide Poisoning, Focus on Motor Vehicles" (2004) **(30311)**

The greatest advance in reducing accidental and suicidal exposure to CO has been the catalytic converter, which converts CO in the exhaust to CO_2 (41503, 41505, 41512, 41514). People are still exposed to CO at service stations (41501), traffic congestion (41502), auto repair shops (41510), US border inspection stations (41515), and car washes (41524). CO is still involved in car crashes and fires (41523, 41525–6). Catalytic converters do not work if the exhaust is plugged by snow (41508) or mud, as in mud bogging (41523) and rusting (41519, 41521). CO poisoning with motor vehicles can even occur outside and not in a garage (41504, 41511, 41520). It is

Reference Number: 41414

shocking that CO poisoning has occurred in school buses (41507), transit vans (41509), and ambulances (41521) due to poorly maintained exhaust systems. It is recommended that school buses should not be parked in large groups bumper to bumper and idling (41507). This section illustrates that we still cannot let down our guard against our ancient, silent enemy for the will o' the wisp will somehow find a way.

Reference Number: 41501

MAYERS, M.R. "Carbon Monoxide Poisoning in Industry." *American Journal of Public Health*, 17: 108–113, 1927

Abstract: The ambient CO concentration was measured in an automobile service garage. The blood COHb was also determined in 20 male employees and two visitors to the service station. The ambient CO concentration was found in a range up to 1,100 ppm. The men were generally exposed to the air for about two to four hours. The blood COHb concentrations ranged from 2.3% to 40%.

> All of the men examined were followed for several hours, and later questioned as to symptoms following their return home from work. Without exception, every one developed a typical headache of considerable severity and duration. The carbon monoxide headache is typical in that it usual begins from 2 to 3 hours after exposure to fresh air and continues for many hours after all the carbon monoxide has apparently been removed from the blood. While it was not possible to make additional determinations of the carbon monoxide saturation of the blood of these men during the last hours of the headache, because of its extraordinary duration by every calculation of the rate of carbon monoxide elimination from the body it would seem proper to assume that the headache outlasts the presence of carbon monoxide in the blood by a great many hours.

Reference Number: 41502

BUTKA, H.E. "Chronic Carbon Monoxide Poisoning." *California and Western Medicine*, June 1927, 809.

Abstract: A police officer directing traffic in a congested intersection of Los Angeles in the 1920s was so affected by the automobile gases that he appeared somewhat intoxicated by the end of the day. Tests have shown that a patrolman at the end of an eight-hour day may have up to 30% COHb.

Chronic poisoning results in an increased red blood cell count running from six to nine million with 95 to 125 percent hemoglobin. The subjective symptoms are dizziness, headache, blurring of vision, weakness, and palpitation. Usually, the symptoms disappear quite rapidly upon removal from the influence of the gas.

Reference Number: 41503

HAMPSON, N.B., AND J.R. HOLM. "Suicidal Carbon Monoxide Poisoning Has Decreased with Controls on Automobile Emissions." *Undersea and Hyperbaric Medicine*, 42: 159–164, 2015 (4 figures, 14 references)

Abstract: The catalytic converter on cars built in the United States since 1975 oxidizes CO to CO_2 and has largely been responsible for the decrease in ambient air concentrations since that time. Prior to 1970, automobiles CO exhaust could be 7–12% or greater. Now, most automobile exhausts have an upper limit of 1% CO, and some new cars as low as 0.1%. The mean blood COHb concentration for 461 patients treated in a HBOT centre for intentional CO motor vehicle poisoning has decreased from 29.1% in 1981 to 19.1% in 2013. Non-fatal motor-vehicle-related intentional CO poisoning has decreased by 63% over 33 years.

In summary, the incidence of both fatal and nonfatal intentional CO poisoning related to motor vehicle exhaust have decreased since the 1980s. These declines correlated temporally and quantitatively with reductions in vehicle CO emissions and are in large part due to the US Clean Air Act of 1970, and the widespread introduction of automotive catalytic converters in 1975.

Reference Number: 41504

STRUTTMANN, T., A. SCHEERER, T.S. PRINCE, AND L.A. GOLDSTEIN. "Unintentional Carbon Monoxide Poisoning from an Unlikely Source." *Journal of the American Board of Family Practice*, 11: 481–484, 1998 (1 table, 1 Figure, 19 references)

Abstract: In 1995, CO poisoning was the cause of 42 occupational fatalities and, in 1993, for 862 non-fatal poisonings, but it is estimated that one-third of CO poisoning are undetected. A 37-year-old female farmer began planting tobacco in a 4.5-acre field at 1:00 p.m. The temperature exceeded 90 degrees Fahrenheit, and it was very humid with no breeze. The victim continued to operator the gas-powered tractor to plant tobacco in spite of severe headaches, drowsiness, and dizziness. At 6:30

p.m., she collapsed on a tray of tobacco plants in front of her on the setter. She was admitted to hospital and administered 100% oxygen for 36 minutes, and her initial COHb saturation was 23.3%. Three hours later, her COHb was 7.3%, and she was released. Her gas-powered tractor was found to emit 447 ppm of CO, which was above the 200 ppm limit recommended. Four additional gas-powered tractors built between 1947 and 1979 also emitted CO above the 200 ppm limit.

This report suggests that workers might not be aware of the risks of carbon monoxide poisoning associated with the use of gasoline powered engines, especially outdoors. Such risks can be reduced by using a tractor with an upward exhaust system to vent fumes upward. Reconfiguring a tractor in this manner would cost about $600 plus installation. If this alternative is financially prohibitive, workers can limit their exposure to carbon monoxide by taking frequent breaks, rotating their positions on the setter, and getting off the setter if any symptoms of carbon monoxide intoxication appear. Primary prevention also includes proper engine maintenance, exhaust system inspections, and emissions testing.

Reference Number: 41505

VOSSBERG, B., AND J. SKOLNICK. "The Role of Catalytic Converters in Automobile Carbon Monoxide Poisoning. A Case Report." *Chest*, 115: 580–581, 1999 (8 references)

Abstract: A 56-year-old man with an orthopaedic prosthesis attempted to commit suicide by running a hose from the exhaust pipe of a running car into the cab, where he sat for 8–10 hours. The car was a 1995 model with a catalytic converter. When the emergency medical personnel arrived, the patient was alert and talking, although mildly confused. After being administered 100% oxygen for just over one hour, his blood COHb was 4.8%. Three and one-half hours later, his blood COHb was 3.0%. It is estimated that his blood COHb was probably less than 10% when he was discovered by emergency medical personnel. Catalytic converters remove CO, hydrocarbons, and NO_2 from the exhaust of the car and produces CO_2 and water. The patient underwent a urgent fasciotomy of his lower right extremity but died five days later.

Inhaling motor vehicle exhaust fumes is a common method used by people attempting to commit suicides; however, the decreased carbon monoxide concentration found in the exhaust of late-model automobiles equipped with catalytic converters are changing the clinical presentation of exhaust inhalation.

Reference Number: 41506

MCAULEY, F., AND J.G. WIGMORE. "Case Report: An Unusual Case of Carbon Monoxide Poisoning." *Canadian Society of Forensic Science Journal*, 20: 83–86, 1987 (3 tables, 5 references)

Abstract: An 18-year-old woman who went missing the previous day was found dead in the trunk of her car at her residence. She had emotional problems and had attempted suicide a year earlier. She was found inside the closed trunk with the keys to the car in her hand and attempts to open the trunk from the inside were obvious. The only significant toxicological finding was a blood carboxyhemoglobin saturation of 55%. The cause of death was due to aspiration of stomach contents and CO poisoning. A reconstruction experiment showed the car, which did not have a catalytic converter, emitted a high concentration of CO (5% volume by volume) and that the trunk once closed (unlike the garage) maintained a high CO concentration.

Table: Concentration of Ambient CO in the Driver's Seat/Garage and in the Closed Trunk after the Engine Was Stopped

Time after Engine Stopped (minutes)	Driver's Seat/Garage (ppm)	Closed trunk (ppm)
4	-	3500
5	2000	-
10	1500	3000
20	1000	2800
25	-	2500

Source: Adapted from McAuley and Wigmore 1987.

While the exact sequence of events leading to this death could not be determined, the reconstruction showed that the trunk could retain a potentially lethal concentration of carbon monoxide for a considerable time. Death from carbon monoxide could therefore occur in the trunk of the automobile without the automobile running.

Reference Number: 41507

JOHNSON, C.J., J. MORAN, AND R. PEKICH. "Carbon Monoxide in School Buses." *American Journal of Public Health*, 65: 1327–1329, 1975 (2 tables, 8 references)

Abstract: On 16 December 1971, eight children became ill on a school bus carrying a group of private school students in the Seattle area. The symptoms included headache, drowsiness, and nausea. Four of the children (including one who was unconscious) were so ill that they were admitted to hospital. All eight children sat at the rear of the bus, which had a rear-mounted engine and a defective exhaust system. All five buses operated by the private school were tested and had ambient CO concentrations of 10–25 ppm (mean 15 ppm). A CO check of 33 buses after a one to two hours running showed that four had levels of 9–35 ppm. The CO concentration of 65 buses idling at lunch time showed 24 buses with an ambient concentration of 9–100 ppm. It was recommended that buses should not park bumper to bumper with the engine idling and to stop idling once the bus was warm. Regular maintenance and regular checking of the exhaust system is also important.

Another interesting point observed was that lower tolerance levels of CO should be recommended for higher altitudes. This is because the CO detection devices measure an absolute and not a relative amount of CO, and a nominal value of 35 ppm probably represent an actual concentration of CO that is greater in proportion to the density of air at a higher altitude. In addition, persons from a low elevation going to a high altitude for skiing or mountaineering may have a higher uptake of CO at sea level.

Reference Number: 41508

HENRETIG, F., F. DEROSS, K. HARDY, D. CONTOSTAVLOS, R. LEVENSON, J. LOGUE, M. TOUGER, AND B.A. MOJICA, "Carbon Monoxide Poisonings Associated with Snow-Obstructed Vehicle Exhaust Systems, Philadelphia and New York City, January 1996." *Morbidity and Mortality Weekly Report*, 45: 1–3, 1996 (6 references)

Abstract: After a blizzard in the northeastern United States during January 1996, 21 persons in New York City were admitted to Jacobi Medical Centre for HBOT due to CO poisoning as a result of car exhausts being blocked with snow. In addition, there were three other reported cases in Philadelphia in which one person died due to CO poisoning.

The epidemiology of CO poisonings in the United States is characterized by seasonal increases during the winter months—particularly because of the risks for exposure to the exhaust from vehicles and combustion appliances during periods when heating appliances are in use and ventilation is more likely to be inadequate. The findings in this report also underscore that heavy snowfalls are associated with particularly hazardous conditions in areas where vehicles are parked outdoors. Following heavy snowfalls, the public should be reminded to inspect vehicles to ensure that exhaust pipes are cleared of snow before engines are started. Other precautions to prevent CO poisoning include avoiding running automobile engines in enclosed spaces (e.g. garages), inspecting furnaces each year, using space heaters only in well-ventilated rooms, and inspecting exhaust systems of all combustion appliances that vent to the outside to ensure that vents have not been damage or blocked with snow.

Reference Number: 41509

SMITH, R.A., AND R.J. BALL. "Carbon Monoxide Poisoning in Two Children Riding at the Back of a Van." *Archives of Disease in Childhood*, 71 (5): 482, 1994 (3 references)

Abstract: A brother and sister ages nine and 10 years respectively became drowsy, had a headache, and vomited after riding in the back of a transit van for 40 minutes. They were taken to hospital, and their initial blood COHb concentrations were 24.5% and 19.7% respectively. There were treated with 100% oxygen in a tightly fitted mask and showed great improvement after one hour. After six hours, their blood COHb concentrations had decreased to less than 1%.

Treatment with hyperbaric oxygen is recommended if COHb concentrations exceeded 40% or if there is a history of loss of consciousness, persisting neurological or intellectual deficits, or cardiovascular abnormalities. None of these features was present in these two cases and both made a complete recovery. However, had the journey been of any longer duration or the symptoms not recognized, a serious or fatal poisoning may have occurred.

Reference Number: 41510

BOL, O., S. KOYUNCU, AND N. GUNAY. "Prevalence of Hidden Carbon Monoxide Poisoning in Auto Service Workers: A Prospective Cohort Study." *Journal of Occupational Medicine and Toxicology*, 13 (1): 35, 2018 (3 tables, 2 figures, 16 references)

Abstract: The COHb concentration was determined (using fingertip Masimo Rad-57 CO-oximeter) in 99 workers from six different auto repair shops on four different occasions at two-hour intervals and compared to the COHb concentrations of a control group of 100 people in the morning. The COHb ranged between 0% and 18% in the auto repair group and 0% and 4% in the control group. The highest mean COHb concentration in the auto repair group was 7.04% at 12:00 p.m. compared to 1.61% for the control group at 8:00 a.m. The highest COHb concentrations were 11%, 13%, 15%, 15%, 15%, and 18% and were all obtained from the auto repair workers.

> We believe that CO poisoning cases other than acute poisoning are even more remarkable because of the difficulty of detecting such cases and the impossibility of finding them without definitive proof. Diagnosis of chronic CO poisoning in patients admitted to emergency services is still not easy as the presence of similar clinical presentations cannot be excluded. CO gas may harm any cell in the body, which is dependent on oxygen regardless of its amount, causing a range of conditions from acute coronary syndrome to detectable damages in the brain.

Reference Number: 41511

NNOLI, M., N.L. LEBGHOSI, I.I. CHUCKWUONYE, AND C.O. NWABUKO. "Toxicological Investigations of Acute Carbon Monoxide Poisoning in Four Occupants of a Fuming Sport Utility Vehicle." *Iranian Journal of Toxicology*, 7: 973–976, 2014 (2 tables, 10 references)

Abstract: Four male friends were travelling in a second-hand sport utility vehicle (SUV) in Nigeria when they were stopped in a major traffic jam. They opened the back hatch of the SUV to allow fresh air into the motor vehicle, but, instead, CO was produced by the exhaust fumes of the idling SUV. Then, the traffic finally started moving again, the SUV would not move, as all the occupants were dead. At autopsy, there was generalized cyanosis of the bodies with sub-conjunctival hemorrhage. There was also marked larygno-trachea edema hyperemia with frothy discharges. Postmortem peripheral blood COHb was measured using a Hitachi ultraviolet spectrophotometer. The COHb concentrations were elevated and ranged between 480 and 650 ppm.

> We suggest that the government be more responsive in removing heavy broken trucks that could create traffic jams, build more good and accessible roads to ensure free flow of traffic, close potholes and remove gallops that could break exhaust pipes producing fuming vehicle. Health education of the public on sources and dangers of CO can also help them stem the increasing reports of CO poisoning in Nigeria.

Reference Number: 41512

GALATSIS, K., W. WLODARSKI, B. WELLS, AND S. MCDONALD. "Vehicle Cabin Air Quality Monitor for Fatigue and Suicide Prevention." *SAE Technical Papers*, March 2000, 55–59 (4 tables, 3 figures, 11 references)

Abstract: The CO concentration emitted by older automobiles was typically 7–12%, but, with the catalytic converter, the CO concentration emitted was usually less than 0.1% while idling; however, some cars may still emit between 2% and 4% CO. The CO, oxygen, and temperature were monitored in the cabins of four models of new cars (1995–7) in city traffic or a motorway and while the air into the cabin was recycled or fresh. The rate of fresh air entering the vehicle cabin is dependent mainly on the speed of the vehicle, on the efficiency of the door/window seals, and on the number of occupants. In city driving with recycled air, the ambient oxygen concentration decreased to, on average, 19.8%. The maximum ambient CO concentration detected was 16 ppm in city traffic with fresh air. In a simulated suicide attempt, the exhaust was directed back into the vehicle cabin, the oxygen level decreased to 7% as it was displaced by the vehicle exhaust, and, within 30 minutes, the CO level in the cabin had increased to over 600 ppm.

Table: Oxygen and CO Concentrations in the Cab of Four Different Car Models in City Traffic in the Recycle or Fresh Air Mode

Car model and engine size	Oxygen level, city, recycle (%)	Oxygen level, city, fresh (%)	CO level city recycle	CO level city fresh
Ford Falcon (4.0 L)	20.5	20.5	13 ppm	7 ppm
Holden Commodore (3.8 L)	20.2	20.4	7 ppm	7 ppm
Mitsubishi Magna (3.0 L)	19.2	20	8 ppm	18 ppm
Toyota Camry (2.2 L)	19.1	20.2	< 1 ppm	< 1 ppm

Source: Adapted from Galatsis et al. 2000.

Low oxygen, high carbon monoxide and carbon dioxide concentrations can typically exist within a vehicle cabin. Such poor air quality may cause drowsiness, fatigue, impairment of judgment, and poor coordination to vehicle occupants. Also, many deaths are caused by motor vehicle exhaust gas suicides from carbon monoxide poisoning. The introduction of an air quality monitor within the vehicle cabin can alarm occupants preventing any adverse health effects. A vehicle air quality monitor was designed and developed.

Reference Number: 41513

KUMAZAWA, T., K. WATANABE-SUZUKI, H. SSENO, A. ISHI, AND O. SUZUKI. "A Curious Autopsy Case of Accidental Carbon Monoxide Poisoning in a Motor Vehicle." *Legal Medicine*, 2: 181–185, 2000 (4 figures, 23 references)

Abstract: A 26-year-old man was found dead in his car with all the doors and windows closed and locked in the parking lot of a bookstore. The ignition was in the on position, but the gas tank was empty. His post-mortem lividity was cherry pink. His post-mortem blood cardiac COHb concentration was 46.6% and also contained traces of alcohol due to putrefaction. His blood amphetamine and methamphetamine concentrations were 3.25 and 0.84 micrograms per mL respectively, which, while not fatal, was toxic. No suicide note was found, and the car was idling with the air conditioning on, but exhaust gas leaked from holes in the exhaust pipe into the driver's compartment.

From the autopsy and the police investigation, the possibility of homicide or suicide was excluded in this case. We finally concluded that the cause of his death was accidental CO poisoning; the cardiomyopathy and the presence of stimulants in blood might facilitate his death.

Reference Number: 41514

STUDDERT, D.M., L.C. GURRIN, U. JATKAR, AND J. PIRKIS. "Relationship between Vehicle Emissions Laws and Incidence of Suicide by Motor Vehicle Exhaust Gas in Australia, 2001–06: An Ecological Analysis." *PLoS Medicine*, 7 (1): e1000210, 2010 (3 tables, 1 figure, 41 references)

Abstract: In Australia, the incidence of motor vehicle exhaust gas (MVEG) suicides decreased by 67% between 1997 and 2006. In 1986, new passenger vehicles had to have a catalytic converter, and, in 1997–9, the CO emission level was further reduced. Between 2001 and 2006, there were 1,872 MVEG suicides in Australia, and 83% of the victims were male. If the proportion of pre-1986 (pre-1999) vehicles had remained at the same level, it is estimated that there would have been 62 (366) more MVEG suicides.

Table: Number and Percent of MVEG Suicides in Australia between 2001 and 2006

Notification year	Number	Percent (%)
2001	498	22
2002	427	19
2003	393	17
2004	382	17
2005	324	14
2006	231	10

Source: Adapted from Studdent et al. 2010.

Areas of Australia with fewer predating stringent CO emission laws experience lower rates of MVEG suicides. Although those emission laws were introduced primarily for environmental reasons, countries that lack them may miss the benefits of a serendipitous suicide prevention strategy.

Reference Number: 41515

COHEN, S.I., G. DORION, J.R. GOLDSMITH, AND S. PERMUTT. "Carbon Monoxide Uptake by Inspectors at a United States-Mexico Border Station." *Archives of Environmental Health*, 22: 47–54, 1971 (2 tables, 4 figures, 23 references)

Abstract: Breath and blood COHb concentrations were determined in 20 US border inspectors exposed to automobile exhaust at the Mexico-US order and six controls before and after a shift. The hourly ambient CO levels ranged from 5–170 ppm. The mean COHb concentration increased by a mean of 1.5–3.6% in non-smokers and 4.8–6.4% in smokers. There was no increase in COHb in the controls.

> We also have some concern for the effects of CO exposures on drivers crossing such congested inspection facilities. The data on the effect of CO on visual sensitivity, psychological tests and the accurate estimation of time intervals suggest mechanisms by which CO exposures could influence driving performance. As noted, these effects may be associated with as little as 2% COHb. Laboratory studies and available epidemiologic information are consistent with the possibility that such an increase in COHb might have an influence on the frequency of accidents. It should be pointed out that the risk of such accidents may still be present for a considerable time after the CO exposure has been discontinued because it takes approximately two hours to reduce the COHb level by one half when breathing air containing no CO once the increased level has been established.

Reference Number: 41516

HAMPSON, N.B., AND D.M. NORKOOL. "Carbon Monoxide Poisoning in Children Riding in the Back of Pickup Trucks." *Journal of the American Medical Association*, 267: 538–540, 1992 (2 tables, 13 references)

Abstract: A study was conducted of 20 children (ages 4–16 years) who suffered CO poisoning while riding in the back of a pickup truck and treated with HBOT in Seattle, Washington, between 1986 and 1991. The mean COHb at the ER was 18.2% (range 1.6–37.0%). In 17 cases, the children were riding under a rigid closed canopy at the rear of the truck, and three had rode beneath a tarpaulin. Loss of consciousness occurred in 15 of the 20 children (75%). One died of cerebral edema, and one had permanent neurologic deficits. In all cases, the truck exhaust system had a previously known break in the tail pipe, and so the exhaust vented towards the rear rather than the side of the truck.

> In the present series, specific configurations of the vehicle exhaust systems were associated with the injury in every instance. A leaking exhaust system or a tail pipe exiting beneath the rear bumper of the pickup truck was always present. We hypothesize that exhaust emitted at the rear of the moving pickup truck is drawn upward via the Venturi effect. In the

case of canopy-related incidents, fumes presumably enter the bed of the vehicle through the space between the bed and the tailgate, or between the tailgate and canopy door. We have seen no cases involving passengers in campers on the rear of pickup trucks or in pickups with intact exhaust systems exiting the side of the vehicle. Most pickup trucks are currently manufactured with tail pipes that exit the side of the vehicle, typically at the rear wheel well. When exhaust systems are worn or damaged, however, they may be replaced with systems that exit the rear of the vehicle. This practice should be avoided.

Reference Number: 41517

PRATHER, E.C., J.F. MC GARRY, K. FOULTON, AND J. ALLEY. "Carbon Monoxide Poisoning, Florida." *Morbidity and Mortality Weekly Report*, 19: 272–273, 1970

Abstract: On 20 June 1970, a 63-year-old woman awoke at 1:30 a.m. with nausea and a burning sensation in her upper chest. Her husband also felt sick. She crawled into the bathroom and vomited. Several hours later, she woke up, and her husband was dead in the bedroom, and her 80-year-old mother was unconscious. They were given supportive therapy at hospital and recovered. The husband had a post-mortem COHb of 40%. The car had been left idling in the adjoining garage where the air-conditioning intake was located. An ambient CO of 400 ppm was detected with the car idling and the air-conditioning on after four minutes.

The family had been for a drive in the car the evening of June 19 and had become drowsy and logy during the drive. It is postulated that they became slightly CO intoxicated and as a result neglected to turn off the car's motor after their return. With the garage door closed and the car and air conditioning running, toxic and lethal levels of CO were then developed within the house. When the car stopped due to overheating or battery failure, the CO in the house was apparently cleared, leading to the regaining of consciousness in the women. The initial negative carboxyhemoglobin levels on the two women were found to be positive at a low level on retesting.

Reference Number: 41518

HAYS, P., AND R.A. BORNSTEIN. "Failed Suicide Attempt by Emission Gas Poisoning." *American Journal of Psychiatry*, 141: 592–593, 1984 (3 references)

Abstract: In North America, due to emission standards, the amount of CO emitted by cars is now one-tenth that emitted by cars made before 1967. A 31-year-old man drank 12 ounces of whiskey late one evening and connected a tube from the exhaust of his car to the closed rear window. He was discovered six hours later in a comatose state. He was administered oxygen and regained consciousness. His skin was pink, and he had a blood COHb of 23%. He was discharged three days later.

> When cars had no emission control systems, fatal carbon monoxide concentrations were achieved within a few minutes in a closed, single-car garage. Our patient, a trained engineer, directed the carbon monoxide into the interior of his car and left the motor running all night. His recovery and freedom from gross sequelae imply that he was not severely intoxicated by 6 hours of exposure. The most likely explanation for his survival is that the carbon monoxide levels produced by his modern car were low. We predict a reduction in fatalities when emission gas is used as a method of attempted suicide.

Reference Number: 41519

FISHER-HUBBARD, A.O., C. APPLEFORD, K. KESHA, C.J. SCHMIDT, AND A. GUPTA. "Accidental Carbon Monoxide Poisoning While Driving: A Case Report with Review of the Literature." *American Journal of Forensic Medicine and Pathology*, 39: 270–272, 2018 (3 figures, 9 references)

Abstract: A female driver crashed into a house in the early hours of the morning. She had minor damage to her car and complained of car trouble. At hospital, she had a BAC of 104 mg/dL and a COHb of 44.9%. There were several rusted holes in the undercarriage of the car that allowed exhaust to enter the passenger compartment.

> Although most automobile CO poisoning cases are suicides, it is important to remember that accidental CO poisoning may occur while in an automobile. The case that we report adds to the existing literature showing that defective exhaust systems can allow CO to enter the cabin of a vehicle.

Reference Number: 41520

ZABA, C., P. SWIDERSKI, K. SOMMERFELD-KLATTA, Z. ZABA, K. PLUTA-HADAS, AND M. URBANIAK. "Suicidal Carbon Monoxide Poisoning Using Motor

Vehicle Exhaust in an Open Space." *Medical Principles and Practice*, 28: 490–492, 2019 (1 figure, 9 references)

Abstract: A 60-year-old man was found dead outside in a forest beneath the exhaust pipe of his car. A suicide note was found. The engine was turned on but had stopped running. His post-mortem blood COHb as determined with an CO-oximeter was 80%. In a properly functioning car with a catalytic converter, the average ambient CO concentration is 12 ppm inside and 2–5 ppm outside. In this car with the engine running, the exhaust near the victim's head was 45–200 ppm with momentary peaks (depending on the wind) of up to 790 ppm.

This study shows that long exposure to carbon monoxide in an open space can cause acute carbon monoxide poisoning. This should be considered by forensic, emergency, and occupational professionals for diagnosing poisoning by unknown agents.

Reference Number: 41521

BILTAJI, M.P. "Carbon Monoxide: An Occupational Hazard to Health Care Workers." *Journal of Occupational Medicine*, 30 (11): 852, 1988 (1 reference)

Abstract: A 30-year-old female nurse accompanied a cardiac patient who was wearing an oxygen mask at the back of an ambulance for a 90-minute round trip. She developed classic symptoms of CO poisoning, and her blood COHb was 25.9% before 100% oxygen therapy. The CO levels of 42 ambulances were determined and ranged from 10 ppm to 200 ppm while idling and 5 ppm to 130 ppm while travelling.

Carbon monoxide-laden exhaust gases appear to enter the ambulance through holes in the side body undercarriage, rear door seals of through the air conditioning/ventilation system.

Reference Number: 41522

WIRTHWEIN, D.P., AND J.E. PIESS. "Carboxyhemoglobin Levels in a Series of Automobile Fires: Death Due to Crash or Fire?" *American Journal of Forensic Medicine and Pathology*, 17: 117–123, 1996 (1 table, 7 references)

Abstract: The post-mortem blood COHb concentrations were determined in 23 male and five female victims (ages 17–53 years) found dead in an automobile involving a fire in Indianapolis. All victims had some degree of body burn, and 25 had severe charring. The COHb concentrations ranged from less than 10% to 92%. In seven cases, there was no motor vehicle collision (MVC), and six of these victims had a blood COHb greater than 47%. In all cases with a COHb less than 10%, a MVC occurred, and, in 12 of the 16 victims, the blunt force trauma was sufficient to cause death. A blood COHb concentration greater than 30% strongly indicates inhalation of combustion products, including CO as a cause of death.

> The data presented in this article illustrate how auto-related fire and trauma can result in death through a variety of mechanisms. Although the fire ravaged body presents a formidable challenge to the pathologist, COHb levels can be useful in differentiating fire and trauma as possible causes of death. Due to extensive charring and incineration in most cases, the significance of thermal body injury is usually impossible to ascertain. Three groups are discernible in the present study: (a) those who die as a result of smoke inhalation and/or CO intoxication, (b) those who die as a result of thermal airway injury, and (c) those who die as a result of trauma. Our data indicate that a COHb level of > 30% strongly implicates inhalation of combustion products as cause of death. A level of < 20% should prompt an exhaustive search for other causes. In light of the exceptions presented in this article, one must resist the temptation to ascribe cause of death to a single cause.

Reference Number: 41523

MENKE, N.B., K. FITZPATRICK, M.J. LYNCH, AND A.M. KING. "Carbon Monoxide Poisoning Due to Mud Bogging." *New England Journal of Medicine*, 373 (11): 1082, 2015 (1 reference)

Abstract: Mud bogging is an off-road motorsport in which vehicles are driven over a muddy track or pit. The vehicles frequently get stuck, and the mud can clog the exhaust pipes increasing CO concentrations in the vehicle. An 18-year-old male driver and a 16-year-old female passenger lost consciousness while mud bogging. Their hospital COHb concentrations were 18.5% and 14.0% respectively. They were administered HBOT and recovered. In another incident, two male passengers (ages 16 and 19 years) lost consciousness in the back of a mud-bogging vehicle. The hospital COHb concentrations were 39.6 and 47.5% respectively. They recovered after being administered HBOT. A cast member of the MTV series *Buckwild* died of CO poisoning while mud bogging.

Mud Bogging may place participants at risk for carbon monoxide poisoning. Clinicians should increase their suspicion of carbon monoxide poisoning when assessing patients with otherwise unexplained symptoms who have recently participated in this activity. Although there are no available supportive data, it is reasonable to expect that maximizing ventilation in the passenger cabin by keeping windows open at all times and using portable carbon monoxide detectors may reduce exposure and prevent toxic effects.

Reference Number: 41524

CARSON, H.J. "Four Deaths Due to Carbon Monoxide Poisoning in Car Washes." *American Journal of Forensic Medicine and Pathology*, 20: 274–276, 1999 (1 table, 11 references)

Abstract: Four men (aged 20–36 years) were found dead in self-car-washing facilities with the vehicle running and the bay doors closed over a period of 13 months. All were found outside their running vehicle. Their post-mortem blood COHb ranged from 78% to 83%. All four had mind-altering drugs such as tetrahydrocannabinol, amphetamine, and alcohol detected, which may impair judgment, perception of outside conditions, and self-awareness. Three cases were determined to be accidental, and one was suicidal.

All four died in winter months. For three men, the deaths were ruled accidental, and for the remaining man, the previous deaths appear to have provided a model for suicide. Warning signs may not be effective to prevent future CO deaths in car washes because of the possible role of intoxication. Mechanical or electronic methods to prevent a bay door from closing completely may be preferable.

Reference Number: 41525

WESTERMAN, R.A., AND A.L. BENJAMIN. "Carbon Monoxide and the Impaired Driver." *Medicolegal Society Proceedings, Royal Australasian College of Surgeons*, April 1978, 133–152 (45 references)

Abstract: A detailed paper regarding chronic CO poisoning and the impaired driver was presented at a Medicolegal Society meeting in

Melbourne, Australia, on 8 April 1978. In one case, a law student drove an old model Volkswagen sedan with a defective heating circulating system. The police stopped the car due to erratic driving, and the driver showed intoxicated behaviour such as incoherent speech. He admitted to consuming some alcohol hours before and refused a Breathalyzer test. Later testing in the Volkswagen sedan showed an ambient CO of 0.26% inside the car with the windows closed and the heater on. The CO concentration decreased to 0.19% when moving and to 0.01% with the heater off and the windows open. He was acquitted of driving under the influence of alcohol.

Incidents in Canada and the USA of deaths from CO in automobiles are cited including one where the medical examiner judged that the driver in a fatal accident may have breathed the gas coming into his car from vehicles ahead of himself in heavy traffic, since blood tests indicated a COHb saturation of thirty-five per cent. Individuals respond in various ways to CO, sometimes acting drunkenly or hilariously. The author suggests that the blood of such drivers be tested for both alcohol and CO content. We would concur.

Reference Number: 41526

FISHER-HUBBARD, A.O., C. APPLEFORD, K. KESHA, C.J. SCHMIDT, AND A. GUPTA. "Accidental Carbon Monoxide Poisoning While Driving." *American Journal of Forensic Medicine and Pathology*, 39: 270–272, 2018 (3 figures, 9 references)

Abstract: A 23-year-old woman who reported having car trouble crashed into a house in the early hours of the morning. She suffered non-fatal trauma. At hospital, her blood COHb was 44.9%, and her BAC was 104 mg/dL. She suffered anoxic brain injury and died five days later. There were rusted defects in the top of the muffler as well as a large, rusted defect in the trunk directly above the muffler. This allowed CO to enter the passenger compartment.

Although most automobile CO poisoning cases are suicides, it is important to remember that accidental CO poisoning may occur while in an automobile. The case we report adds to the existing literature showing that defective exhaust systems can allow CO to enter the cabin of a vehicle.

4.16 PAINT STRIPPER (METHYLENE CHLORIDE)

> Dichloromethane (DCM) (methylene chloride or methylene bichloride) is an organochlorine compound with the formula (CH_2Cl_2). This colorless, volatile liquid with a chloroform-like sweet odor is widely used as a solvent.
>
> —*Wikipedia*

The problem with DCM is that it is readily absorbed into the body via the skin and lungs, and, once in the blood, it is converted eventually to formyl chloride by the liver.

CH_2Cl_2 →	$CH(OH)Cl_2$ →	CO →	COHb
DCM	formyl chloride	carbon monoxide	carboxyhemoglobin

It then decomposes to CO, which then binds to hemoglobin forming carboxyhemoglobin. It takes about 48 hours for DCM to completely leave the body, and, during that time, it is being converted to CO so oxygen must be applied for a longer period of time, or the CO will recur until all the DCM is gone (41601–6).

Reference Number: 41601

RUDGE, F.W. "Treatment of Methylene Chloride Induced Carbon Monoxide Poisoning with Hyperbaric Oxygenation." *Military Medicine*, 115: 570–572, 1990 (15 references)

Abstract: Methylene chloride (CH_2Cl_2) is a colourless, highly volatile solvent used widely as a paint stripper, degreaser, and solvent for cement and plastic. CH_2Cl_2 is highly lipid soluble and can be slowly released from the various tissues and converted into CO by the liver. A 25-year-old man presented at hospital with a severe headache, nausea, and incoordination. The day earlier, he was cleaning computer equipment with CH_2Cl_2 in a poorly ventilated room without wearing gloves. His initial COHb was 20.1%. He was administered 100% oxygen and then HBOT.

A very important difference exists between the treatment of poisoning from inhaling CO and poisoning from the endogenous production from methylene chloride exposure. In the former, tissue CO levels continue to rise after the exposure is terminated as endogenous production by the liver continues. As a result, patients adequately treated soon after exposure may develop a recurrence of symptoms which may require treatment.

A period of observation (12–24 hours after exposure) is required after methylene chloride poisoning to ensure complete resolution.

Reference Number: 41602

MAHMUD, M., AND S.N. KALES. "Methylene Chloride Poisoning in a Cabinet Worker." *Environmental Health Perspectives*, 107: 769–772, 1999 (1 table, 39 references)

Abstract: A 26-year-old male cabinet worker complained of recurrent headache for a month. He was diagnosed with a stress headache. His job involved spraying contact cement, containing 70% CH_2Cl_2 in an ill-ventilated spraying booth, which was heated in the winter with a gas-powered heating fan with doors closed during winter. There was also a gas-powered forklift at the plant. When the doors were open, his post-shift COHb was 6.4%, and he had a mild headache. Four days later, when the doors were closed, he developed nausea and vomiting. His COHb on this occasion was 21%, and he was administered NBOT. The company substituted a water-based product for the CH_2Cl_2.

Adverse health effects due to $MeCl_2$ [methylene chloride] can be avoided by substituting safer products or processes for those using $MeCl_2$, such as alternative process adopted by the manufacturer in this case. When $MeCl_2$ is use, adequate ventilation is essential to keep exposure at low levels, $MeCl_2$ should never be used in enclosed or poorly ventilated areas because of the unacceptable risk of loss of consciousness and death. It is important to educate those using $MeCl_2$ containing products about safety hazards and monitoring of airborne concentrations.

Reference Number: 41603

NAGER, E.C., AND R.E. O'CONNOR, "Carbon Monoxide Poisoning from Spray Paint Inhalation." *Academic Emergency Medicine*, 5: 84–86, 1998 (14 references)

Abstract: A 52-year-old woman presented to the emergency department (ED) with sporadic complaints of shortness of breath, nausea, weakness, and left-sided parenthesis after painting her house outdoors with Rust-Oleum spray for four days. She smoked one pack of cigarettes per day. Her COHb was 11.7%. She was administered 100% oxygen via a mask until her COHb had decreased to 3.9%. She was advised to stop smoking. She returned to the ED the next day with the same symptoms, and her COHb was 9.4%. She was administered 100% oxygen for 24 hours until her blood COHb was 3.1%. The paint spray was found to contain 31% methylene chloride, also known as DCM. The half-life of CO derived from DCM is about 2.5 times longer than inhaled CO (that is, up to 13 hours).

It is generally accepted to discharge the patient to home if he or she is symptomatic, the COHb level is < 10%, and there is no risk of re-exposure at home. What is particularly disturbing is that this patient met these criteria, was discharged home, yet had rebound symptoms and elevated COHb levels the next day. This case is suggestive of the storage of DCM in body tissues with CO release over a period of days.

Reference Number: 41604

FAGIN, J., AND J. BRADLEY. "Carbon Monoxide Poisoning Secondary to Inhaling Methylene Chloride." *British Medical Journal*, 281 (6253): 1461, 1980 (5 references)

Abstract: A 20-year-old female art student developed nausea and a severe throbbing headache while using a commercial paint remover (methylene chloride) in a poorly ventilated area. She was admitted to hospital with a cherry red skin colour and had a blood COHb of 50%. She was administered 60% oxygen (4L per minute), and, after 12 hours, her blood COHb was 20%. She was discharged from hospital three days later.

Carbon monoxide poisoning is a serious hazard to patients with coronary artery disease, since only marginal increments in carboxyhemoglobin concentration necessitates greatly increased coronary blood flow to prevent ischemia. Though our patients responded to 60% oxygen, higher concentrations (ideally 100%) are mandatory, especially in patients with coronary artery disease. Methylene chloride is present in several paint removers and any patient with coronary artery disease likely to be exposed to this agent should be warned.

Reference Number: 41605

BONFIGLIOLI, R., L. CARNEVALI, M. DI LELLO, AND F.S. VIOLANTE. "Bilateral Hearing Loss after Dichloromethane Poisoning: A Case Report." *American Journal of Industrial Medicine*, 57: 254–257, 2014 (3 figures, 19 references)

Abstract: DCM is also known as methylene chloride and is an industrial solvent widely used as a degreaser, aerosol propellant, and lacquer/paint remover. Approximately, 90% of the DCM absorbed into the body is eliminated in the breath, and the rest is converted by the liver to CO and CO_2. A 37-year-old female worker spilled a large tank of DCM at her workplace, which she cleaned up by using blotting paper, which she then threw into the garbage bin near her workplace. She lost consciousness, and then, when she woke up, she experienced nausea, vomiting, vertigo, and headache. At the ED, she was administered HBOT as her blood COHb was 6.9%. The next day, her COHb was still elevated at 5.1%. By day 4, her COHb was 1.8%. She experienced bilateral hearing loss, which regressed by day 25 of her HBOT.

> Here we report a case of sudden hearing loss after acute DCM exposure, however the patient had complete recovery of hearing function, which tends to exclude auditory damage due to other exposure. It is hard to establish whether the reported hearing loss is the consequence of DCM direct toxic effect, or of its metabolite CO. The latter can induce sensorineural hearing loss after both acute and chronic poisoning, although the exact incidence of this is not known.

4.17 PORTABLE ELECTRIC GENERATORS/GAS-POWERED ENGINES

> Carbon monoxide poisoning does not occur only in poorly ventilated areas and enclosed spaces. As this case illustrates, toxicity may occur even in the outdoors. Exposure to internal-combustion engines and compatible symptoms should suggest potential carbon monoxide toxicity, even if the patient has not been in an enclosed space.
>
> —J.S. Huff and E. Kardon, "Carbon Monoxide Toxicity in a Man Working Outdoors with a Gasoline-Powered Hydraulic Machine" (1989) **(41705)**

Gas-powered generators are a potent source of CO, even when the generators are placed in the basement, garage, or back porch (41701–2). Gas-power washers are the next most common source of CO poisoning and are used in indoor swimming pools, indoor parking lots, and barns (41703–4). Gas-powered floor burnishers have caused CO poisoning (41708). CO poisoning even occurred in a person who was using a gas-powered wood splitter outdoors (41705).

Reference Number: 41605

Reference Number: 41701

MARC, B., A. BOUCHEZ-BUVRY, J.-L. WEPIERRE, L. BONIL, P. VAQUERO, AND M. GARNIER. "Carbon Monoxide Poisoning in Young Drug Addicts Due to Indoor Use of a Gasoline Powered Generator." *Journal of Clinical Medicine*, 8: 54–56, 2001 (1 table, 15 references)

Abstract: The bodies of five male and one female drug addicts (ages 19–24 years) were found dead on their mattresses in a rented house, due to CO poisoning. The source of the CO was a gasoline-powered generator recently moved to the basement from outdoors as the neighbours complained about the noise. The generator was used since the electricity to the house was cut off due to unpaid bills. The post-mortem blood COHb concentrations ranged from 68.7% to 82.7%. Urine was screened positive for cocaine, cannabis, and amphetamines.

> The very high carboxyhemoglobin levels observed in the six victims explains the severity of symptoms and the death since the gasoline powered generator operated inside of a non-ventilated house and since the exposure to a CO environment was several hours long. A study has established that a gasoline powered, 5 hp, 4-stroke engine could generate CO concentrations exceeding the 200 ppm ceiling and 1200 ppm immediate danger to life and health concentration when used inside buildings. The study determined that a level of ventilation able to produce 120 air changes per hour would be required to prevent the CO concentration from exceeding the 200 ppm ceiling for short periods. Obviously, the danger awareness of CO was certainly low or non-existent in this group at greater risk.

Reference Number: 41702

HAMPSON, N.B., AND J.L. ZMAEFF. "Carbon Monoxide Poisoning from Portable Electric Generators." *American Journal of Preventative Medicine*, 28: 123–125, 2005 (12 references)

Abstract: Between 1978 and 2004, 34 male and 29 female patients were treated for CO poisoning using HBOT due to a portable gas electric generator in 37 separate incidents. The ages of the victims ranged from 2 to 85 years (mean 39 years). Loss of consciousness was reported in 34 cases (54%). Blood COHb concentrations ranged from 6.6% to 49.7% (mean 23%). Sixty patients spoke English, and three spoke only Spanish. Poisonings occurred mainly in the autumn and winter, and 81% between

October and March. Sixteen incidents occurred when the generator was used as a result of a power blackout.

Table: Locations of Gas-Powered Electric Generators in the Home Resulting in CO Poisoning

Location	Number of cases
Garage	15
Basement	4
Living room	2
Back porch	1

Source: Adapted from Hampson and Zmaeff 2005.

To develop effective prevention schemes such as these requires knowledge of the reasons that generators are used in poorly ventilated spaces, information that has not been previously reported. Based on the current study, it would appear that people poisoned commonly lack awareness of CO risk. Among those who are aware, there is a lack of understanding of ventilation requirements. Fear of theft, inadequate extension cord length and concern for the generator getting wet were reported only infrequently.

Reference Number: 41703

NESSEL-STEPHENS, L., S. WELCH, J.L. WEEKS, K. HUNTING, AND J. CARDENAS-AMAYA. "Carbon Monoxide Poisoning from Use of Gasoline-Fueled Power Washers in a Underground Parking Garage, District of Columbia, 1994." *Morbidity and Mortality Weekly Report*, 44: 356–357, 1995 (7 references)

Abstract: At 11:00 p.m. on 11 June 1994, four male labourers and a foreman (ages 22–39 years) began preparing the floor of an underground garage for resurfacing. The underground garage door was closed, and the ventilation fans were not working. At 12:30 p.m., two eight-hp gasoline-power washers were started. At 3:30 a.m., a worker collapsed, and, shortly later, a second worker collapsed. All five men had symptoms of dizziness, confusion, headache, and nervousness. The two men who collapsed had COHb of 20% and 17% and were treated with HBOT later that day. The other three men had COHb of 10.3%, 13.4%, and 7.9% and were treated with 100% oxygen and later released. A CO concentration of 648 ppm was detected in the parking garage, one hour after the power washers were turned off.

The risk of CO exposure to workers can be reduced through improved ventilation. In addition, however, risk-reduction efforts must include air monitoring for CO levels. Reliable air monitoring includes the requirement for persons who have been trained to perform monitoring and for equipment that has been properly calibrated and maintained. Training and warning labels can increase awareness among contractors and workers about the risk associated with use of gasoline-fueled equipment in enclosed spaces.

Reference Number: 41704

HUSTON, B., V. FROLOFF, K. MILLS, AND M. MCGEE. "Carbon Monoxide Poisoning and Death in a Large Enclosed Ventilated Area." *Journal of Forensic Sciences*, 58: 1651–1653, 2013 (2 figures, 14 references)

Abstract: A 55-year-old male smoker power-washed an indoor swimming pool and cement deck with a Honda GC-190 3000 psi gas-powered washer. The dimension of the pool room was 40 feet by 100 feet with 14-foot-high ceilings. Although the ventilation system was not in operation, two Ridgid industrial fans were operating pushing air across the pool to the outdoors. He power-washed for two hours, then had lunch, and showed no signs of CO intoxication. He returned and continued to power-wash a corner of the room, away from the flow of the fans, where he collapsed. Resuscitation was conducted but was stopped after 18 minutes. A post-mortem examination showed a cherry pink livor mortis and 80–95% narrowing of the arteries. Post-mortem blood COHb was 27%, and the hospital admission sample was 49%. The fire department detected a 250-ppm ambient CO concentration. The gas power washer was not defective.

In this case, the movement of air from the hallway door to the outside door resulted in a slightly diagonal direction of air movement that excluded the corner when the hot tub was located and where the decedent was working prior to his collapse. This would have resulted in a buildup of CO in this area of the pool room despite the attempt to ventilate the room with industrial air movers resulting in CO toxicity. The decedent's atherosclerotic heart disease and tobacco use would have made him more susceptible to the myocardial toxic effects of CO and were considered contributing factors in his death. In this case, warning labels were present on the engine clearly stating the dangers of CO. However, the decedent and his co-worker felt a false sense of security due to the large size of the pool room and the presence of industrial blowers for ventilation.

Reference Number: 41705

HUFF, J.S., AND E. KARDON. "Carbon Monoxide Toxicity in a Man Working Outdoors with a Gasoline-Powered Hydraulic Machine." *New England Journal of Medicine*, 8 June 1989, 1564 (7 references)

Abstract: A non-smoking teenager was splitting wood with a hydraulic wood splitter outdoors. He gradually became fatigued and nauseous and had a headache. He was seen by a physician and had a blood COHb concentration of 13.7% one hour after he stopped working.

Carbon monoxide poisoning does not occur only in poorly ventilated areas and enclosed spaces. As this case illustrates, toxicity may occur even in the outdoors. Exposure to internal-combustion engines and compatible symptoms should suggest potential carbon monoxide toxicity, even if the patient has not been in an enclosed space.

Reference Number: 41706

KAHLER, M., W. KUHSE, AND L.A. WINTERMEYER, "Unintentional Carbon Monoxide Poisoning from Indoor Use of Pressure Washers, Iowa, January 1992–January 1993." *Morbidity and Mortality Weekly Report*, 42: 777–779, 1993 (6 references)

Abstract: Six farmers in Iowa were exposed to CO fumes while operating gasoline-powered pressure washers indoors, while cleaning out barns between 1992 and 1993. One farmer died after having replaced his electric pressure washer with a gasoline-powered one. His post-mortem blood COHb was 75.6%. The other survivors had a blood COHb at the time of medical treatment that ranged between 9.2% and 50%. It is estimated that a 3–11 hp pressure washer operating in a 4,700 cubic foot unventilated space will produce dangerous ambient CO concentrations within minutes.

In addition to gasoline-powered pressure washers, CO poisonings among persons on farms have been associated with unvented or inadequately vented space heaters and indoor tractor maintenance, underscoring that gasoline engines, irrespective of size, should not be operated indoors. Although warning labels and operator's manuals often advise against operating gasoline-powered equipment without adequate ventilation, adequate ventilation cannot be safely determined. Therefore, these labels and manuals should clearly indicate the CO hazard associated with indoor operation and prohibit any indoor use.

Reference Number: 41707

ERDOGAN, M.S., S.S. ISLAM, A. CHAUDHARI, AND A.M. DUCATMAN. "Occupational Carbon Monoxide Poisoning among West Virginia Workers' Compensation Claims: Diagnosis, Treatment Duration, and Utilization." *Journal of Occupational and Environmental Medicine*, 46: 577–583, 2004 (5 tables, 1 figure, 36 references)

CO is involved in approximately 35% of work-related fatal inhalations in the United States. This study was conducted of 182 workplace-related CO poisonings that occurred in West Virginia between 1992 and 2000. The most frequent sources of CO were gas-powered engines (28%) and furnaces (20%). Treatment duration ranged from one day to eight years. The average cost for CO claims was $2,130.00, and the median loss time was 45 days.

Table: Mean Incident Rate per 100,000 Workers of Workplace-Related CO Poisonings in West Virginia between 1995 and 2000

Year	Mean incidence rate for female workers	Mean incidence rate for male workers
1995	4.7	5.4
1996	5.8	7.3
1997	2.4	2.7
1998	8.8	3.5
1999	2.6	1.6
2000	3.6	3.4

Source: Adapted from Erdogan et al. 2004.

Affective, cognitive, or neurologic and neuropsychiatric sequelae such as apathy, memory disturbances, depression anxiety, focal neurologic abnormalities, Parkinson-like syndromes, ataxia, fecal or urinary incontinence, dementia and psychosis could follow CO intoxication. In our series, 12 workers sustained therapy for more than 1 year. The distinguishing characteristics of those treated more than 1 year is their affective and neuropsychiatric conditions, which can either be sequelae of CO poisoning or pre-existing conditions. Insidious or development of delayed neurologic symptoms could explain why some patients in our study were late in seeking treatment, although we cannot rule out incorrect diagnosis. The impact on costs of chronic neurologic problems is substantial as evidenced by prolonged treatment duration.

Reference Number: 41708

URANECK, K., R. MCCANDLESS, S. MEYER, R. HOUSEKNECHT, AND L. PAULOZZI. "Carbon Monoxide Poisoning Associated with a Propane-Powered Floor Burnisher, Vermont 1992." *Morbidity and Mortality Weekly Report*, 42: 726–728, 1993 (7 references)

Abstract: A case of CO poisoning is defined as a COHb of 2% or more in non-smokers and 9% or more in smokers. On 24 and 28 July 1992, a pharmacy store's floor was cleaned with a propane-powered floor burnisher. Nine of the ten pharmacy employees had an elevated COHb, which ranged from 6.7% to 25.3% (mean 16.6%) within 1.5 to 3 hours after exposure. Three employees were administered HBOT. The burnisher's exhaust pipe emitted 2,000 ppm CO after one minute, 3,000 ppm while idling, and 50,000 ppm at full throttle.

Table: Symptoms Experienced by Nine Employees Exposed to CO Fumes from a Gas-Powered Floor Burnisher

Symptom	Number of employees
Headache	9
Light-headedness	7
Tunnel vision	5
Nausea/vomiting	4
Syncope	2
Difficulty breathing	2
Chest pain	2
Decreased hearing	1

Source: Adapted from Uraneck et al. 1993.

As a result of this investigation, the pharmacy and cleaning contractor and subcontractor were fined. VDH [Vermont Health Department] recommended that liquid propane-powered burnishers be replaced with electric-powered burnishers and that CO alarms be installed if use of liquid propane-powered machines continue.

4.18 STORM/DISASTER-ASSOCIATED CO POISONING

Although the fumes of charcoal have, probably from the earliest times when this form of carbon was used as fuel, been acknowledged to possess specific and very remarkable toxic properties, it has only very recently been shewn that carbonic acid plays an important part in their production.

—A. Gamgee, "On Poisoning by Carbonic Acid Gas, and by Charcoal Fumes" (1867) **(40401)**

CO-poisoning epidemics are common yet preventable causes of morbidity and mortality after disasters. Use of emission-control devices on generators might decrease severe CO poisonings. CO alarms use should be encouraged and educational messages should focus on safe generator use and avoidance of unsafe heating and cooking practices.

—E. Lutterloh et al., "Carbon Monoxide Poisoning After an Ice Storm in Kentucky, 2009" (2011) **(41814)**

Sometimes more people die or are hospitalized as a result of the clean-up and lack of power after a natural disaster than as a result of the natural disaster due to CO poisoning. Exposure to CO has occurred after ice/snowstorms (41801, 41806, 41808–9, 41811, 41814–15), floods (41804, 41812), earthquakes (41803, 41810), hurricanes (41802), and power outages (41807). These poisonings have been due to gas-powered generators/tools, propane/kerosene heaters, and indoor use of charcoal grills, among others (41801). Extreme weather events are expected to increase with climate change, and more post-disaster CO poisonings are to be expected (41807).

Reference Number: 41801

DALEY, W.R., A. SMITH, E. PAZ-ARGANDONA, J. MALILAY, AND M. MCGEEHIN. "An Outbreak of Carbon Monoxide Poisoning after a Major Ice Storm in Maine." *Journal of Emergency Medicine*, 18: 87–93, 2000 (5 tables, 1 figure, 25 references)

Abstract: In early January 1998, a severe ice storm struck the northeastern United States and southeastern Canada, disrupting electrical power to over 600,00 people. Forty-nine female and 41 male patients (mean age 38 years), including 22 children attended one of four hospitals in Maine due to CO poisoning within two weeks after the storm. The mean COHb was 14.2% (range 2.1–49.8%). Eleven were admitted to hospital, five were provided HBOT, and one died. Headache, nausea, dizziness, and vomiting were the most common symptoms. Nine had chest pains. Loss of consciousness occurred in eight patients (age 21–95 years) and was the strongest indicator of CO-poisoning severity (mean COHb was 25.8%). Victims who were smokers had a higher mean blood COHb than non-smokers. A comparison group of 522 households was selected by random telephone dialing to determine risk factors for CO poisoning. The most common source of CO was a gas-powered generator, a kerosene heater, a propane heater, and a charcoal grill. The strongest risk factors for CO poisoning included locating a gas generator in the basement (OR = 312.8 times) or in a garage or attached structure (19 times) or a lack of awareness of warnings in the owner's manual (1.8 times).

Table: Source of CO Poisonings Due to the 1998 Ice Storm in Maine, Number of Cases, and Mean COHb in Non-Smokers and Smokers

Source	Number of cases	Mean COHb (non-smoker) (%)	Mean COHb (smoker) (%)
Generator	74	14.9	15.0
Kerosene heater	5	2.9	11.5
Generator (propane heater)	9	12.7	18.0
Charcoal grill	3	23.6	26.3
Portable gas stove	2	-	21.2
Generator (kerosene heater)	2	6.4	16.3
Propane heater	1	3.4	-
Cement cutter	1	-	10.7
Kerosene heater-portable stove	1	3.2	-
Kerosene heater-propane heater	1	2.1%	-
Kerosene heater-fireplace	1	-	12.1%
TOTAL	100	13.8%	16.0%

Source: Adapted from Daley et al. 2000.

Environmental CO measurements were available for six incidents. The highest recorded measurement was 1,060 ppm in an incident involving both a generator and a kerosene heater, taken in the garage where the generator was located. Three CO measurements were obtained from incidents involving generators alone: 700 ppm from a workshop above a garage where the generator was operating, 550 ppm from a home with the generator in the basement, and 420 ppm from a home with the generator in the garage. The remaining environmental CO measurements were 15 ppm from an incident involving a kerosene heater alone, and 6 ppm from one involving a kerosene heater and a fireplace.

Reference Number: 41802

HAMPSON, N.B., M.W. LAI, M. MCNEIL, P. BYERS, R. RATARD, ET AL. "Carbon Monoxide Poisoning after Hurricane Katrina, Alabama, Louisiana, and Mississippi, August–September 2005." *Morbidity and Mortality Weekly Report*, 54: 996–998, 2005 (1 table, 1 figure, 8 references)

Abstract: After Hurricane Katrina made landfall on 29 August 2005, causing massive damage, loss of life and power outages, some residents used gasoline-powered generators and other devices for electrical power and clean-up. Between 29 August and 24 September 2005, a total of 51 cases of CO poisoning were reported by HBO_2 facilities in Alabama, Louisiana, and Mississippi. Five of the 51 cases were fatal, and all involved the use of a gasoline-powered generator in the house or garage. The improper placement of a portable generator was responsible for 98% of the CO poisonings.

CO poisoning is easily preventable by reducing exposure to CO. CO detectors can alert users when dangerous levels of CO are present. Educating the public on the proper placement and operation of generators and pressure washers will reduce exposure to CO. In anticipation of CO poisonings after the hurricane, CDC [Centers for Disease Control and Prevention] developed CO-poisoning prevention information sheets and collaborated with the New Orleans major's office to distribute them. In addition, CDC collaborated with local organizations to work with retailers and rental companies of generators and gasoline powered tools to alert buyers and renters of the risk for CO poisoning with improper use of this equipment.

Reference Number: 41803

ISEKI, K., A. HAYASHIDA, Y. SHIKAMA, K. GOTO, AND C. TASE. "An Outbreak of Carbon Monoxide Poisoning in Yamagata Prefecture Following the Great East Japan Earthquake." *Asia Pacific Journal of Medical Toxicology* 2: 37–41, 2013 (3 figures, 3 tables, 19 references)

Abstract: On 11 March 2011, a 9.0 magnitude earthquake struck northeastern Japan knocking out electrical power for 24 hours. The temperature was below 0 degrees Celsius with snowfall that night. Some people used alternative fuel sources to provide power and warmth. A total of 18 male and 33 female patients (mean age 49.8 years) attended hospital due to disaster-related CO poisoning. The COHb levels ranged from 0.5% to 41.6% (mean 22.0%). Most patients were treated with normobaric oxygen (NBO_2), but five were treated with HBO_2. Three of the patients experienced delayed neuropsychiatric syndrome (two had mild dementia, and one had severe cognitive impairment). The sources of CO were charcoal briquettes (44%), gas electric generators (35%), electric generators and oil stoves (15%), oil stoves (4%), and motor vehicles (2%).

> Gasoline powered electric generators are also a major source of storm-related CO poisoning. The indoor use of generators can lead to CO poisoning in many storms but even the outdoor use of generators can result in CO poisoning when placed less than 7 feet from home. Our study showed that most of the patients aged 19–64 years were exposed to CO through the use of generators occurred for the most part in Yamagata prefecture. Yamagata prefecture is famous for rice and various kinds of fruits, and the farmers use generators in plastic greenhouses. In this disaster they suffered from CO poisoning while trying to obtain electric power.

Reference Number: 41804

DALEY, W.R., L. SHIRELEY, AND R. GILMORE. "A Flood-Related Outbreak of Carbon Monoxide Poisoning, Grand Forks, North Dakota." *Journal of Emergency Medicine*, 21: 249–253, 2001 (1 table, 2 figures, 18 references)

Abstract: A record flood occurred in the city of Grand Forks, North Dakota, on 18 April 1997. The dike failed, and 75% of Grand Forks was underwater, and 50,000 residents were evacuated. Between 2 May and 23 June 1997, 27 men and 6 women (ages 7–67 years) attended the local ED due to CO poisoning. Dizziness was the most reported symptom. One patient lost consciousness, had a blood COHb of 29.4%, and was the only patient hospitalized. The initial mean blood COHb was 20.4% (range 10.3–32.4%). Every incident involved gas-powered pressure washers used in basements.

Table: Symptoms in Flood-Related CO Poisoning Cases from the Flood in Grand Forks, North Dakota

Symptom	Number of cases
Dizziness	25
Headache	21
Nausea	16
Fatigue	7
Tachycardia	5
Vomiting	4
Altered Mental status	3

Source: Adapted from Daley, Shireley, and Gilmore 2001.

The need for gasoline-powered equipment to provide a temporary supply of electricity or rehabilitate water-damaged homes increases after flooding. This equipment includes generators, pumps, and pressure washers. Pressure washers have caused CO poisoning previously, when used in an enclosed space such as a barn or parking garage. Environmental models and field simulations have demonstrated that CO will reach dangerous levels within minutes when small gasoline-powered engines run inside buildings. Hazardous accumulation can result even if open doors and exhaust fans provide ventilation. Nearly every patient in this outbreak reported that the basement was ventilated when the pressure washer was in operation. Still, these devices caused CO poisoning after running a few hours or less.

Reference Number: 41805

HEIDARI, M., S.S. MIRESMAEELI, AND N. ESKANDARY. "Carbon Monoxide Poisoning Post Disaster: Incident after the Incident." *Disaster Medicine and Public Health Preparedness*, 14: 429–430, 2020 (12 references)

Abstract: The role of portable power generators and heating appliances in CO poisoning after a disaster such as an earthquake, hurricane, and flood is discussed.

Given that CO gas is invisible, odorless, and tasteless which is very hard to detect and is known as the silent killer, the symptoms are nonspecific, especially in the early stages, which are similar to common cold and make diagnosis difficult. The lack of awareness regarding the dangers of incomplete combustion of heating appliances, as well as the symptoms of CO poisoning and the severe mental stresses of survivors after the incident, causes people to use heating appliances and generators irrespective of the safety issues. After disaster, survivors may pay less attention to the safety issues and self-protection items and focus on other issues, due to the imposed stress and tension. Survivors of the incident may be exposed to another incident due to carelessness. Unlike the first incident, this incident may occur slowly and without noise, has no obvious sign, and people will not even have the chance to escape.

Reference Number: 41806

WRENN, K., AND G.P. CONNERS. "Carbon Monoxide Poisoning during Ice Storms: A Tale of Two Cities." *Journal of Emergency Medicine*, 15: 465–467, 1997 (1 table, 7 references)

Abstract: An analysis was conducted of 55 CO-poisoning patients seen in Rochester, New York, over a 10-day period in March 1991 and 13 CO-poisoned patients over four days in February 1994 seen in a Nashville ER due to ice storms, which had caused prolonged power outages. The mean source of CO poisoning was a gas generator (67%) in Rochester and a charcoal grill (46%) in Nashville. The mean COHb was 16% in Rochester and 26.5% in Nashville. HBOT was used in 2% of the Rochester patients and 38% of the Nashville patients.

Accidental CO exposures are potentially preventable. If ice storms are expected to occur, early alerting of the public to common indoor misuse of gas generators, charcoal or gas grills, and kerosene or propane heaters and stoves should be done by responsible public officials. It is the duty of emergency physicians and public health officials to plan ahead on how to inform the media and disseminate this information.

Reference Number: 41807

WORSHAM, C.M., J. WOO, M.J. KEARNEY, C.F. BRAY, AND A.B. JEA. "Carbon Monoxide Poisoning during Major US Power Outages." *New England Journal of Medicine*, 386: 191–192, 2022 (1 figure, 5 references)

Abstract: The ER visits due to CO poisoning were determined before and after power outages using the MarketScan database between 2007 and 2018. There were 581 major power outages with a mean duration of 47.9 hours during that time period. The rates of CO-related ER visits increased significantly during power outages that lasted more than one day and peaked on the third day of the outage. Among persons aged 18 years of age or younger, there was a 13.5-times increase and, overall, a 9.5-times increase in CO-related ER visits during outages. Similar patterns were not observed after extreme weather unless accompanied by a power outage.

> Many of the outages included in our study resulted from severe weather events, which are expected to increase in coming years. Our findings help to quantify the magnitude of the risk of this underrecognized harm associated with power outages, such as investment in grid infrastructure.

Reference Number: 41808

HOUCK, P.M., AND N.B. HAMPSON. "Epidemic Carbon Monoxide Poisoning Following a Winter Storm." *Journal of Emergency Medicine*, 15: 469–473, 1997 (1 tables, 2 figures, 13 references)

Abstract: A severe winter storm hit western Washington State during the morning of 20 January 1993. Cases in which the blood COHb concentration was at least 2% for non-smokers and 9% for smokers and occurred on 20–3 January 1993 were deemed to be storm-related CO poisonings, resulting in 81 patients (mean age 21 years) being treated at hospital. The median COHb upon presentation was 17% (range 2–46%). Eleven patients (14%) lost consciousness and were treated with HBOT along with 24 other patients. The other 46 patients received oxygen via a mask. The main sources of CO poisoning were charcoal briquettes (44 victims), portable generators (32 victims), and space heater (6 victims).

> This is the largest epidemic of storm-related CO poisoning to be documented in the United States. Previous reports have described up to 22 cases in each of several large storms. The 81 cases represent the minimum number of storm-related poisonings and may be a substantial underestimate of the true CO morbidity. We identified 6 symptomatic persons who were treated for CO poisonings and were epidemiologically linked to case patients, but who did not meet our case definition because COHb levels were not measured. We have missed additional cases if patients were treated in clinics; if physicians did not recognize nonspecific symptoms, such as headache and nausea, as being caused by CO; if COHb levels had dropped to the normal range prior to testing; or if coroners and medical

examiners neither sought nor found evidence of CO poisoning during death investigations. This even was notable because of the prominence of charcoal briquettes as the source of CO, especially for persons of Asian ancestry.

Reference Number: 41809

HARTLING, L., W. PICKETT, AND R.J. BRISON. "The Injury Experience Observed in Two Emergency Departments in Kingston, Ontario, during the Ice Storm 98." *Canadian Journal of Public Health*, 90: 95–98, 1999 (3 tables, 1 figure, 13 references)

Abstract: The ice storm hit the Kingston, Ontario, area on 7 January 1998. A state of emergency was declared on 9 January as almost 80% of Kingston was without power. The state of emergency was lifted on 19 January 1998. A total of 669 injuries were presented to two Kingston ERs during that time period, of which 254 (38%) were directly attributable to the storm. Two hospital deaths occurred, one due to CO poisoning and the other to burns. The most common injuries were superficial abrasions/lacerations (31%), fractures/dislocations (19%), and open wounds (15%). CO poisoning occurred in 9% of the cases.

Table: Agent of Injury for Hospitalization for Ice Storm in Kingston, Ontario, 1998

Agent of injury	Number of cases
Ice	146
Clean-up of trees	39
CO	22
Motor vehicle gas	11
Darkness due to power outage	10
Fire	8
Other/Unknown	18

Source: Adapted from Hartling, Pickett, and Brison 1999.

During the 13-day period beginning the day of the ice storm, there were 254 storm-related injuries that presented to the two emergency departments in Kingston, Ontario. The patterns of injury observed were clearly related to the unusual environmental conditions posed by the storm. This analysis provides useful information for public officials to review disaster plans and generate recommendations for future occurrences.

Reference Number: 41810

NAKAJIMA, M., S. ASO, H. MATSUI, K. FUSHIMI, Y. YAMAGUCHI, AND H. YASUNAGA. "Disaster-Related Carbon Monoxide Poisoning after the Great East Japan Earthquake, 2011: A Nationwide Observational Study." *Acute Medicine and Surgery*, 6: 294–300, 2019 (1 table, 4 figures, 20 references)

Abstract: A study was conducted of adult hospital admissions from CO poisoning in Japan between 2010 and 2017, which were divided into 988 patients in the earthquake disaster area (Tohoku prefecture) and 6,826 patients in non-disaster areas. The earthquake occurred on 11 March 2011. CO poisoning was 7.50 times greater in the disaster area post-earthquake than before.

Table: Characteristics of CO-Poisoning Patients Related to Great East Japan Earthquake and CO-Poisoning Patients Not Related to This Disaster

Characteristic	Earthquake related	Non-earthquake related
Median age	56 years	50 years
Male (%)	55	62.3
Inhalational injury (%)	0.7	8.4
ICU admission (%)	4.4	21.7
In-hospital mortality	0.7	7.0

Source: Adapted from Nakajima et al. 2019.

This study indicates that the number of CO poisoning cases significantly increased after the Great East Japan Earthquake in the affected area. For this reason, we must continue to provide information to the population regarding the hazard of earthquake related CO poisoning.

Reference Number: 41811

RAO, R., W. TOUGER, P. GENNIS, J. TYRELL, J. ROCHE, AND E.J. GALLAGHER. "Epidemic of Accidental Carbon Monoxide Poisonings Caused by Snow-Obstructed Exhaust Systems." *Annals of Emergency Medicine*, 29 (2): 290–292, 1997 (1 table, 10 references)

Abstract: On 8 January 1996, the New York City area experienced a blizzard that resulted in an accumulation of 24 inches of snow. During the next five days, 25 patients received HBOT from CO poisoning.

Eighteen patients were treated within 24 hours after the blizzard ended. All patients were exposed to CO in stationary automobiles with the engine running with the exhaust system blocked with snow. CO poisoning occurred even with cars with the windows or doors slightly opened. The mean blood COHb (interquartile range) was 34.3% (24.5–39.5%).

Public education about the dangers of CO intoxication is warranted during heavy snow accumulation. Practical recommendations should emphasize visual inspection of the vehicle exhaust system before the engine is started, the importance of an unobstructed tailpipe for vehicles caught in snowdrifts, recognition of unusual odors or fumes in the car as an indication of possible exposure, warning that an open window or door may not be sufficient to prevent CO toxicity, and alerts to parents to protect children from unintentional CO exposure.

Reference Number: 41812

DUCLOS, P., O. VIDOONE, P. BEUF, P. PERRAY, AND A. STOEBNER. "Flash Flood Disaster, Nimes, France, 1988." *European Journal of Epidemiology*, 7: 365–371, 1991 (6 tables, 2 figures, 4 references)

Abstract: On Monday, 3 October 1988, 263 millimetres (mm) of rain fell within nine hours, which resulted in a flash flood. Although 45,000 houses were damaged and 1,100 vehicles destroyed, there were only nine deaths reported. There were 12 cases of CO poisoning among firefighters, civilians, and members of the military who were involved with pumping water from flooded basements.

The limited death toll might be attributed to (1) the disaster occurred early in the morning when people were still home, (2) traffic heading into the city was stopped by flood water on some access roads, (3) the rescue operations were set up within the framework of the French disaster plan, and (4) the civilian response. No outbreaks of infectious disease were determined after the disaster.

Reference Number: 41813

HAWKES, A.P., J.B. MCCAMMON, AND R.E. HOFFMAN. "Indoor Use of Concrete Saws and Other Gas-Powered Equipment." *Journal of Occupational and Environmental Medicine*, 40: 49–54, 1998 (1 table, 3 figures, 15 references)

Abstract: A study was conducted of 349 workplace-related unintentional CO poisonings in Colorado between 1985 and 1994. Twenty-four workers died, and, of those, 14 were firefighters who died in a forest fire. Gasoline-powered engines accounted for the largest number of CO poisonings (40%), followed by the automobile (26%) and furnaces (12%). In contrast to non-occupational poisonings, the occupational CO poisonings are fairly well distributed throughout the year. Of the gas-powered engines, concrete saws were the most frequent source of poisonings. Concrete saw operators had a higher mean COHb concentration (26.8%) compared to other gas-powered engines (19.5%).

Table: Gas-Powered Equipment Most Frequently Involved in Unintentional Work-Related CO Poisonings

Type of equipment	Percent involved (%)
Concrete saws	20.9
Power trowel	10.1
Spray washers	8.6
Compressor	7.2
Propane buffer	6.5
Welder/Gas torch	5.8
Propane forklift	5.8
Forklift	5.0

Source: Adapted from Hawkes, McCammon, and Hoffman 1998.

The reason for the higher exposure to CO in users of concrete saws is unclear. It is possible that concrete saws were used in smaller spaces than other equipment, thus increasing the concentration of CO in the area. The output of CO in the exhaust of concrete saws, or the load on the saw engine, may also have played a role in the higher exposure. The finding that approximately 30% of the cases did not immediately remove themselves from the environment or seek treatment once they developed symptoms of CO poisoning underscores the need to educate workers on the symptoms of CO poisoning. CO poisoning not only has acute effects (the most severe of which are coma and death) but can also result in neurologic sequelae. In addition, with continued exposure to CO, a worker may become confused, which will limit his/her ability to make judgments about his/her environment or to seek help. One of the workers in this study, a drywall texturizer, was unable to identify an escape route from the building because of confusion.

Reference Number: 41813

Reference Number: 41814

LUTTERLOH, E., S. IQBAL, J.H. CLOWER, H.A. SPILLER, M.A. RIGGS, T.J. SUGG, K.E. HUMBAUGH, B.L. CADWELL, AND D.A. THOROUGHMAN. "Carbon Monoxide Poisoning after an Ice Storm in Kentucky, 2009." *Public Health Reports*, 126 (S1): 108–115, 2011 (2 tables, 3 figures, 22 references)

Abstract: A severe ice storm occurred in Kentucky on 26–27 January 2009, causing an extended power outage. During the two weeks after the ice storm, there were 144 cases of CO poisoning reported, 202 ED visits (26 admissions), 10 deaths (8 due to improperly placed generators), and 28 people received HBOT. The main source of the CO poisonings were kerosene heaters, generators, propane heaters, charcoal, gas fireplaces, and gas ovens.

CO-poisoning epidemics are common yet preventable causes of morbidity and mortality after disasters. Use of emission-control devices on generators might decrease severe CO poisonings. CO alarms use should be encouraged and educational messages should focus on safe generator use and avoidance of unsafe heating and cooking practices. Vendors should be encouraged to display CO alarms and other safety equipment alongside generators. Prevention messaging should begin before storms and should include components directed toward minority populations.

Reference Number: 41815

STYLES, T., P. PRZYSIECKI, G. ARCHAMBAULT, L. SOSA, B. TOAL, J. MAGRI, AND M. CARTTER. "Two Storm-Related Carbon Monoxide Poisoning Outbreaks, Connecticut, October 2011, and October 2012." *Archives of Environmental and Occupational Health*, 70: 291–296, 2015 (1 table, 3 figures, 26 references)

Abstract: Two storm-related CO mass poisonings occurred in Connecticut in 2011 and 2012 involving 133 cases and 30 cases respectively. This included three deaths in 2011. In Connecticut, a blood COHb of 9% or greater is a physician and laboratory reportable to the Department of Public Health. Most of the victims (60%) were male. The most common CO sources were power generators and charcoal burners. The most common CO symptoms were headache, dizziness, nausea, and fatigue.

Table: Common Sources of CO after Winter Ice Storms, Hurricanes, and Floods

Winter ice storms	Hurricanes	Floods
Gasoline-power generators Charcoal heaters	Gasoline-power generators	Gasoline-power washers Other cleaning equipment

Source: Adapted from Styles et al 2015.

In 2011, after an early snowstorm and massive power outages, Connecticut experienced one of the largest recorded post-disaster CO-poisoning outbreaks in the United States, a similar, although smaller outbreak occurred in 2012 after Hurricane Sandy. In both outbreaks, persons other than non-Hispanic whites were disproportionately affected; generators and charcoal were the primary CO sources identified; unsafe use of CO-producing equipment and lack of functional household CO alarms were common; and effectiveness of mass media warnings and product warning labels were questionable. A multifaceted approach is needed to decrease postdisaster CO poisonings.

4.19 SCUBA DIVING

The next morning, we sampled the cylinders. The analysis showed 1/2000 of carbon monoxide. At a depth of one hundred and sixty feet, the effect of carbon monoxide is sixfold. The amount we were breathing may kill a man in twenty minutes. We started our new Diesel-powered free-piston air compressor. We saw the compressor sucking in its own exhaust fumes. We had all been breathing lethal doses of carbon monoxide.

—Jacques Cousteau describing a near fatal dive in the 1940s[6]

CO is a greater hazard underwater than on land. The main problem is that the gas/diesel-powered compressors can suck in CO from any nearby exhaust and pressurize the scuba tank with not only air but also toxic levels of CO (41901–6). The compressor may produce clean air when cool but CO-contaminated air when heated (41903). Activated charcoal does not absorb CO (41903).

6 Jacques Cousteau, with Federic Dumas, *The Silent World* (New York: Harper Publishing, 1953).

Reference Number: 41901

HAMPSON, N.B. "Carbon Monoxide Poisoning While Scuba Diving: A Rare Event?" *Undersea and Hyperbaric Medicine*, 47: 487–490, 2020 (9 references)

Abstract: A 42-year-old woman was diving in salt water with a buddy to a depth of 121 feet but, during ascent, developed a sudden change in mental status and dropped the regulator from her mouth. Her diving companion was unable to reinsert her regulator and inflated their emergency vests and rose rapidly to the surface. They reached the shore within five minutes, and cardiopulmonary resuscitation (CPR) was administered. She was intubated by paramedics and received 100% oxygen. The COHb measured at hospital two hours after the incident was 15% (back-estimated to 40–50%, as the half-life of CO is 58 minutes with 100% oxygen, intubated). She was also treated with HBO_2 but was brain dead. Her post-mortem COHb was 5%. The CO concentration in her scuba tank was 1,320 ppm. Other CO-poisoning events related to scuba diving are also reported, including a case in 1946 with Jacques Cousteau while breathing compressed air filled by a new diesel-powered compressor.

There are two ways in which CO may contaminate scuba tanks. First, the compressor air intake hose can be placed too close to the exhaust from a combustion-powered compressor or generator allowing entrainment of CO. In addition to distance, the intake should be upwind from the exhaust. This was felt to be the problem in one of the scuba cases and a surface supplied air case. The second way that scuba tanks may become contaminated with CO is production of CO within the compressor itself. Incomplete combustion of lubricating fluid or debris resulting from defective piston washers or rings was found to be the problem in three of the scuba cases. Note that this can occur even in electric compressors.

Reference Number: 41902

MCDERMOTT, J.H., C. REYNARD, J. PERRY, J.W. DEAR, F. CHILD, AND R. JENNER. "Acute Carbon Monoxide Toxicity in a Pediatric Cohort: Analysis of 10 Boys Poisoned during a Scuba Diving Lesson." *Clinical Toxicology*, 56: 856–859, 2018 (1 table, 1 figure, 13 references)

Abstract: Ten boys (ages 14–16 years) participated in a diving lesson in an indoor swimming pool. One boy became unconscious in the swimming pool. The boy was unresponsive for about 20 seconds before being given 100% oxygen in a facemask. A venous blood sample at hospital showed a COHb of 32.8%. Five other boys had symptoms of CO poisoning, and their initial blood COHb concentrations ranged from 6.9% to 20.5%. All were provided with oxygen therapy and recovered uneventfully. Multiple samples were collected, and the mean half-life of CO was 74.2 minutes. Five boys were asymptomatic, and the blood COHb was less than 1% in three of the boys and less than 12.5% in one of them. Of the 15 cylinders around the pool that were analyzed, eight had dangerously high CO concentrations of greater than 1,000 ppm. The high-pressure bank had CO levels of 2,550 ppm, and the compressor that supplied the high pressure bank had a level of 30 ppm.

> This work described an uncommon clinical presentation, representing the largest single cohort of its kind. This work exemplifies the variable symptomatology of CO toxicity, of which clinicians should be alert to if patients fall ill after scuba diving.

Reference Number: 41903

MILLAR, I.L., AND P.G. MOULDEY. "Compressed Breathing Air: The Potential for Evil from Within." *Diving and Hyperbaric Medicine*, 38: 145–151, 2008 (1 table, 19 references)

Abstract: Scuba diving requires an adequate supply of breathable air without contamination. In oil-lubricated, high-pressure compressors, over-heating is a common problem that may produce CO and/or hydrocarbons. The compressor may produce clean air while cold but CO-contaminated air when heated. Where scuba tanks are randomly tested, this contamination may escape detection. Activated charcoal filters do not absorb CO, and so a manganese/cooper oxide combination (hopcalite) is required, which oxidizes the CO into CO_2. A slight increase in CO_2 in the tank usually does not affect the diver.

> Even relatively low levels of CO inhalation therefore remain undesirable for older or unfit drivers as uptake will increase with depth and exertion, bringing with it an increased risk of cardiac ischemia and/or arrhythmia. Also critical to the case for avoiding low-level CO inhalation is the significant reduction in exercise tolerance that results.

Reference Number: 41904

ALLEN, H. "Carbon Monoxide Poisoning in a Diver." *Archives of Emergency Medicine*, 9: 65–66, 1992 (5 references)

Abstract: A 25-year-old commercial driver was working on a dock underwater using an air line for a surface compressor. The air inlet was situated upwind from the compressor inlet to prevent exhaust fumes from entering the air line. He dived to 10 metres for 30 minutes and, upon surfacing, was confused, disorientated, and weak. He was taken to an ER where he was alert and orientated but complained of a headache and nausea. He also had tachycardia (120 bpm). His COHb was 37%. He was administered oxygen by a face mask and received HBOT. He recovered uneventfully.

Subsequent examination of the air compressor revealed a worn washer on the piston ring, with carbon particles in the first air filter, indicating leakage, and partial combustion of the lubricating fluid. This would produce carbon monoxide contamination of the delivered air supply.

Reference Number: 41905

HOLT, J., AND L.K. WEAVER. "Carbon Monoxide Poisoning Mimicking Arterial Gas Embolism in a Commercial Diver." *Undersea and Hyperbaric Medicine*, 39: 687–690, 2012 (15 references)

Abstract: A 37-year-old male commercial diver breathed compressed air while spending two hours cleaning a municipal water tank at 7,000-foot altitude at a depth of 27 feet (equivalent to 40 feet at sea level). The diver felt weak all over his body, could barely hold on to the ladder, and had a headache. He was initially diagnosed with an arterial gas embolism and flown by pressurized aircraft to a regional medical centre and treated with HBOT. His admission blood COHb was 8.8% (six hours after surfacing and after four hours of NBO_2). It was estimated that his blood COHb upon surfacing would have been about 45%. The air compressor that was used to charge the diver's tanks was run by a gas-powered generator with its exhaust approximately three feet from the air intake of the compressor.

Finally, this commercial diver was performing work of varying intensity throughout the dive, increasing his uptake of CO compared to individuals at rest. The diver was able to function normally during the dive not because his CO exposure was negligible, but because of the increased partial pressure of oxygen at depth, with his symptoms manifesting during decompression.

Reference Number: 41906

CHIN, W., O. HUCHIM-LARA, AND S. SALAS. "Decreasing Carbon Monoxide in the Diving Air of Artisanal Fishermen in the Yucatan Peninsula by Separation of Engine Exhaust from Compressor Intake." *Undersea and Hyperbaric Medicine*, 43: 411–419, 2016 (7 figures, 2 tables, 32 references)

Abstract: Artisanal fishermen in the Yucatan Peninsula of Mexico use the hookah dive system, which uses an air gas-powered compressor in the boat to pump air underwater. The intake of the compressor typically is on the same level as the exhaust of the compressor. It is recommended that the compressed air for breathing underwater should not contain more than 10 ppm. The mean CO concentration detected in the compressed air for seven fishermen was 29 ppm (range, 9–69 ppm). When a one-inch diameter hose was attached to the compressor intake to make it at least five feet away from the exhaust, the mean CO concentration dropped to eight ppm (range 3–17 ppm).

The simple intervention in this study allows for continued collaboration with the fishermen divers. This intervention allows us to further grasp how learning will occur among this population of fishermen and gauge how willing the rest of the cooperative might be in implementing this intervention. This evidence should be applicable to other health and management issues among the fishermen that involve behavioural changes to promote occupational health.

4.20 TENTS/CAMPING

One does not expect death due to toxic fumes in a pristine wilderness setting, yet this case illustrates the extreme hazard that can occur when CO is allowed to accumulate in any closed space.

—R.G. Foutch and W. Henrichs, "Carbon Monoxide Poisoning at High Altitudes" (1988)

> Opening tents flaps, doors, or windows is insufficient to prevent build-up of CO concentrations from these devices.
>
> —R. Wheeler et al., "Carbon Monoxide Poisoning Deaths Associated with Camping- Georgia, March 1999" (1999) **(42004)**

The risk of CO poisoning in tents is often under-recognized by campers (42005, 42009). Camping at high altitudes produces a greater risk as more CO is produced by camping stoves, and there is less ambient oxygen to detoxify the CO (42002, 42007). Cold winter conditions can also increase the risk of CO poisoning, especially when the tent is covered with snow or ice (42003, 42008). Using a large diameter pan to melt snow for water produces more CO than a small diameter one (42001). Ventilation may be impaired with a lack of wind if the tent walls are covered in condensation, blocked openings, or if the walls are made of less permeable material (42010).

Reference Number: 42001

LEIGH-SMITH S., R. STEVENSON, M. WATT, I. WATT, AND A. MCFAYDEN. "Does Pan Diameter Influence Carbon Monoxide Levels during Heating of Water to Boiling Point with a Camping Stove?" *Wilderness and Environmental Medicine*, 15: 171–174, 2004 (1 table, 1 figure, 11 references)

Abstract: The amount of CO produced by boiling 500 mL of water on a Coleman Dual Fuel 533 stove was determined using a small pan diameter (16.5 cm) and a large pan diameter (22 cm) in a partially ventilated 200 L cardboard box model. The large pan produced a higher CO production than the small pan from 60 seconds onward with a mean difference of 185 ppm higher.

> This study has shown that there is a significantly higher CO production with a large-diameter pan compared with a small-diameter pan. These findings were evident with a camping stove used to heat water to a boiling point when a maximum blue flame was present throughout. This increase has a potential to enhance the risk of CO poisoning to tent occupants. Thus, in enclosed environments it is recommended that small-diameter pans be used in an attempt to prevent high CO levels. The safest small-diameter pan should be used in tents bearing in mind the added instability with a taller but smaller-diameter pan.

Reference Number: 42002

FOUTCH, R.G., AND W. HENRICHS. "Carbon Monoxide Poisoning at High Altitudes." *American Journal of Emergency Medicine*, 6: 596–598, 1988 (1 figure, 25 references)

Abstract: A Swiss six-man climbing party began their ascent of Mount McKinley after flying to the Kahiltna Glacier at 7,200 feet on 29 May 1986. A 23-year-old and 17-year-old who shared a tent experienced some mild symptoms of acute mountain sickness (that is, headache and malaise). These two climbers went to their tent at 9:20 p.m. to cook dinner using a small butane cook stove and sleep for the night. It snowed lightly all evening and morning. In the afternoon of the next day, they were found dead in their sleeping bags. They seemed peaceful in their semi-rigid position and, although some blood pooling was noted, the skin was not cherry red. The butane stove was still on 1/4 turn (low simmer), and the fuel cannister was empty. Their blood COHb concentrations were 56.9 and 65.6% respectively.

Despite existent warnings concerning the dangers of stove use in tents, CO may be a greater hazard than is generally appreciated by mountaineers, backpackers, and others who use tightly sealed tents and portable cook stoves in pursuit of outdoor adventure. In light of this risk, it would be prudent for tent and stove manufacturers to more clearly print a warning regarding the potential for CO poisoning on their merchandise.

Reference Number: 42003

PUGH, L.G.C.E. "Carbon Monoxide Hazard in Antarctica." *British Medical Journal*, 1 (5116): 192–196, 1959 (2 tables, 2 figures, 26 references).

Abstract: CO is an important hazard in polar exploration as Roald Amundsen, Richard E. Bryd, and Vilhjalmur Stefansson all had narrow escapes from fatal CO asphyxia. Melting snow with the Nansen and primus portable stoves can produce high ambient CO concentrations. Motorized vehicles also pose a CO hazard, and high altitudes can cause a greater blood COHb saturation and greater risk of poisoning than at sea level.

Table: CO at Sea Level and Blood COHb Produced at Sea Level and 10,000 Feet (3,050 Metres)

CO percent at sea level (%)	Resultant COHb at sea level (%)	Resultant COHb at 10,000 ft (3,050 m)
0.005	8.5	11.6
0.010	15.6	20.7
0.020	27.0	35.2
0.030	36.1	46.2
0.040	44.8	52.4

Source: Adapted from Pugh 1959.

Carbon monoxide poisoning on the Commonwealth Trans-Antarctic Expedition is described. The symptoms occurred mainly at 7,000–10,000 feet (2,135–3050 metres) and represented a combination of the characteristic effects of altitude and CO. In the British party symptoms were associated with long hours of travelling in motorized vehicles with the cabin windows closed. In the New Zealand party symptoms were associated with shorter exposures in a small caravan provided with an exhaust heater and two primus stoves for cooking. Symptoms of CO poisoning were also experienced by the T.A.E. advance party who spend the winter of 1956–7 in a crate with three primus stoves for cooking and heating purposes.

Reference Number: 42004

WHEELER, R., M.A. KOPONEN, A.B. JOHNSON, P.J. MEEHAN, S.E. LANCE-PARKER, AND K.E. POWELL. "Carbon Monoxide Poisoning Deaths Associated with Camping, Georgia, March 1999." *Morbidity and Mortality Weekly Report*, 48: 705–706, 1999 (7 references)

Abstract: On the afternoon of 14 March 1999, a 51-year-old man, his 10-year-old son, a 9-year-old boy, and a 7-year-old girl were found dead inside a 10-by-14-foot two-room zipped-up tent at a campsite in southeast Georgia in the United States. A propane stove was found still burning inside. The post-mortem blood COHb concentrations were 50%, 63%, 69%, and 63% respectively. Their dog also died of CO poisoning. On 27 March 1999, a 34-year-old man and his 7-year-old son were found dead in a zipped-up tent. A charcoal grill, which had been used for cooking outdoors, had been brought inside the tent for warmth.

Evening temperatures often drop unexpectedly, even during warmer months of the year. Campers who are unprepared for colder weather may overlook the danger of operating fuel-burning camper heaters, portable gas stoves or charcoal grills inside tents and campers. Camping stoves and heaters are not designed to be used indoors and can emit hazardous amounts of CO, and smoldering charcoal emits large amounts of CO. Inside a tent or camper, these sources produce dangerous concentrations of CO, which becomes even more dangerous to sleeping persons who are unable to recognize the early symptoms of CO poisoning. To avoid hazardous CO exposures, fuel-burning equipment such as camping stoves, camping heaters, lanterns, and charcoal grills should never be used inside a tent, camper, or other enclosed shelters.

Reference Number: 42005

KIM, Y.-J., C.H. SOHN, B.J. BIN, K.S. LIM, AND W.Y. KIM. "Carbon Monoxide Poisoning during Camping in Korea." *Inhalation Toxicology*, 28: 719–723, 2016 (2 tables, 2 figures, 15 references)

Abstract: In the last five years, family camping has become more popular in Korea. Between 2010 and 2014, 72 CO poisonings were analyzed that involved 29 patients admitted to an ED in Seoul, Korea, and 43 victims were reported in the media who had suffered from CO poisoning related to camping. The CO poisoning patients due to camping accounted for 15.3% of all patients treated for CO poisoning at the ED. Their mean initial COHb concentration was 21.2% (range 9.5–35.4%). Nineteen of the patients were treated with HBO_2. The mean CO exposure time was 90 minutes (range 37–690 minutes). The CO exposure was the result of burning charcoal for heating in the tent in all cases except for one, which involved a gas stove. The camping incidents occurred most frequently in May and January. All patients survived, and the majority of patients (83.4%) were not aware of the potential danger of charcoal for camping.

It is noteworthy that all of the incidents occurred in popular campsite. In Korea, there is no legal requirement for safety regulations for campsites and this leads to increasing incidents of accidental CO poisoning. CO poisoning related to camping could be easily prevented if camping stoves, camping heaters, and charcoal grills were not used inside or near a tent. In the United States, a warning label with a nonverbal pictogram has been required on all bags of charcoal briquettes since 1998 by U.S.

Consumer Product Safety Commission. Modification of the label has been associated with a 60% decrease in charcoal-related deaths. Among other features, this includes a picture of a tent with a charcoal grill inside with a slash through it. This label showed dramatic effectiveness and should be required in Korea. Continuous public education and campaigns about the potential danger of CO are necessary. Posters and leaflets on campsites as well as a warning label on all bags of charcoal briquettes would be also effective. Handheld CO detectors are another effective solution for preventing accidental CO poisoning.

Reference Number: 42006

THOMASSEN, O., G. BRATTEBO, AND M. ROSTRUP. "Carbon Monoxide Poisoning While Using a Small Cooking Stove in a Tent." *American Journal of Emergency Medicine*, 22: 204–206, 2004 (1 figure, 12 references)

Abstract: Seven men (ages 18–38 years) burned a small kerosene cooking camp stove inside an igloo-type polyester/cotton tent for two hours with the ventilation canals open but with the zip fasteners closed. The tents were placed on the snow at a five-to-six-degrees-Celsius ambient temperature with no wind. The CO concentration in the ambient air inside the tent was above 150 ppm after 15 minutes after the stove was lit and increased to 350–500 ppm after 30 minutes and obtained a peak ambient CO concentration of 600 ppm. The mean initial blood COHb was 2.9% and increased linearly to a mean of 21.5% after two hours after exposure. The mean heart rate increased from 63 bpm to 90 bpm after two hours. Most of the volunteers experienced a slight headache.

The half-life of CO in blood is between 4 and 6 hours. Because the COHb concentration is cumulative over time, prolonged exposure to low concentrations of CO can result in considerable poisoning. In a tent during winter, one must use the stove in the morning as well as in the evening for cooking, heating, and the melting of snow for water. A hike or military manoeuvre over a period of days could therefore lead to a dangerous rise in COHb and chronic poisoning.

Reference Number: 42007

MURANO, D.B., E.P. SAVAGE, AND R.M. BUCHAN. "Altitude Affects Carbon Monoxide Formation in Pickup Campers." *Journal of Environmental Health*, 40: 198–202, 1978 (1 table, 5 figures, 10 references)

Abstract: Each year, thousands of tourists travel in their pickup campers to high-altitude recreational areas in the Rocky Mountains above 8,500 feet. The camper vans contain a variety of gas appliances/heaters/lights. The lower oxygen concentration at the higher altitudes causes an increase in the CO production in these combustion products. In addition, the lower ambient oxygen at high altitudes increases the toxicity of CO. The CO concentration inside 30 pickup camper vans was determined at altitudes about 8,500 feet and then below 5,500 feet after the gas appliances were turned on for 15 minutes. The mean ambient CO concentration detected in the camper vans were 57 ppm at high altitude and 38 ppm at low altitudes.

Table: The Top Ten Altitude Differences and Ambient CO Concentration Change from Low to High Altitudes

Altitude difference in feet	CO concentration (ppm) at low altitudes	CO concentration (ppm) at high altitudes
5,300	6	8
5,100	12	14
5,050	4	14
5,050	14	15
5,050	66	172
5,000	10	14
4,963	15	25
4,963	10	17
4,900	20	25
4,900	24	45

Source: Adapted from Murano, Savage, and Buchan 1978.

Three factors have a significant effect upon the variability in CO concentration among pickup campers at high altitude. These factors are: the presence or absence of gas lights, the presence or absence of gas ovens, and the degree of ventilation over the gas range. At high altitudes many pickup campers can contain CO concentrations of sufficient magnitude to constitute a potential health hazard to the occupants.

Reference Number: 42008

LEIGH-SMITH, S. "Carbon Monoxide Poisoning in Tents: A Review." *Wilderness and Environmental Medicine*, 15: 157–163, 2004 (5 tables, 37 references)

Abstract: There have been frequent narrow escapes of fatal CO poisonings by early Arctic and Antarctic explorers, including those reported by Roald Amundsen. Dangerous levels of CO can be obtained inside a tent within 30 minutes, while cooking using a Primus stove. A review of all accidental CO poisoning deaths in California between 1979 and 1988 showed that 10 of the 136 deaths were associated with camping equipment, seven from lanterns or lamps, and three from stoves. A propane gas stove inside a large tent in the US state of Georgia killed an adult and three children. On military manoeuvres in Norway, two royal marines died, and one survived CO poisoning in their tent pitched tactically (that is, dug into the snow with camouflage netting placed on top) after cooking breakfast inside on a camping stove. CO diffusion through tent walls may be prevented by snow and ice accumulation or by condensation. In poorly ventilated environments, two phases of CO production were observed, an initial slow linear phase followed by an exponential increase before the flame is extinguished. Partial ventilation can therefore cause higher CO levels than no ventilation, which can extinguish the flame quickly. Fatal CO levels can be produced even by setting camp stoves or lamps at a low setting. CO production is markedly increased when the previously blue flame flares yellow when heating pans. Kerosene produces the most CO in camping stoves.

Table: Some Risk Factors for CO Poisoning in Tents and Preventative Measures

Risk factor for CO poisoning	Recommendations for avoidance
Cooking	Avoid prolonged simmering, use a maximum blue flame, and avoid low flames; use small diameter pans
Yellow flame	Turn stove off, ventilate tent for a few minutes
Inadequate ventilation	Ventilation CO egress post as high as possible, ventilation oxygen ingress port as low as possible
Insidious onset if sedentary	Beware headache and tachycardia
Duration of CO exposure	Regular trips outside to unmask symptoms
Dehydration	Good hydration
Tent icing and snow cover	Keep tent fabric porous by regular cleaning

Source: Adapted from Leigh-Smith 2004.

Reference Number: 42008

Case reports verify that CO poisoning within tents and snow caves is a real and probably overlooked problem. It is potentially an even greater problem at altitude because of the multiplicity of risk factors for CO poisoning. Despite multiple anecdotal reports of climbers perishing from CO poisoning on Himalayan peaks circulating in climbing circles, the danger does not appear to be widely recognized. Diagnosing CO poisoning in the early stages may be difficult because of the nonspecific nature of symptoms and (at altitude) their similarity to AMS [acute mountain sickness]. The masking of symptoms when subjects are sedentary exacerbates the problem, and these are likely to be the occasions when individuals are subjected to the highest CO levels, such as resting and cooking in tents for hours during inclement weather.

Reference Number: 42009

ALDOBEABAN, S.M., A.I. ALYAHYA, S.F. IBRAHIM, S.S. ALRUSSAYES, AND O.S. ELSERAFY. "Camping-Related Carbon Monoxide Poisoning: Knowledge and Risk Behaviours among Young Saudi Campers." *Journal of Pakistan Medical Association*, 72 (10): 2025–2029, 2022 (2 tables, 1 figure, 22 references)

Abstract: During winter camping trips to the desert or at one of Saudi Arabia's parks that are popular among youth, where night temperatures can reach below zero degrees Celsius. A survey was conducted of 235 male Saudi campers from Riyadh in 2018–19 regarding the risk of CO poisoning and camping. Most of the campers (58.3%) had a good knowledge score, and 68.8% had a good practice score. Having civil defence training, a monthly income greater than 10,000 Saudi riyals, and a good knowledge score all related to CO detector use.

The campers had relatively high knowledge regarding carbon monoxide poisoning, but this knowledge was not fully translated into good health practice.

Reference Number: 42010

LONG, I.J., AND G.T. FLAHERTY. "Silent Killer: The Dangers of Carbon Monoxide Poisoning during International Travel." *International Society of Travel Medicine*, 24: 1–3, 2017 (1 table, 20 references)

Abstract: A young Irish man died from CO poisoning from a lit stove in his room while travelling in Argentina. A Canadian tourist died from CO leaking from a natural gas water heater in an illegal hostel in

Taipei. This article lists the dangers of CO poisoning while travelling internationally.

> Many extreme athletes travel around the world to participate in adventure sports such as caving, mountaineering, and diving. These adventurous travellers should pay particular attention to the risk of CO poisoning. One of the many items of equipment carried by cavers is an acetylene lamp. This lamp uses calcium carbide to produce acetylene, which then burns to emit light. Acetylene does not always burn completely, increasing the risk of CO being produced. A common practice among mountaineers is the use of gas stoves inside tents or snow caves, for the purposes of heating, cooking, or melting snow. This can potentially generate an environment of toxic gases, due to poor ventilation or incomplete combustion. A tent's ventilation can be worsened by a lack of wind, if its openings are closed, it is produced from a less permeable material, or if its walls are covered by snow, ice, or condensation. These are important considerations for the snow or high altitude tourist.

4.21 WAREHOUSE WORKERS HEADACHES

> The exposure of warehouse workers to CO poisoning has been so common that it has its own name: Warehouse Workers Headache Syndrome.
>
> —James Wigmore, "This Headache Cannot Be Treated with Aspirin" (2022)[7]

CO poisoning of warehouse workers operating gas or propane forklifts is so common that it has a name (42101–3, 42105–7). Catalytic converters are only efficient at normal-operating engine temperatures (42104). Even propane-powered forklifts that are thought to be cleaner than gas, can cause highly dangerous blood COHb of 28.7% in warehouse workers (42108). The safest approach is electric forklifts (42104).

Reference Number: 42101

ELY, E.W., B. MOOREHEAD, AND E. HAPONIK. "Warehouse Workers' Headache: Emergency Evaluation and Management of 30 Patients with Carbon Monoxide Poisoning." *American Journal of Medicine*, 98: 145–154, 1995 (3 tables, 4 figures, 83 references)

7 James Wigmore, "This Headache Cannot Be Treated with Aspirin: Beware of Warehouse Workers' Headache!," *LinkedIn*, 20 November 2022.

Abstract: Gasoline, wood, and coal are known for their incomplete combustion and resultant production of CO, whereas cleaner fuels such as methane, propane, and butane are thought to be able to be used to operate machinery such as forklifts safely indoors. In a small storage warehouse, a single forklift was converted from gasoline to propane due to concerns about CO, but the forklift was left idling near the air duct, which cased CO poisoning in three male and 27 female workers. The room CO concentration was later measured at 386 ppm, but when the forklift was turned off and the doors and windows opened, it dropped to three ppm within minutes. Five workers from the inspection and packing area were the most severely ill and had the highest expired CO levels—21.1% compared to 4.7% of the other workers with less direct exposure. All workers were treated with high-flow 100% oxygen. Four workers were transferred to a tertiary care centre and received HBOT. Two years later, 44% reported persistent symptoms, and 20% continued to undergo neuropsychiatric testing and counseling. CO is eliminated mostly as unchanged gas through the lungs, and 1% is oxidized to CO_2. The company switched to an electric forklift and had no further CO-poisoning incidents.

Table: Follow-Up Symptoms Two Years after CO Exposure

Symptom	Percentage of total exposed (%)
Numbness in arms or legs	30
Restlessness	30
Persistent headaches	27
Irritable or violent behaviour	13
Confusion	13
Difficulty holding your water	13
Difficulty walking	13
Memory loss	13
Difficulty speaking	3
Seizures	0

Source: Adapted from Ely, Moorehead, and Haponik 1995.

CO poisoning is an important cause of accidental morbidity and mortality, but mass exposure to CO is reported infrequently. Ongoing poisoning may go unrecognized for a considerable interval until a catastrophic single event occurs. Because of the prevalence of this medical problem, all physicians should be aware of the many domestic and industrial sources of CO, the multiple nonspecific manifestations of CO poisoning and principles of timely medical evaluation and management. Because of the gravity of acute and chronic CO intoxication, prevention remains the primary approach and requires the awareness of the myriad of settings in which serious exposures occur and correction of predisposition to this injury.

Reference Number: 42102

SINKS, T., P.R. KEMUT, AND K.M. WALLINGFORTH. "Two Episodes of Acute Illness in a Machine Shop." *American Journal of Public Health*, 79: 1024–1028, 1989 (2 tables, 2 figures, 15 references)

Abstract: An explosion occurred in the boiler of a machine shop due to a malfunctioning valve. Fortunately, no one was injured. The plant opened seven days later, and 10 employees became ill with headache, dizziness, vomiting, and chest pains. Arterial blood COHb at hospital ranged from 13.3% to 14.8%. The machine shop also used trichloroethylene as a degreaser. Propane forklifts, open flames in the welding and brazing operations, and a portable heater were identified as possible CO sources. Two of the seven propane forklifts were taken out of service due to excessive CO emissions. Two weeks later, workers again experienced symptoms, but the mean COHb was 2.3% (range 0– 8.0%), and the ambient CO concentration at the machine shop was less than 15 ppm. It is thought that the second set of illnesses was due to the collective stress reaction.

Whenever workers are exposed to toxic substances, both they and their unexposed co-workers are subjected to psychological stress and the situation may produce collective stress reactions above and beyond the primary symptoms. This, in turn can precipitate additional illness episodes, such as the one illustrated in this paper. This emphasizes the importance of a thorough and sensitive debriefing of the workers following any environmental or medical assessment. If worker fears and suspicions are not sufficiently allayed and episodes of illness do recur, both the investigators and the management can lose credibility, and the situation may escalate.

Reference Number: 42103

MCCAMMON, J.B., L.E. MCKENZIE, AND M. HEINZMAN. "Carbon Monoxide Poisoning Related to the Indoor Use of Propane-Fueled Forklifts in Colorado Workplaces." *Applied Occupational Environmental Hygiene*, 11: 192–198, 1996 (2 tables, 12 references)

Abstract: Three workplaces in which employees attended hospital as a result of CO poisoning were investigated by the Colorado Department of Public Health and Environment. The blood COHb of the 15 people affected ranged from 15% to 24.3%. All incidence involved propane-powered forklifts in a warehouse, and one involved a portable gas heater as well. The tailpipe CO emission was as high as 9% before maintenance.

> The use of propane-fueled equipment as a substitute for equipment fueled by gasoline does not solve the problem of carbon monoxide emissions. Although burning propane typically results in lower carbon monoxide emission concentrations than burning gasoline, use of propane equipment indoors must be accompanied by diligent equipment maintenance and testing of carbon monoxide emissions and resulting employee exposures. If carbon monoxide emissions cannot be sufficiently reduced thorough routine maintenance, careful adjustment of the fuel-to-air ratio, and regular monitoring of carbon monoxide emissions, then the use of catalytic converters or an alternative power supply should be considered. If carbon monoxide emissions are reduced sufficiently to control the hazard, tailpipe emissions should be regularly tested to ensure that carbon monoxide emissions remain low.

Reference Number: 42104

GALLAGHER, F., AND H.J. MASON. "Carbon Monoxide Poisoning in Two Workers Using an LPG Forklift Truck within a Coldstore." *Occupational Medicine*, 54: 483–488, 2004 (2 tables, 2 figures, 26 references)

Abstract: One 50-year-old non-smoking male worker and one 45-year-old non-smoking male worker developed CO poisoning symptoms over three days after using a rented LPG forklift in a cold storage warehouse. The first worker developed a grand mal seizure on the first day and was admitted to hospital. The second worker (driver) had persistent tiredness, headache, and gastrointestinal upset over the next two weeks. No blood COHbs were measured. Their usual forklift, which was removed for maintenance, produced 737 ppm of CO, whereas the rented forklift produced 42,250 ppm (4.25%) of CO. Using mathematical modeling, the first worker (material handler) could obtain a COHb of 17.8–38.5%, whereas the forklift operator (worker 2) could obtain a COHb between 8.8% and 19.5% while working in the cold storage facility with the rental forklift.

While regular maintenance can help to reduce toxic exhaust emissions in forklift trucks, the fitting of catalytic converters substantially reduces CO emission. However, catalytic converters are only efficient at normal engine operating temperatures, so reductions in CO emission may depend on usage of the forklift truck. Therefore, the safest approach would be to use only electric-powered forklift trucks in enclosed spaces, although such trucks may have power limitations and need charging facilities. The delay in diagnosing CO poisoning, as in this case, is not uncommon. This is largely due to the non-specific symptoms that are presented; often CO poisoning is described as a flu-like viral illness.

Reference Number: 42105

HIRSCH, A.E., R.L. LANGLEY, AND J.S. MCDANIEL. "Carbon Monoxide Poisonings from Forklift Use during Produce Packing Operations." *Journal of Agromedicine*, 23: 132–135, 2016 (23 references)

Abstract: During August 2013, the use of a propane-powered forklift to load pallets of blueberries from a farm warehouse into a 53-foot refrigerated trailer caused two employees to lose unconsciousness. The employee who was found unconscious sitting on the forklift, which was still running, died and had a post-mortem blood COHb of 60% and also had underlying heart disease. The other employee was successfully treated with HBO_2. Sixteen people, including first responders, were exposed to CO during the rescue, of whom 13 showed symptoms such as headache and dizziness and were treated with 100% oxygen. Firefighters measured a CO concentration in the warehouse of 2,214 ppm. Tests conducted in the trailer found that the ambient CO concentration exceeded 1,000 ppm within 15 seconds after starting the forklift.

In addition to forklifts, agricultural workers may be exposed to CO while working with or near other combustion equipment such as pressure washers, small gasoline engines, portable generators, heaters, tractors, and other vehicles. As this incident illustrates, CO can accumulate quickly and overcome workers when combustion equipment is operated indoors without adequate ventilation. Animals can also be affected by exposure to CO. To prevent serious injury and death from CO poisoning, employers should consider replacing gas-powered equipment with electric-powered equipment, which does not produce CO. After the incident described above, the employer replaced their propane-powered forklift with an electric-powered forklift to prevent future CO exposures. If employers are unable to eliminate combustion equipment, they can reduce CO exposure by limiting the amount of time combustion equipment is used indoors, providing adequate ventilation, installing CO alarms, keeping combustion equipment properly maintained, and educating employees on the symptoms of CO poisoning.

Reference Number: 42106

KUCERA, H.L., J.R. MEWES, AND P.H. ORR. "Carbon Monoxide from Engines in Warehouse Operations." *American Potato Journal*, 49: 93–97, 1972 (3 tables, 1 figure)

Abstract: Forklifts powered by internal combustion engines are used extensively in potato warehouses, and headaches and dizziness have been reported by workers in closed potato storage facilities in Minnesota when potatoes were being loaded by the forklifts. The production of CO was determined in gasoline and LPG (that is, propane and/or butane) engines at normal ratios and 10% rich air fuel ratios and with or without catalytic mufflers.

Table: Ambient CO Concentration in Test Room (10,000 Cubic Feet) after One-Hour Running of Forklift Powered by Gasoline or LPG

Engine type	Stock muffler (%)	Catalytic muffler (%)
Gasoline (properly tuned)	1.00	0.73
Gasoline (10% rich)	1.21	1.01
Butane/propane (properly tuned)	0.55	0.09
Butane/propane (10% rich)	0.89	0.61

Source: Adapted from Kucera, Mewes, and Orr 1972.

Caution (i) Beware of carbon monoxide. CO is never completely eliminated in any of these tests! (ii) Beware of the possible reduction of irritants by catalytic mufflers. Irritants can be removed but CO may still be present.

Reference Number: 42107

COMSTOCK, R.D., R.W. CURRIER, K.V. MARKIEWICZ, R.L. WELKE, M.P. QUINLISK, AND T.H. GREINER. "Carbon Monoxide Poisoning Associated with Use of LPG-Powered (Propane) Forklifts in Industrial Settings, Iowa 1998." *Morbidity and Mortality Weekly Report*, 48: 1121–1124, 1999 (7 references)

Abstract: Employees at two warehouses and a light industrial plant in Iowa, experienced CO poisoning symptoms (mainly headaches) during work due to the use of propane-powered forklifts at their facilities. The forklifts were found to emit between 40,000 and 75,000 ppm of CO.

> The events described in this report illustrate factors that result in failure to adequately prevent CO poisoning and to promptly recognize such incidents when they occur. Timely and correct clinical diagnosis of acute CO poisoning remains elusive because of the non-specific and protean nature of its signs and symptoms (i.e., headache, nausea, lethargy, weakness, abdominal discomfort/pain, confusion, dizziness, visual disturbances [including blurred vision], numbness and tingling, ataxia, irritability, agitation, chest pain, dyspnea on exertion, palpations, seizures, and loss of consciousness). In incident 1, failure to diagnose illness correctly in the first employees evaluated resulted in some CO-intoxicated employees being sent back to work and further exposure and in continued exposures to other workers at the plant. Correct diagnosis can be achieved by determining COHb levels in the patient. However, screening can be performed by breath analyzer instruments. Pulse oximeter testing does not reflect tissue hypoxia and cannot be used to screen or diagnose.

Reference Number: 42108

FAWCETT, T.A., R.E. MOON, P.J. FRACICA, G.Y. MEBANE, D.R. THEIL, AND C.A. PIANTADOSI. "Warehouse Workers' Headache Carbon Monoxide Poisoning from Propane-Fueled Forklifts." *Journal of Occupational Medicine*, 34: 12–15, 1992 (7 references, 2 tables)

Abstract: Of 220 patients treated for acute CO poisoning at the Duke University Hospital in North Carolina, 17 patients had confirmed CO exposure due to propane-fueled forklifts. Fifteen of the 17 cases occurred during the winter months in closed warehouses. The average time of exposure to the exhaust fumes was 4.4 hours. The most common symptoms were global headache, nausea, and light headedness. The average COHb at the time of the medical examination was 21.5% (range 4.2–28.7%). The 10 smokers had a slightly higher average COHb of 23.3%. The CO concentration in the exhaust of 12 propane-powered forklifts was 35,600 ppm (range 220–88,770 ppm), which decreased to 30,500 ppm (range 861–67,940 ppm) at working speed. An average forklift in a large (10 x 60 x 100 metres) unventilated warehouse would produce an ambient CO concentration of one ppm per minute of operation.

It is a common belief that propane fueled equipment is very efficient and can be used indoors. Our report of warehouse workers presenting with acute CO poisoning and this analysis of forklift exhaust should help correct this serious misconception. Awareness of the problem should lead to reduced exposure to warehouse workers and a heightened index of suspicion by heath care providers. Patients who present with unexplained headache and constitutional symptoms related to working around propane-fueled machinery in closed spaces warrant routine measurement of HbCO and appropriate treatment with oxygen if indicated.

4.22 WATER PIPE SMOKING

Emergency physicians should remember CO poisoning as a differential diagnosis when water pipe users present with syncope.

—S. Türkmen et al., "Carbon Monoxide Poisoning Associated with Water Pipe Smoking" (2011)[8]

Water pipes are also known as hookah, shisha, narghile, or ghalyun, and they have their tobacco heated by burning charcoal, and the fumes are bubbled through a container of water seemingly filtering the smoke (42002–3). But that is not the case as water pipes produce substantially more CO than cigars or cigarettes (42214) and have caused numerous CO poisoning incidents (42204–12). As such, water pipes have been given a section of their own.

Reference Number: 42201

LIM, B.L., G.H. LIM, AND E. SEOW. "Case of Carbon Monoxide Poisoning after Smoking Shisha." *International Journal of Emergency Medicine*, 2: 121–122, 2009 (5 references)

Abstract: A 19-year-old man smoked a water pipe (shisha) with three other friends at a restaurant and hit his head after a fall. He was admitted to hospital where the fall was thought to be due to cardiac arrhythmia, transient hypoglycemia, viral illness, or substance overdose. A COHb analysis of his blood showed a 27.8% saturation. He was immediately placed on

8 S. Türkmen, U. Eryigit, A. Sahin, S. Yeniocak, and S. Turedi, "Carbon Monoxide Poisoning Associated with Water Pipe Smoking," *Clinical Toxicology*, 49 (7): 697–698, 2011.

100% oxygen for the next 4–6 hours, and his blood COHb dropped to 1%. Smoke exposure with a water pipe can equal 100–200 cigarettes.

> CO poisoning is difficult to diagnose in the ED as its presenting symptoms are usually non-specific. Undoubtedly, a history of potential CO exposure is the most reliable indicator of poisonings. EPs should have a high index of suspicion, even though the exposure can be exotic as in this case. It is important to realize that when dizziness occurs in CO poisoning, it usually corresponds to a carboxyhemoglobin level of 20% and above, making a level considered to be at least moderately severe. Making a diagnosis in a timely fashion can expedite the management of such patients.

Reference Number: 42202

ARZIMAN, J., Y.A. ACAR, A.O. YILDIRIM, O. CINAR, E. CEVIK, Y.E. EYI, AND U. LALDIRIM. "Five Cases of Carbon Monoxide Poisoning due to Narghile (Shisha)." *Hong Kong Journal of Emergency Medicine*, 18: 254–257, 2011 (2 figures, 17 references)

Abstract: Narghile is a type of water pipe in which sweetened tobacco (tumbeki) is burned using charcoal. Three female and two male narghile smokers (ages 17–27 years) were admitted to an ED in Turkey as a result of five separate CO poisonings. Their COHb saturations ranged from 11.4% to 21.3%.

> Our cases were not life threatening or with persistent neurological sequel. It could be related to the short duration of narghile smoking. However, in elderly people especially those with co-morbidities, life threatening outcome can occur directly or indirectly because of trauma arouse out of substance effect. We want to remind that CO poisoning is not easy to diagnose and ED physicians must be alert of this diagnosis especially when encountering patients with non-specific neurological symptoms.

Reference Number: 42203

FAZLZADEH, M., R. ROSTANI, S. HAZRATI, AND A. RASTGU. "Concentrations of Carbon Monoxide in Indoor and Outdoor Air of Ghalyun Cafes." *Atmospheric Pollution Research*, 5: 550–555, 2015 (2 tables, 3 figures, 37 references)

Abstract: Ghalyun, which is also known as a hookah or water pipe, is tobacco heated by burning charcoal and then passing it through a bowl of water and inhaling through a hose. The indoor and outdoor CO concentrations were measured in 68 Ghalyun cafes in Ardabil City, Iran. Two types of tobacco are smoked, traditional and fruit flavoured. Most of the cafes were located in basements.

Table: Mean (and Range) of CO Concentration in Parts per Million under Different Conditions

Condition	CO concentration of indoor air, mean (range) (ppm)	CO concentration of outdoor air, mean (range) (ppm)
Traditional tobacco	8.8 (6.4–13.4)	3.3 (1–5)
Flavored tobacco	28.5 (6.1–112.4)	2.5 (1–6)
Basement location	30.4 (8.4–112.4)	2.5 (1–6)
Ground floor	12.9 (6.1–35.2)	3 (1–5)
Mechanical ventilation	22.4 (8.1–36.3)	2.9 (1–6)
No mechanical ventilation	25.9 (6.1–112.4)	2.5 (1–5)

Source: Adapted from Fazlzadeh et al. 2015.

Indoor CO concentrations are much higher than the recommended exposure limits at many sampling locations. Within the variable studied; type of tobacco smoked, location of the cafe inside the building and utilizing mechanical ventilators, respectively were recognized as the most influential factors controlling CO concentrations in Ghalyun cafes. Therefore, enhancing air exchange rates through local exhaust and general ventilating systems along with applying restrictions on operating the cafes located in the basement may significantly reduce risk of both occupational and second hand exposure to tobacco smoke in Ghalyun cafes.

Reference Number: 42204

TORREY, C.M., K.A. MOON, D'A.L. WILLIAMS, T. GREEN, J.E. COHEN, A. NAVAS-ACOEN, AND P.N. BREYSSE. "Waterpipe Cafes in Baltimore, Maryland; Carbon Monoxide, Particulate Matter, and Nicotine Exposure." *Journal of Exposure Science and Environmental Epidemiology*, 25: 405–410, 2015 (2 tables, 3 figures, 24 references)

Abstract: CO, particulate matter ($PM_{2.5}$), and airborne nicotine concentrations were measured in seven water pipe cafes in the Baltimore area between 2011 and 2012. They were measured in the main seating areas one to two metres above the ground. The overall median CO concentration was nine ppm, and the highest mean CO was 53 ppm. The maximum CO concentration measured was, on average, 39 ppm with a range from seven ppm to 115 ppm. Previous testing has shown that the side-stream smoking from a single water pipe session produces up to 30 times the CO as a smoked cigarette.

> Indoor concentration of both $PM_{2.5}$ and CO were markedly elevated in waterpipe cafes and exceeded concentrations that were observed in cigarette smoking bars. Air nicotine concentrations, although not as high as in venues that allow cigarette smoking were markedly higher than in smoke-free bars and restaurants. Concentrations of PM approached occupation exposure limits and CO exceeded occupation exposure guidelines suggesting that worker protection measures need to be considered. This study adds to the literature indicating that both employees and patrons of waterpipe venues are at increased risk from complex exposures to secondhand waterpipe smoke.

Reference Number: 42205

MISEK, R., AND C. PATTE. "Carbon Monoxide Toxicity after Lighting Coals at a Hookah Bar." *Journal of Medical Toxicology*, 10: 295–298, 2014 (3 figures, 25 references)

Abstract: A 24-year-old man who worked at a hookah bar and lit 30 to 40 hookahs filled with flavoured tobacco during his shift was admitted to hospital after falling unconscious at work. Upon waking at hospital, he reported nausea, headache, and generalized weakness. His arterial blood COHb was 33.8%, and he was administered 100% oxygen and HBOT. The fire department obtained normal ambient airborne CO concentration at the hookah bar, suggesting that the patient was primarily exposed by inhaling the hookah to set it up.

> The case demonstrates the potential dangers of hookah smoke as well as the danger posed to employees at hookah bars who light coals by inhaling through the pipe while torching the coal. Further research is needed to delineate the dangers posed to employees at hookah bars required to light coals. The practitioner should maintain a high index of suspicion for CO poisoning in patients with a history of hookah use and early HBO therapy should be carefully considered in all symptomatic patients found to have CO poisoning.

Reference Number: 42206

MAALEM, R., A. ALALI, AND S. ALGAHTANI. "Tobacco Hookah Smoking-Induced Carbon Monoxide Poisoning: A Case Report of Non-Ambient Exposure." *Clinical Case Reports*, 7: 1178–1180, 2019 (8 references)

Abstract: A 43-year-old man who smoked hookah outdoors experienced a gradual diffused headache, them unilateral left side weakness, confusion, and vomiting. The patient's initial COHb at hospital attendance was 38.8%. The patient refused HBOT and so was administered NBOT. He was discharged 10 hours later.

Table: Decrease of COHb Percent in a Hookah Smoker Undergoing NBOT

Duration of treatment	Blood COHb (%)
Initial	38.8
1.5 hours	23.2
3 hours	14.7
6 hours	6.4
10 hours	2.8

Source: Adapted from Maalem, Alali, and Algahtani 2019.

Carbon monoxide (CO) toxicity should be considered in patients presented to the emergency department with any acute neurological manifestations or any vague symptoms after hookah smoking. The CO exposure among hookah smoking is significant and can occur through the inhaled smoke from within the water pipe circuit rather than the ambient CO exposure. Therefore, smoking in a non-closed space environment does not eliminate the risk of CO toxicity. If CO toxicity is suspected; high-flow supplemental oxygen should be provided immediately while further investigations and management options are considered.

Reference Number: 42207

VERWEIJ, B.G.F., P.P.M. ROOD, S.C.E. SCHUIT, AND M.G. BOUEHUIS. "Waterpipe Smoking: Not as Innocent as It May Seem." *Netherlands Journal of Medicine*, 77: 156–159, 2019 (2 tables, 18 references)

Abstract: Two male and one female hookah smokers (ages 17, 18, and 21 years respectively) were treated at hospital as a result of loss of consciousness, dizziness, and weakness. The blood COHb concentrations were 26%, 19%, and 26% respectively. All patients were treated with NBOT.

> Waterpipe smoking and using hot coals as heat source is popular but more harmful than one might consider due to the risk of CO intoxication. A broader awareness of this risk in both waterpipe users and primary and emergency care providers is essential. In patients presenting with altered mental status or atypical neurological complaints, CO intoxication should be included in the differential diagnosis. Informing about water pipe use and VBG [venous blood gas] analysis are rapid and simple but essential in diagnosing CO intoxication. We advise that patients should be treated with high flow oxygen until symptoms resolve and HbCO normalizes. Standard treatment duration is not strongly supported by studies and treatment with high-flow oxygen may well have harmful effects.

Reference Number: 42208

NGUYEN, V., M. SALAMA, D. FERNANDEZ, J.D. SPERLING, A. REGINA, R. RIVERA, J. WANG, B.W. FRIEDMAN, AND S.W. SMITH. "Comparison between Carbon Monoxide Poisoning from Hookah Smoking versus Other Sources." *Clinical Toxicology*, 58: 1320–1325, 2020 (1 table, 3 figures, 27 references)

Abstract: Of the 265 adult patients treated with HBOT, 58 patients (22%) had hookah-associated CO toxicity (HACO). The non-HACO patients initially had higher COHb concentrations (24.6%) than the HACO patients (20.1%), although the HACO patients had greater neurological symptoms. The proportion of HACO patients has increased from 9.5% in 2015 to 41.6% in 2018.

Table: Characteristics of Hookah-Smoking and Non-Hookah-Smoking CO Exposure Victims

Characteristics	Hookah	Non-hookah
Mean age	28.2 years	45.0 years
Smoker (%)	42	18
Syncope (%)	93.1	70.5
Seizures (%)	10.3	2.9
Shortness of breath (%)	0	8.7

Source: Adapted from Nguyen et al, 2020

In our retrospective review, within a short time span, hookah smoking became a substantial contributor to CO toxicity. Its rapid increase is of particular concern, outnumbering traditional sources of CO toxicity. Frontline physicians should ask about hookah exposures in patients with CO toxicity, and individuals who smoke hookah should be warned about its health risks. We urge public health authorities to undertake the necessary prospective investigations and risk mitigation measures.

Reference Number: 42209

OZKAN, S., T. OZTURK, Y. OZMEN, AND P. DURUKAN. "Syncope Associated with Carbon Monoxide Poisoning due to Narghile Smoking." *Case Reports in Emergency Medicine*, 2013 (1): 796857, 2013 (1 figure, 5 refences)

Abstract: A 19-year-old male narghile smoker (two over four to five hours) attended the ER due to dizziness, syncope, and blurred vision. His initial COHb was 32.7%. He was administered 100% NBO_2 in a non-rebreather mask, and, four hours later, his COHb was 9.5% and was 0.3% at discharge.

As a conclusion, increased popularity of narghile smoking in recent years in especially young population will increase the number of cases of CO poisoning cases with nonspecific symptoms. So, it must be kept in mind that patients with nonspecific symptoms should remind us to CO poisoning, even narghile smoking.

Reference Number: 42210

RAAIJMAKERS, N.J., A.H.F. BOKMA, C.G. EGGEN, G. PRINS, AND J. ALSMA "An Unhealthy Blush: Secondary Erythrocytosis due to Waterpipe Smoking." *Netherlands Journal of Medicine*, 78: 202–205, 2020 (1 table, 1 figure, 11 references)

Abstract: A 22-year-old man was referred to an outpatient clinic for evaluation of polycythemia. He had been under treatment by a dermatologist for 18 months due to his facial plethora (that is, swelling and redness). He had an elevated erythropoietin level of 15.9 units per L, and, after an extensive diagnostic evaluation, a COHb concentration of 21% was detected. He denied smoking or using other drugs. He was administered NBOT, which lowered his COHb to 6% within hours. On inspection of his house by the fire department, a water pipe was found in his bedroom, which his father said he used every day.

To our knowledge, this is one of the first reported cases of a relevant secondary erythrocytosis as a result of a chronic symptomatic CO intoxication due to frequent waterpipe smoking. In patients with a normal or high EPO [erythropoietin] level, we propose to test for the presence of an elevated COHb. The test is not expensive and may quickly reveal CO intoxication. As illustrated with this case, CO intoxication has potentially disastrous consequences and patients with chronic CO poisoning may be virtually asymptomatic. Not all individuals consider using a waterpipe as dangerous as smoking or drugs, and therefore physicians need to specifically ask for its use during history.

Reference Number: 42211

EICHHORN, L., D. MICHAELIS, M. KEMMERER, B. JUTTNER, AND K. TETZLAFF "Carbon Monoxide Poisoning from Waterpipe Smoking: A Retrospective Cohort." *Clinical Toxicology*, 56: 264–272, 2018 (1 table, 4 figures, 67 references)

Abstract: Forty-one male and 20 female water pipe smokers (mean age 23 years) were admitted to an ER due to CO poisoning between 2013 and 2016 in Bonn, Germany. The mean initial COHb was 26.93%, which declined to 18.14% after a mean interval of 74 minutes. Dizziness was reported in about half of all the patients, and syncope was associated with a COHb concentration greater than 30%. The patients were treated with HBOT.

Table: Initial Symptoms Presented by Water Pipe Smokers with CO Poisoning

Symptom	Percent (%)
Syncope	80
Dizziness	67
Headache	58
Nausea	48
General malaise	23
Ataxia	19

Source: Adapted from Eichhorn et al. 2018.

Due to the high prevalence of waterpipe smoking, especially young patients with unspecific symptoms and waterpipe smoking should be evaluated for CO poisoning.

Reference Number: 42212

MATINASEK, M.P., K.D. WARD, AND A.V. CALVANESE. "Change in Carbon Monoxide Exposure among Waterpipe Bar Patrons." *Nicotine and Tobacco Research*, 16: 1014–1019, 2014 (2 tables, 44 references)

Abstract: The breath CO concentration was measured in 166 patrons of six water pipe bars in Tampa, Florida, using a piCO device before and after hookah smoking. The mean breath CO increased from 6.5 ppm to 8.2 ppm (an 80% increase) after hookah smoking. Dual smokers (hookah and cigarettes) had the greatest breath CO increase compared to hookah-only smokers. Breath CO concentration was related to the frequency of water pipe use, the amount of charcoal, the number of tobacco bowls, and the time spent in the bar.

> US waterpipe bar patrons are exposed to considerable amounts of CO, which could put them at risk of acute illness and chronic heart and lung diseases. Environmental and policy controls are needed to curb this increasingly popular tobacco use method in the United States.

Reference Number: 42213

BONADIES, N., A. TICHELLI, AND A. ROVO. "When Water Does Not Clear the Smut from Smoke." *BMJ Case Reports*, 2013200665: 1–3, 2013. (1 figure, 13 references)

Abstract: One to two hours of water pipe smoking is equivalent to smoking 100–200 cigarettes in terms of CO, tar, nicotine, and polycyclic aromatic hydrocarbons inhaled. A 25-year-old university student and American football player reported irritability, dizziness, impaired concentration, and reduced physical performance. His hemoglobin, erythrocyte, and thematic blood values were elevated. He denied smoking, but his blood COHb was 6.1%, and, upon more detailed questioning, he admitted to smoking hookah one to two hours daily. Shisha smoking was trendy in his American football player peer group. He stopped smoking the water pipe, and, within six weeks, all of his blood values returned to normal. His blood COHb was still at 6.1%, and so it was assumed he was still smoking, albeit at a much lower level.

> Similar to chronic CO poisoning in heavy cigarette smokers, we found an increased erythrocyte volume with reduced plasma volume. However, the patient credibly denied cigarette smoking. Retrospectively, increased

awareness and a more detailed personal history would have facilitated identifying water-pipe smoking as the possible cause from CO poisoning. The clinical spectrum of CO poisoning is non-specific including headache, dizziness, nausea, confusion, fatigue, chest pain, shortness of breath and loss of consciousness. The correlation with COHb seems to be poor and life threatening symptoms appear at COHb >25% with metabolic acidosis. In heavy cigarette smokers, COHb up to 10% have been observed, which is in the range of our case report.

Reference Number: 42214

ZEIDAN, R.K., S. RACHIDI, S. AWADA, A. EL HAIJE, W. EL BAWAB, J. SALAME, R. BBEJJANY, AND P. SALAMEH. "Carbon Monoxide and Respiratory Symptoms in Young Adult Passive Smokers: A Pilot Study Comparing Waterpipe to Cigarette." *International Journal of Occupational and Environmental Medicine*, 27: 571–581, 2014 (3 tables, 54 references)

Abstract: The exhaled breath CO concentration was measured in 50 young restaurant workers serving water pipes, 48 university students exposed to cigarette SHS and 49 university students not exposed to SHS in Beirut, Lebanon.

Table: Characteristics of Non-Exposed, Water Pipe Smoke Exposures and Cigarette SHS Exposures

Characteristic	Non-exposed	Exposed to water pipe smoke	Exposed to cigarette smoke
Mean breath CO level (ppm)	3.76	26.84	12.25
Has chronic cough (%)	16.3	58.0	54.2
Has chronic sputum production	6.1	6.1	2.1
Has chronic wheezing	2.0	0.0	2.1
Has chronic bronchitis	4.1	4.0	0.0

Source: Adapted from Zeidan et al. 2014.

In summary, young non-smoker subjects demonstrated more chronic cough and elevated carbon monoxide levels when exposed to SHS while the effect of waterpipe was even more evident.

4.23 WOOD BURNING

Wood smoke is one of the most harmful air pollutants on earth.

—G. Oberfeld, World Health Organization's Health and Environment in Sustainable Development (1999)

Cooking fires are a major source of CO in less developed countries (42305–7). CO poisoning has also occurred in wood-fire saunas (42304), fireplaces (42302), and heating systems (42308) and in wood burners (42301). Wood produces CO not only when burning but also when stored as freshly made wood pellets, which has caused several deaths when homeowners assessed the storage area (42309, 42310). Ocean ships carrying wood pellets also develop dangerous CO in the holds during transport. All need to have their ambient CO concentrations determined and ventilated before entering (42311). Outdoor wildfires, which are becoming more common due to climate change, can also cause CO poisoning even in the open air. In one intense wildfire, 37 victims (10 were firefighters) were treated for CO poisoning at hospital with HBOT. Twenty percent unfortunately developed long-term delayed neurological symptoms (42312).

Reference Number: 42301

CUSHEN, R., H. BRUNT, S. JONES, H. LEWIS, P. CALLOW, AND A. KIBBLE. "An Unusual Incident: Carbon Monoxide Poisoning Risk in 540 Homes Due to Faulty Wood Burner Installations." *Public Health*, 173: 17–20, 2019 (14 references)

Abstract: In England and Wales, about 40 people die each year from CO poisoning, 200 are admitted to hospital, and about 4,000 visit the ER. At least 540 homes in Wales had a faulty installation of wood burners over a seven-year period even though it was supervised by a registered engineer. The homeowners were notified of the potential dangers, and CO detectors were provided by the incident management team and the local authority.

> No cases of CO poisoning were confirmed by local general practitioners or emergency departments following the IMT [incident management team] notification. However, given that installations took place over a seven-year period and the non specific symptoms of CO poisoning, it is possible that there were unidentified and/or unreported cases during that time. Also, given that the affected homes covered the whole LA [local authority] area,

cases that were reported would not necessarily have been linked. However, healthcare providers do need to be reminded regularly of the symptoms and signs of CO poisoning so that they include it in their differential diagnosis. But there is also a need to ensure that they refer CO cases to EPH [environmental public health] for action; it is possible that issue could have been identified sooner through reporting of health concerns.

Reference Number: 42302

ABELSOHN, A., M.D. SANBORN, B.J. JESSIMAN, AND E. WEIR. "Identifying and Managing Adverse Environmental Health Effects: 6. Carbon Monoxide Poisoning." *Canadian Medical Association Journal*, 166: 1685–1690, 2002 (2 tables, 43 references)

Abstract: A mother brings her 15-year-old son to the family physician in late November because he had frequent dull, frontal headaches for six weeks. The patient had moved his bedroom to the basement, which was insulated and dry-walled by his father. The house was heated with a gas-burning furnace in the basement and a wood-burning fireplace in the living room. The boy's blood COHb was 5%. The gas company was notified, but the ambient CO in the basement bedroom was normal.

There are 2 important findings: the case subject's COHb level is 5% and the carbon monoxide level in the basement, where the subject's bedroom is located is normal. A COHb level of 5% in a person who does not smoke is definitely considered elevated, although headaches do no usually start until the level is over 16%. The half-life of COHb is short, about 3 hours, so it is likely that the patient's level fell off substantially after he left the basement. A normal level would not have definitely ruled out carbon monoxide poisoning as a cause of his headaches. With regard to the normal carbon monoxide level in the basement, the important clue in this case is the timing of the headaches. They started in the fall, at the beginning of the heating season. On further questioning, a relationship is established between the headaches and the evenings when the fireplace was used. A heating, ventilation and air conditioning contractor conducts a chimney flow test with the fireplace in use and finds a backdraft and concurrent carbon monoxide level of 77 ppm. When the lit fire in the living room consumed oxygen, it caused a negative pressure in the inadequately ventilation house, including the furnace room in the basement, which resulted in the reversed flow of combustion gases including carbon monoxide, down the chimney into the living area in the basement. The renovation that increased the insulation in the basement may also have contributed to increasing the back drafting by reducing air flow and ventilation in the furnace. Reduced ventilation may also have allowed the concentration of carbon monoxide to increase.

Reference Number: 42303

SANTOS, L.R., M. ALVES-CORREIA, M. CAMARA, M. LELIS, C. CALDEIRA, M.D.L. BRAZO, AND J.J. NOBREGA. "Multiple Victims of Carbon Monoxide Poisoning in the Aftermath of a Wildfire: A Case Series." *Acta Médica Portuguese*, 31: 146–151, 2018 (2 figures, 1 table, 20 references)

Abstract: A total of 37 patients (ages 14–71 years) were administered HBOT due to exposure to CO as a result of intense wildfires on Madeira Island during August 2016. Ten were firefighters, four were children, and two were pregnant women, and 96% were exposed to CO outdoors for a median time of six hours. The mean COHb was 2.9% for smokers and 3.7% for non-smokers. The most common symptoms were headache (49%), followed by dizziness. Five patients developed syncope at the wildfire scene. The median time to the HBOT chamber was 4.8 hours. HBOT was performed at 2.5 atmospheres absolute for 90 minutes. The occurrence of DNS occurred in 20% of the patients after 30 days. HBOT should be considered in all cases of serious CO poisoning, loss of consciousness, ECG abnormalities, neurological symptoms, and metabolic acidosis, regardless of COHb.

Wildfires can cause clinically relevant CO poisoning, with typical signs and symptoms. This is a cause of concern as Portugal experiences more and bigger wildfires each year. We present a cohort of cases treated with HBOT that presented acutely to the emergency department in the time span of 72 hours after CO poisoning from wildfires. This is an unheard number in Portugal at least. Availability of a hyperbaric chamber on location eased treatment with HBOT.

Reference Number: 42304

LAM, N., M. NICAS, I. RIZ-MERCADO, L.M. THOMPSON, C. ROMERO, AND K.R. SMITH. "Non-Invasive Measurement of Carbon Monoxide Burden in Guatemalan Children and Adults Following Wood-Fired Temazcal (Sauna Bath) Use." *Journal of Environmental Monitoring,* 13: 2172–2181, 2011 (2 tables, 6 figures, 42 references)

Abstract: Wood fire steam baths or *temazcales* are a potentially dangerous source of CO exposure in the Guatemalan Highland communities where they are regularly used for relaxation and healing purposes. The temazcal is constructed of clay and sticks with the only opening covered by a heavy blanket to keep heat in. About four to five kgs of wood is burned to heat the rocks that are on a rack above the fire. Once the fire is reduced to glowing embers (one to two hours), the users pour water over the rocks to generate steam. The ambient CO concentration was measured as well as the breath CO and pulse CO-oximetry (Masimo Rad-57) in sauna participants, both adult and children. The mean ambient CO concentration during temazcal use was 661 ppm. The COHb concentration in the participants generally ranged from 12% to 14% (with an upper limit of 30% and lower limit of 2%) and did not differ between children and adults, although the children spent 66% less time in the steam bath than adults. This is due to children absorbing CO faster than adults, which puts them at a greater health risk.

Table: The Duration of the Sauna Bath and Mean COHb Concentration in Adult and Child Participants

Participants	Duration of temazcal	Mean COHb concentration (%)
All adults (15 years +)	45 minutes	13
Adult female	48 minutes	14
Adult males	41 minutes	12
All children (<15 years of age)	21 minutes	13
Female children	23 minutes	13
Male children	20 minutes	16

Source: Adapted from Lam et al. 2011.

Temazcal use results in high COHb levels in children and adults based on all three non-invasive measurement methods. Reducing exposure to children and pregnant women is of particular importance due to the adverse health and developmental risks associated with CO exposure in these sensitive populations. The need for culturally appropriate modifications to temazcales and education to change behavioural determinants of exposure (e.g., time of entry, duration of use), or recognizing signs and symptoms of CO toxicity is urgent. Results from this study and the cultural significance of the temazcal, suggests that education alone may not be enough to reduce CO exposure to levels posing minimal health risk. Thus, educational efforts need to be coupled with appropriate engineering modifications to the temazcal.

Reference Number: 42305

THORSSON, S., B. HOLMER, A. ANDJELIC, J. LINDEN, S. CIMERMAN, AND L. BARREGARD. "Carbon Monoxide Concentrations in Outdoor Wood-Fired Kitchens in Ouagadougou, Burkina Faso: Implications for Women's and Children's Health." *Environmental Monitoring and Assessment*, 186: 4479–4492, 2014 (3 tables, 11 figures, 38 references)

Abstract: Approximately 3 billion people (50% of households in developing countries) rely on biomass such as wood or coal to cook. Ambient CO concentrations were determined in 31 households with outdoor open wood fires for cooking in Ouagadougou, Burkina Faso. In eight households, CO was measured using passive sampling. A majority of the households (64%) cooked once a day, 26% cooked twice a day, and 10% cooked three times a day. The average CO concentration in the kitchen during cooking was 4.3 ppm (range 0.3–65.3 ppm) with the higher levels occurring during the evening cooking times. The World Health Organization's guidelines for exposure to CO is 52 ppm for 30 minutes, 26 ppm for one hour, and 9 ppm for eight hours. Eye irritation and coughing were reported by about 30% of the households.

A clear daily pattern was observed with relatively low concentrations during the day and higher levels during the evening. CO concentrations were on average 43% higher in the kitchens located in closed yards than in kitchens located in open yards, showing that the location of the fire affected the levels of CO that was measured in the kitchens. The CO concentrations were generally higher in the households with the highest fuel costs, indicating that fuel costs could be used as a simple and useful tool for assessing biomass smoke emissions. However, further studies are needed. It is shown that outdoor kitchen areas contribute considerably to the air pollution exposure among women and children in Ouagadougou. Considering the generally stable nocturnal weather conditions in the region, burning should to the greatest extent possible be limited between sunset and dawn and in areas with limited ventilation. By restricting the biomass burning to daytime hours and to well ventilated areas, pollution levels can be reduced and with it, women's and children's exposure to biomass smoke.

Reference Number: 42306

JOON, V., K. KUMAR, M. BHATTACHARYA, AND A. CHANDRA. "Non-Invasive Measurement of Carbon Monoxide in Rural Indian Woman Exposed to Different Cooking Fuel Smoke." *Aerosol and Air Quality Research*, 14: 1789–1797, 2014 (4 tables, 4 figures, 38 references)

Abstract: More than 70% of the population in India use biomass fuels for cooking, which impact women the most. The exhaled breath CO concentrations were measured by a Smokerlyzer in 80 non-smoking Indian cooks before and after cooking. The Smokerlyzer has an electrochemical sensor that creates an electrical charge in proportion to the CO in the breath. The device is calibrated twice a year using 50 ppm of CO gas. The subject inhales deeply and holds the breath for about 15 seconds before exhaling slowly into the mouthpiece. The ambient CO concentration was measured in the breathing zone of the cook by placing a sensor (Testo 350XL) 2.5 feet about the ground and 3 feet away from the stove. The type of biomass burned were LPG, wood, crop residues, and dung cakes, indoor and outdoor. The highest breath and ambient CO concentrations were highest with the use of dung cakes indoor. The highest incidence of self-report CO poisoning symptoms, especially headache also occurred with the use of dung cakes indoor.

Table: Mean CO Levels in the Breathing Zone Area of Cooks

Fuel/cooking location	Mean ambient CO (ppm)
LPG	5.2
Wood/outdoor	108.7
Crop residue/semi outdoor	143.4
Crop residue/indoor	195.5
Dung cakes/outdoors	195.6
Dung cakes/indoors	277.9

Source: Adapted from Joon et al 2014.

The present study reveals that the use of biomass fuels lead to very high levels of CO in the breathing zone atmosphere of the cooks among rural women in the Delhi-NCR [National Capital Region] region. These levels far exceed the indoor standards prescribed by WHO [World Health Organization]. Measurements of exhaled breath CO after the cooking cycle reveals the prevalence of high exhaled breath CO levels among the biomass fuel users. A very high positive correlation is obtained between average CO levels during the cooking cycle and exhaled breath CO after the cooking cycle.

Reference Number: 42307

LUCARELLI, K., K. WYNE, AND J.E. SVENSON. "Improved Cookstoves and Their Effect on Carbon Monoxide Levels in San Lucas Toliman, Guatemala." *International Journal of Environmental Health Research*, 28: 64–70, 2018 (4 tables, 16 references)

Abstract: The blood CO concentrations were measured using a handheld Masimo Rad-57 pulse CO-oximeter in 122 inhabitants of six villages in the Guatemala highlands. The inhabitants use a traditional unventilated wood stove indoors or outdoors or an improved ventilated wood stove (*plancha abierto*). The average CO level of the 122 participants was 4.6%, which is higher than expected in a non-smoking population. Neither CO levels or CO symptoms (headache, dizziness, confusion) were associated with the type of stove used. It may be that the poor construction of dwellings and the close proximity of neighbours, who cook outdoors cause smoke (CO) exposure to all inhabitants.

Our findings demonstrate that personal CO levels do not correlate with type of stove used, time of day measured, or village inhabited in the rural communities of San Lucas Toliman. We do not dispute the benefits of improved stoves in reducing exposure to indoor air pollution, but propose that some other factor may play a greater role than stove type in determining personal CO levels.

Reference Number: 42308

RICCARDI, A., P. BLENTINESI, M. MONTEVERDI, AND R. LERZA. "Chronic Carbon Monoxide Poisoning: A Report of Two Cases." *Emergency Care Journal*, 17 (2): 9677, 2021 (1 figure, 20 references)

Abstract: A 44-year-old man and his 40-year-old wife attended hospital due to diarrhea, fatigue, mildly defined chest pain, and confusion respectively. The man has extensively reddened skin mainly in his face, chest, and both hands. They both smelled of wood smoke, and their COHb were 3.3% and 8.8% respectively. They were treated with oxygen immediately and HBOT for five days. In their country home, an inspection found an irregular indoor home heating system and a CO leak.

It is possible to think of CO poisoning for all patients with acute ischemic complaint evaluated in an ED? All authors agree that venous CO levels are as diagnostic as arterial ones, but they could lead to an increase in sanitary costs without a real impact on health results. Probably the use of oxygen and carbon monoxide oximeter could make diagnosis of ED poisoning easier, but this is useless in case of chronic, severe exposure with normal or quite to normal CO blood levels.

Reference Number: 42309

GAUTHIER, S., H. GRASS, M. LORY, T. KRAMER, M. THALI, AND C. BARTSCH. "Lethal Carbon Monoxide Poisoning in Wood Pellet Storerooms: Two Cases and a Review of the Literature." *Annals of Occupational Hygiene*, 56: 755–763, 2012 (2 tables, 2 figures, 19 references)

Abstract: In Germany, the number of wood pellet heating systems increased from 2,000 in 2000 to 140,000 in 2010. Wood pellets are made from natural compressed wood waste such as sawdust, woodchips, and shavings. The wood pellets are stored in a dry airtight storeroom from which the pellets are fed by a conveyor belt into a furnace in the next room. Unless absolutely necessary for installation or maintenance, no one should enter the storeroom. It is recommended that the storage room be ventilated for 15 minutes before entering. A 43-year-old engineer died when he opened the pellet bunker door, which contained 155 tons of pellets supplying 700 households with heat. His blood COHb was 60%. A man standing next to him survived and called emergency services and had a blood COHb of 16%. In another case, a 28-year-old pregnant woman was found dead in a pellet storeroom providing heat for 60 households. She had gone into the storeroom to unclog the conveyor system. Her post-mortem blood COHb was 75%. Various tests were conducted to determine the amount of CO produced by wood pellets in a closed space.

Table: Production of CO (and Oxygen in Air) by 30 Kilograms of Wood Pellets Stored in a Sealed 60 Litre Container at 26 Degrees Celsius for 16 Days

Water added (% of pellet weight)	Relative humidity (%)	CO content (ppm)	Oxygen content (%)
0	27	3610	9.3
1	33	4727	6.8
2	46	4313	8.1
3	48	2294	10.3
4	59	3059	11.5

Source: Adapted from Gauthier et al. 2012.

In conclusion, it can be said that life-threatening levels of carbon monoxide can arise from wood pellets, not only at high temperatures but also at average room temperature, with a large headspace, by freshly produced or recently filled pellets, and at the humidity usually found in these storerooms. The danger seems to be greater in the winter when the increased demand for pellets results in the more frequent replenishment of freshly produced pellets and more technical problems with the wood pellet heating system might occur. Entering wood pellet storerooms in residential buildings carries a risk that has previously been underestimated.

Reference Number: 42310

GOLOB, N., D. GRENC, AND M. BRVAR. "Carbon Monoxide Poisoning in Wood Pellet Storerooms." *Occupational Medicine*, 68: 143–145, 2018 (1 figure, 10 references)

Abstract: Paramedics and firefighters were called to an unconscious man found on the floor in a wood pellet storeroom. The victim was immediately evacuated and resuscitated. Three rescuers became dizzy and complained of headache, nausea, and fatigue. High CO concentrations were detected in the pellet storeroom (15,000 ppm). The rescuers were administered 100% oxygen at hospital and had blood COHb concentrations of 7.6%, 4.9%, and 4.7%. The victim died, and a high post-mortem blood COHb concentration was determined.

First responders have to be aware of the dangers of carbon monoxide in wood pellet storerooms. Basic precautions and safety instructions should be followed before entering a wood pellet storeroom. Carbon monoxide should be measured before entering and self-contained breathing apparatus should be used. Wood pellet storerooms require continuous ventilation and should be equipped with carbon monoxide detectors.

Reference Number: 42311

SVEDBERG, U., J. SAMUELSSON, AND S. MELIN. "Hazardous Off-Gassing of Carbon Monoxide and Oxygen Depletion during Ocean Transportation of Wood Pellets." *American Occupational Hygiene*, 52: 259–266, 2008 (1 table, 3 figures, 9 references)

Abstract: Wood pellets are produced by milling wood chips, bark, planer shavings, or sawdust into a fine powder, which, after drying, is compressed into dense wood pellets of 6–8 mm in diameter and has a moisture content of 10% or less. In 2007, about 40% of the world's production of 9 million tons of wood pellets are transported by ships. A study was conducted to determine the CO and other gas contents in the stairways in the cargo holds of vessels carrying wood pellets from British Columbia after arriving in Sweden by Fourier Transform Infrared Spectroscopy. The CO concentration in the hold ranged from 1,090 to 14,650 ppm and the percentage of oxygen ranged from 0.8% to 10.9%. A recommended two-hour self-ventilation prior to entry is not sufficient; forced ventilation is required.

Ocean transportation of wood pellets in confined spaces may rapidly produce lethal levels of CO and an oxygen-deficient atmosphere that may leak into adjacent access spaces. This is the first report to recognize this danger. The oxygen depletion and CO formation are probably caused by oxidative degradation of natural lipids and other organic materials present in wood pellets. The oxygen depletion may have a value in that it reduces the fire hazard associated with this cargo.

Reference Number: 42312

SIMPSON, A.T., M.A. HEMINGWAY, AND C. SEYMOUR. "Dangerous (Toxic) Atmospheres in UK Wood Pellet and Wood Chip Fuel Storage." *Journal of Occupational and Environmental Hygiene*, 13: 699–707, 2016 (2 tables, 2 figures, 13 references)

Abstract: Wood pellets are produced by subjecting waste wood material to high pressure and temperatures resulting in a pellet that is at least partially sterilized. Wood chips are not exposed to this processing and have a greater surface area and more open and porous surface that can result in microbiological activity and emissions. Drying the chips can inhibit but not prevent microbiological activity. CO, CO_2, and oxygen levels were determined at seven sites (both pellet and chips), and no toxic atmospheres were detected. In the laboratory, samples of fresh wood pellets and chips were packed into airtight 60L drums (stored at 21 degrees Celsius) and CO, CO_2, and oxygen levels were monitored over 34 days.

Table: Maximum CO, CO_2, and Oxygen Detected in Wood Pellets, Undried Wood Chips, and Dried Wood Chips after up to 34 Days of Storage

Type of wood product	Maximum CO (ppm)	Maximum CO_2 (%)	Minimum oxygen (%)
Wood pellets	>1,042	15.47	0.1
Undried wood chips	60	35.57	5.7
Dried wood chips	543	0.15	20.9

Source: Adapted from Simpson, Hemingway, and Seymour 2016.

These laboratory tests confirm that the pellets encountered during this work were capable of producing dangerous levels of carbon monoxide given suitable conditions. The hazard would be increased by the elevated carbon dioxide and reduced oxygen levels. Wood chips produce less carbon monoxide however; there is still a hazard from carbon dioxide and low oxygen levels.

4.24 MISCELLANEOUS

The following patients may be encountered during an emergency department (ED) shift: a 7-year-old with a first-time seizure, an 80-year-old with syncope, a family with flulike symptoms, a pregnant patient with vomiting and dizziness, a 45-year-old with chest pain, a comatose patient from a house fire and a factory worker with headache. Although the complaints may sound diverse, carbon monoxide (CO) exposure may account for all of these clinical scenarios.

—L.W. Kao and K.A. Nanagas, "Carbon Monoxide Poisoning" (2004)[9]

9 L.W. Kao and K.A. Nanagas, "Carbon Monoxide Poisoning," *Emergency Medical Clinics of North America*, 22: 985–1018, 2004.

An unusual method of producing CO for suicide is combining formic acid and sulphuric acid together (42401–3). A new source of CO poisoning in the general public occurs as a result of underground utility cable fires as a result of the rubber/plastic coating splitting open due to freezing/thawing/salt, causing a short and intense fire (42404). Overheating of polyvinyl chloride produces CO, hydrogen chloride, and CO_2 and caused CO poisoning in 62 workers (42405). Sterno, which is used to heat dishes in a buffet, caused CO poisoning in one of the servers (42406). Meanwhile, 34 people who were watching soccer on televisions indoors were poisoned by CO due to an indoor butane gas stove used for heat (42407). Arc welders have suffered from CO poisoning while welding (42408–9), and one worker died as a result of CO exposure in a foundry (42410). The company constructing the Hoover Dam ruthlessly ignored CO poisoning that was affecting workers in tunnels and refused to replace the gas-powered trucks operating there in order to finish on time and under budget (42410–11). CO is the second most common molecule found in interstellar space. Comets typically have a CO abundance of up to 30%. The Comet 103P contains less than 0.5% of CO as most of it has been ejected into space (42413).

Reference Number: 42401

ERSHAD, M., A. MELISIOTIC, Z. GASKILL, M. KELLY, AND R. HAMILTON. "An Unusual Case of Carbon Monoxide Poisoning from Formic and Sulfuric Acid Mixture." *Clinical Practice and Cases in Emergency Medicine*, 4: 51–54, 2020 (22 references)

Abstract: CO can be produced by combining formic aid and sulfuric acid. A 27-year-old woman was found in the front seat of her car with an 18.9 L drum on the car floor containing a mixture of sulfuric acid and formic acid. First responders measured the ambient CO in the car at 400 ppm. At hospital, her blood COHb was 15% after two hours of NBOT. She was administered HBOT three times. Using a half-life of 75 minutes for the two hours of NBOT, her blood COHb could have been 45% in the car.

Formic acid, when combined with sulfuric acid, gets dehydrated to produce CO. This method is used in the commercial production of CO in laboratories. In the past few years, this method has been used to commit suicide. One possible explanation could be the availability of a considerable number of blogs, webpages, and online forums that provide information on the production of CO for the purpose of suicide. Interestingly another chemical reaction that has been recommended such as heating calcium carbonate and zinc to produce calcium oxide, zinc oxide and CO. It is important that first responders and emergency physicians be aware of such chemical reactions so that they can deploy appropriate on-scene and personal protection precautions.

Reference Number: 42402

YAND, C.-C., J. GER, AND C.-F. LI. "Formic Acid: A Rare but Deadly Source of Carbon Monoxide Poisoning." *Clinical Toxicology*, 46: 287–289, 2008 (1 figure, 17 references)

Abstract: A 26-year-old man committed suicide by mixing 2.5 L of formic acid and 2.5 L of sulphuric acid in three beakers in his bedroom. He collapsed and was found lying on the floor unconscious by his father. His father performed CPR on his son and took the beakers to the porch. The father lost consciousness 10 minutes later. The mother felt extremely dizzy and called the neighbours and paramedics. All three victims were transported to the hospital within 30 minutes. The son was dead on arrival, but a post-mortem blood COHb was not conducted. The father's blood COHb was 45.8% and his serum alcohol concentration was 0.213 g/100mL. He was intubated and administered oxygen over the next 6.5 hours and later had HBOT. He was discharged uneventfully on day 15. The mother had an initial blood COHb concentration of 23% and was administered oxygen. She was discharged on day 12 due to a severe grief reaction.

Formic acid is a highly fatal source of carbon monoxide poisoning when mixed with sulphuric acid. In addition to the toxicities of carbon monoxide, concomitant inhalation of formic acid fumes can cause severe lung injury, which may complicate the management of carbon monoxide poisoning. Treatment is primarily symptomatic and supportive.

Reference Number: 42403

BAKOVIC, M., M. NESTIC, AND D. MAYER. "Suicidal Chemistry: Combined Intoxication with Carbon Monoxide and Formic Acid." *International Journal of Legal Medicine*, 130: 723–729, 2016 (2 tables, 5 figures, 23 references)

Abstract: A 22-year-old man was found dead in his locked bathroom. With the body, two plastic jugs containing formic acid (85%) and sulphuric acid were found, which when combined produce CO. At autopsy, corrosive burns were located on the face and neck and a cherry-pink liver mortis was observed on the back of the body. The femoral blood concentrations were 0.28 g/L formic acid and 30% COHb.

> Concerning survival time, we believe that no exact answer could be offered. However, the starting point for our consideration was the fact of high concentrations measured in lung tissue, gastric content, and blood, as opposed to lower concentrations in other tissue samples, especially liver and bile. Such distribution indicates that no significant amount of formic acid was distributed to the tissues. Furthermore, chemical reaction between formic and sulphuric acid is very rapid producing carbon monoxide almost instantaneously. Experimental data demonstrate that with concentrations of 3000 ppm of carbon monoxide in the air, it takes approximately 30 min to reach 30% of carboxyhemoglobin in the blood. Histology analyses revealed complete lack of inflammation reaction in the tissues. Initial infiltration of neutrophil granulocytes is expected to start within 20 to 30 min after the onset of noxious stimulus. Taking all listed facts into account, we find reasonable to assume that death has occurred within 30 min from the exposure to the chemicals.

Reference Number: 42404

WILBURN, R.E., W.L. WELLES, D.K. HORTON, K. BERKOWITZ, AND W.E. KAYE. "Carbon Monoxide Releases and Poisonings Attributed to Underground Utility Cable Fires, New York, January 2000 to June 2004." *Morbidity and Mortality Weekly Report*, 53: 920–922, 2004 (5 references)

Abstract: Between January 2000 and June 2004, there were 234 CO burn-out events due to underground utility cable fires in New York City. The underground fires burn when the rubber/plastic coatings of the utility cable crack and split due to normal wear and tear, freezing and thawing, road salt and excavations, causing electrical shorts and underground fires that produce CO and other noxious gases, which migrate into the surrounding buildings. Approximately one-half of the CO burn-out events occurred during the snow season. Of the 234 events, 220 (94%) caused evacuation of the surrounding buildings, involving 3,855 persons.

> Because burning rubber and insulation materials release noxious odors that are highly noticeable, the percentage of CO burnout events with victims probably is lower than that of other tractional CO events with victims. Traditional CO events, particularly those associated with gas appliances

tend to take longer to identify because CO itself has no odor. Although few injuries resulted from the CO burnout events described in this report, thousands of persons were evacuated.

Reference Number: 42405

FRONEBERG, B., P.L. JOHNSTON, AND P.J. LANDRIGAN. "Respiratory Illness Caused by Overheating of Polyvinyl Chloride." *British Journal of Industrial Medicine*, 39: 239–243, 1982 (5 tables, 14 refences)

Abstract: On 9 August 1979, 62 workers out of 201 workers (31%), and one out of 60 management personnel (2%), developed upper and lower respiratory irritation, headache, nausea, and fainting at a polyvinyl chloride (PVC) fabricating plant. All were taken to hospital, but no one died. The illness was due to an overheated PVC extruder. One kilogram of PVC heated to 300 degrees Celsius releases an estimated 12.9 grams of hydrochloric acid and 4.9 grams of CO.

Table: Quantities of Toxic Gases Emitted by One Kilogram of PVC at Various Temperatures

Temperature (°C)	Hydrogen Chloride (g)	Phosgene (g)	CO (g)	CO_2 (g)
300	12.9	0	4.9	6.8
600	15.3	0.001	6.9	14.8
900	15.4	0.003	0.48	28.7

Source: Adapted from Froneberg, Johnston, and Landrigan 1982.

Plant management believed initially that the outbreak was caused by psychogenic factors, a belief that was possibly reinforced by the high proportion of women among the affected workers. The hypothesis was, however, abandoned after examination of the spatiotemporal distribution of cases and with the realization that attack rates did not differ significantly between men and women workers. This experience exemplifies the hazard of too ready acceptance of a psychogenic aetiology for outbreaks of mass industrial illness.

Reference Number: 42406

MURRAY, T.J. "Carbon Monoxide Poisoning from Sterno." *Canadian Medical Association Journal*, 118: 800–802, 1978.

Abstract: A 17-year-old female high school student felt nauseous, weak, and shivering and became unconscious after working in a university dining room at the serving table for one hour. She regained consciousness 30 minutes after admission to the local hospital. Her red face faded into blotchy patches and then disappeared. She was diagnosed with epilepsy, but further investigation showed that it was CO poisoning due to serving food that was heated by three cans of Sterno in a chafing dish that was surrounded by aluminum foil. Several other serving workers had experienced similar symptoms. The hall was well ventilated and, during the reconstruction of the incident, showed a ambient CO of seven ppm. Directly over one can of Sterno showed 50–90 ppm of CO. But, in the area of the chaffing dish, 1,000–3,000 ppm of CO was detected.

> The acute and serious carbon monoxide poisoning in this case resulted from a combination of factors. Three cans of Sterno were burning close together at the same time and the chafing dish was partly surrounded by aluminum foil, which cause insufficient oxygenation and incomplete combustion. The girls stood for long periods over the open area where the carbon monoxide and fumes exited. Although the room concentration of carbon monoxide was only 7 ppm owning to good ventilation, the values where the girls were standing were as high as 3,000 ppm.

Reference Number: 42407

GODOY, P., C. PONT, A. ARTIGUES, AND M. ALSEDA. "Mass Carbon Monoxide Poisoning among Television Viewers of a Football Match." *Revista Clinica Espanola*, 216: 409–413, 2016 (1 figure, 2 tables, 19 references)

Abstract: Twenty-eight men and 11 women (mean age 43.8 years) gathered in a 30-square-metre room in Spain on 2 February 2015 to watch a televised soccer match. A butane gas stove (being intended for exclusive outdoor use) was lit at 7:30 p.m. inside to heat the room. The first symptoms of CO poisoning occurred at 8:15 p.m., and the majority of cases occurred between 9:15 p.m. and 9:30 p.m. The mean exposure time was 52.4 minutes, and the mean distance from the stove was 4.2 metres. The most common symptoms were headache (50%), nausea (20.6%), and weakness (20.6%). Thirty-four of the 39 viewers were affected. The mean

COHb was 13.9%. Thirty-three patients required oxygen, and 13 were administered HBOT.

> In conclusion, we report a mass CO poisoning in which a high percentage of individuals who were confirmed to be poisoned (according to COHb levels) were asymptomatic. Therefore, we recommend assessing COHb levels in all individuals exposed to mass CO poisonings, even if they have no symptoms. The public and health practitioners should be informed of the risk of CO poisoning especially in winter when using heating systems that generate CO.

Reference Number: 42408

ANTONSSON, A.-B., B. CHRISTENSSON, J. BERGE, AND B. SJOGREN. "Fatal Carbon Monoxide Intoxication after Acetylene Gas Welding of Pipes." *Annals of Occupational Hygiene*, 57: 662–666, 2013 (2 tables, 9 references)

Abstract: A 23-year-old welder was found unconscious in a boiler room after welding a district heating pipe that had been closed at both ends. He was administered resuscitation for one hour but was declared dead at hospital. Third degree burns were observed on both legs and left hand. There were signs of aspiration contents in the trachea and bronchial tree. His post-mortem blood COHb was 46%. Various tests were conducted to determine the ambient CO concentration produced by welding.

Table: CO Produced by Welding

Welding method	Average CO (ppm) in 10 m^3 after 15 minutes of operation
Gas welding with oxygen	75
Gas welding with odorox	53
Gas welding with greasy ends joined by welding	70
Gas welding with the ends of the pipe locked and with odorox	620
Torch cutting using acetylene and oxygen	225
Soldering of copper pipe	200

Source: Adapted from Antonsson et al. 2013.

During gas welding, CO is generated. Oxidization of CO will not take place if there is a deficit of oxygen and the result will be high concentrations of CO. Welding of district heating pipes and evacuation of newly installed pipes are expanding operations where high and fatal concentrations may occur. This risk has not been recognized previously and consequently information is the first step along the path of prevention.

Reference Number: 42409

MA. S., H. SEO, D.J. PARK, B. CHOI, AND S. YE. "Carbon Monoxide Poisoning-Induced Encephalopathy in a Carbon Dioxide Arc Welder: A Case Report." *Annals of Occupational and Environmental Medicine*, 34: p.e19, 2022 (2 tables, 3 figures, 35 references)

Abstract: CO_2 arc welding produces significant amounts of CO due to the CO_2 gas used as an inert material, or the carbonates of the flux core used can decompose into CO. In the Korean shipbuilding industry, the CO concentration in poorly ventilated areas have been measured at 23.9 ppm. A 40-year-old male welder had intermittent headache and memory loss for the last five years. He used to be a heavy smoker (10 pack-years) and was a heavy drinker. His blood COHb concentration measured between 2011 and 2016 showed a mean of 6%, which is normal for smokers. His brain MRI showed CO-poisoning-induced encephalopathy. He received HBOT and showed some improvement. Ambient CO concentrations of where he worked detected up to 818.5 ppm.

After considering the results of the brain MRI imaging, neurological and psychiatric symptoms, response to hyperbaric oxygen therapy, and work environmental conditions, we concluded that delayed encephalopathy after acute CO poisoning of patient was caused by exposure to CO during CO_2 arc welding, in the course of his occupation. Our study makes a significant contribution to the protection of welders' health because although the risk of exposure with CO_2 arc welding is well-established, no studies have yet reported the incidence of CO poisoning induced encephalopathy in a CO_2 arc welder.

Reference Number: 42410

FAIRFAX, R., AND D. BLACK. "A Foundry Fatality from Carbon Monoxide." *Applied Occupational and Environmental Hygiene*, 14: 285–288, 1999 (3 references)

Abstract: A male worker employed as a new materials-unloading operator was found dead outside of the lever room of the ferrous foundry where he worked. His post-mortem blood COHb was 75%. The foundry produced mainly pipes and used two cupola furnaces, which are well known for generating CO. The employee mainly worked outside around the rail cars helping to unload coke and limestone. He spent three to four hours in the melt area inside checking the conveyer belt. It is proposed, based on the evidence, that the source of the CO was the access doors of the combustion chamber, which had become open, releasing CO into the plant. It is estimated that the worker had been exposed to CO for up to two hours before discovery.

Nevertheless, this is a fatality that perhaps need not have occurred and could have been prevented. The facility had never done monitoring to determine the levels of carbon monoxide. If a stationary monitoring system had been in place capable of sounding an audible alarm the worker may not have entered the danger area. The worker was also working alone in a potentially dangerous area. As we saw, it is likely that he had been dead for several hours before he was discovered. Perhaps if he had been found earlier, CPR may have saved his life. Lastly, the company made available dust-mist respirators for protection against dust and smoke. The workers had not been informed of the limitations of their respirators (that these air purifying respirators were not effective against carbon monoxide or oxygen deficient atmospheres). Because of this, the workers believed that their dust-mist respirators protected them against carbon monoxide. Supplied air respirators were not provided to any workers.

Reference Number: 42411

REUSS, M. "*Hoover Dam: An American Adventure*, By J.E. Stevens, Norman: University of Oklahoma Press, 326 pp, 1988." *Journal of American History*, 76: 295–296, 1989.

Abstract: A book about the construction of the Hoover Dam was reviewed. The Bureau of Reclamation and the Six Companies that oversaw the construction of the dam only wanted to hire single men for the most dangerous construction tasks in order to lower payment to dependents in case of accidental deaths. The workers were at the mercy of the Bureau of Reclamation and Six Companies that built the company town—Boulder City—without a library or school and tried to get the workers to vote for Herbert Hoover in 1932.

The company doctor listed pneumonia as the cause of death of tunnel workers who had died of carbon monoxide poisoning. When civil suits were filed seeking damages for gas poisoning, Six Companies tried in every underhanded way possible to discredit both plaintiffs and their lawyers. To some, Six Companies epitomized the worst of capitalism. To others the fact that the Six Companies completed Hoover Dam over two years ahead of time and at a handsome profit testified to the strength of capitalism.

Reference Number: 42412

STEVENS, J.E. "Hoover Dam: An American Adventure (Reviewed by D.J. Pisani)." *Southern California Quarterly*, 72: 194–198, 1990.

Abstract: This book review is about the construction of the Hoover Dam and its associated health and safety problems.

Tunnel workers often suffered symptoms of carbon monoxide poisoning as the fleet of gasoline powered muck trucks poured heavy fumes into close and poorly ventilated work areas. At times the dust and smoke obscured the strong artificial lights. Headaches, nausea, and muscular weakness were just three of the problems that prompted 48 separate damage suits against the construction consortium responsible for building the dam, the Six Companies. Six Companies knew that carbon monoxide was a potentially lethal hazard, Stevens concludes, but to protect a $300,000 investment in gasoline powered trucks and avoid a costly delay while electric motors were installed, it was willing to circumvent state legal restrictions on the use of internal-combustion equipment underground and put its employees at risk. This was the single major blemish on a safety record that was otherwise commendable, but it also reflected an unwillingness to tolerate any delays in construction, for whatever reason and sheer corporate greed.

Reference Number: 42413

WEAVER, H.A., P.D. FELDMAN, M.F. A'HEARN, N.D. RUSSO, AND S.A. STERN. "The Carbon Monoxide Abundance in Comet 103P/Hartley 2 during the EPOXI Flyby." *Astrophysical Journal Letters*, 734: L5, 2011 (2 tables, 3 figures, 25 references)

Abstract: The 103P comet was discovered in 1984 as a Jupiter family comet with an orbital period of 6.4 years. Using spectral emissions, the CO, CO_2, and water (H_2O) abundance was determined. The CO/H_2O ratio is

0.15–0.45%, which is one of the most CO- depleted comets. Comets can have a CO abundance of up to 30%. Thus, CO is probably a minor part in the comet's current activity and is not responsible for ejecting icy grains into the comet.

> CO is predicted to be the dominant C-bearing gas in the outer parts of the solar nebula where comets formed. However, CO is extremely volatile with a sublimation temperature of ~20 K under solar nebula conditions. Thus, the CO content of ices incorporated into comets may have been very sensitive to the local conditions where the nucleus formed or where the coldest part of the nucleus formed assuming there was significant radial mixing of materials within the nebula as suggested by the results from the Stardust mission.

CHAPTER 5

Post-Mortem Carbon Monoxide

Coal fumes lead to heavy head and death.

—Aristotle, 300 BCE, cited by L.K. Weaver, "Carbon Monoxide Poisoning" (1999)[1]

When someone dies of suspected carbon monoxide (CO) poisoning, post-mortem carboxyhemoglobin (COHb) concentrations need to be analyzed to determine the cause of death. Possible post-mortem changes in COHb concentrations should be considered. Other tissues containing hemo/myoglobin may be analyzed for CO when blood is not available.

5.01 METHODS OF ANALYSIS

Post-mortem blood COHb determinations are more difficult to conduct than ante-mortem blood due to clotting, hemolysis, and putrefaction (50101). The four main methods of analysis for CO are spectrophotometric (that is, ultraviolet [UV]), micro-diffusion, headspace analysis with various detectors, and blood gas analysis (50103). The best method is headspace analysis as it involves the actual CO gas (50102–5, 50107). In one case, the spectrophotometric method using UV found no COHb detected but, by the more accurate headspace method using gas chromatograph (GC) / mass spectrometer (MS), found 52.9%. The CO-oximeter using more than four wavelengths may be acceptable for CO analysis (50106)

Chapter 5

1 L.K. Weaver, "Carbon Monoxide Poisoning." *Critical Care Clinics*, 15: 297–317, 1999.

Reference Number: 50101

CONSTANTINO, A.G., J. PARK, AND Y.H. CAPLAN. "Carbon Monoxide Analysis: A Comparison of Two Co-Oximeters and Headspace Gas Chromatography." *Journal of Analytical Toxicology*, 10: 190–193, 1986 (5 tables, 3 figures, 16 references)

A comparison was conducted of the IL-182 CO-oximeter, IL-282 CO-oximeter, and headspace GC to analyze CO in post-mortem blood that was determined in 62 samples. The CO-oximeters are differential spectrophotometers that use 578, 568, and 548-nanometre (nm) wavelengths. The correlation coefficient for all three methods exceeded 0.99. All three methods are acceptable for the analysis of CO in post-mortem blood.

For decomposed or severely clotted blood, special care must be taken in treating the sample prior to analysis. Homogenizing such samples prior to adding the detergent may be helpful. Filtering the supernatant will also help to avoid clogging the tubing of the CO-oximeters with blood clots.

Reference Number: 50102

ORITANI, S., B.-L. ZHU, K. ISHIDA, K. SHIMOTOUGE, L. QUAN, M.Q. FUJITA, AND H. MAEDA. "Automated Determination of Carboxyhemoglobin Contents in Autopsy Materials Using Head-Space Gas Chromatography/Mass Spectrometry." *Forensic Science International*, 113: 375–379, 2000 (2 figures, 18 references)

Abstract: An automatic headspace GC/MS method was developed for the routine analysis of CO in post-mortem samples, including those samples that have undergone thermo-coagulation, putrefaction, and contamination. Two samples were analyzed. One was saturated with CO after the addition of 0.2 millilitres (mL) of 20% hydrosulfate, and the excess CO (not bound to the heme proteins in the sample) was flushed out with nitrogen. Then, 20% potassium ferricyanide (0.5 mL) and t-butanol internal standard were added to both samples and sealed. The CO peak area was measured by headspace GC/MS. The CO saturation step is responsible for the improved accuracy of this method.

In conclusion, it was established that the automated head-space GC/MS was practical not only for analyses of blood and bone marrow aspirates to confirm findings using the CO-oximeter system but was also useful with thermo-coagulated and putrefied blood (possible some other blood containing material as well) which were not analyzed using Co-oximeters.

Reference Number: 50103

ZANABONI, M., G. RODA, S. ARNOLDI, E. CASAGNI., V. GUMBARO, AND M. DEI CAS. "Comparison of Different Analytical Methods for the Determination of Carbon Monoxide in Postmortem Blood." *Journal of Forensic Sciences*, 65: 636–640, 2020 (2 tables, 3 figures, 15 references)

Abstract: Four different methods to determine the post-mortem blood COHb were compared. The methods were UV-visible spectrophotometry (Beutler+West modification at 420 and 430 nm wavelength), micro-diffusion Conway cell method using 522 nm wavelength, HS-GC/TCD and CO-oximetry (blood gas analysis). Although all 4 methods produced comparable COHb results in 3 postmortem blood samples, the headspace GC/thermal conductivity detector (TCD) was the preferred method in this forensic laboratory as analysis for CO could be conducted in other post-mortem samples other than blood. The spectrophotometric and blood gas analysis are indirect methods of analysis. HS-GC/TCD and micro-diffusion methods are direct methods of analysis.

Table: COHb Results of Three Post-Mortem Blood Samples Using the Four Methods

Method of analysis	Case 1 (%)	Case 2 (%)	Case 3 (%)
Spectrophotometric	57.2	61.7	38.2
Micro-diffusion	55.7	59.2	36.9
Headspace GC/TCD	57.8	62.1	39.0
Blood Gas Analysis	56.3	60.8	37.2

Source: Adapted from Zanaboni et al. 2020.

The detection and quantitative determination of CO in biological samples can be carried out in two ways, either provoking the release of CO from the COHb complex and analyzing it directly, or indirectly or determining COHb using its absorption bands. The best matrix is blood; in cases in which blood is not available, COHb can be determined on tissues rich in hemoglobin such as bone marrow can be suitable for the purpose. The determination of CO and COHb is of utmost importance in forensic toxicology to determine the cause of death in cases of CO poisoning, fire and explosions. To this end, reliable and updated analytical methods are required.

Reference Number: 50104

HAO, H., H. ZHOU, X. LIU, Z. ZHANG, AND Z. YU. "An Accurate Method for Microanalysis of Carbon Monoxide in Putrid Postmortem Blood by Head-Space Gas Chromatography: Mass Spectrometry (HS/GC/MS)." *Forensic Science International*, 229: 116–121, 2013 (6 tables, 5 figures, 29 references)

Abstract: A headspace GC/MS method is described for the analysis of CO in small volumes of putrid post-mortem blood. A 0.25 mL sample of blood is added to a headspace vial with 0.25 mL of lactic acid (25%) to release the CO. The vials are sealed and heated to 50 degrees Celsius for 30 minutes, from which 200 micromoles per litre of headspace is removed and injected onto a GC/MS. The regression coefficient (r^2) of the method was 0.995, and a relative standard deviation of less than 1% was obtained. In one case in which five workers were found dead in their dormitory, the use of this method on blood samples that were four months old showed that the COHb was greater than 50% in all five victims compared to less than 0–29% for a UV method. In another case in which blood samples were stored at room temperature for 15 days, during which the blood turned black red and smelled terrible, the headspace GC/MS method showed COHb concentrations of 65.1% and 68.7% respectively, whereas the UV method showed that COHb was not detected.

Table: A Comparison of the Headspace GC/MS and UV Methods on Putrid Blood Samples for Five Victims of CO Poisoning

Case	COHb percent (headspace GC/MS) (%)	COHb percent (UV method) (%)
A	56.8	19
B	73.9	29
C	52.9	not detected
D	64.5	18
E	71	23

Source: Adapted from Hao et al. 2013.

In China, death due to CO poisoning was identified by elevated COHb levels (>40%) in postmortem blood. In contrast, use of UV spectrometry provided values of COHb% that were well below 30% and in three cases above, not even detectable, which would not have supported CO poisoning as the cause of death in these individuals. The UV method is often used to detect CO concentrations in fresh blood samples, but may not be suitable for spoiled blood, where it may give values lower than the true levels. According to our method it was shown that the effect of preserved temperature was negligible. The % COHb levels in the putrefied blood samples showed good consistency with the % COHb levels determined in the fresh blood. Thus, the HS/GC/MS technique may have significant advantages over the UV method, particularly when blood samples are not fresh.

Reference Number: 50105

COSTANTINO, A.G., J. PARK, AND Y.H. CAPLAN. "Carbon Monoxide Analysis: A Comparison of Two CO-Oximeters and Headspace Gas Chromatography." *Journal of Analytical Toxicology*, 10: 190–193, 1986 (5 tables, 3 figures, 16 references)

Abstract: The post-mortem blood COHb concentration were determined in 62 post-mortem blood samples using the CO-oximeter model IL-182 (578, 568, and 548 wavelengths) and IL-282. These results were compared to a headspace GC method that used saturated and unsaturated blood samples, which were analyzed by a molecular sieve column after the CO in the blood was liberated using Ke3Fe(CN)6, and the CO was converted to methane when passed over a catalyst to be analyzed by the detector. The correlation between the methods ranged from 0.981 to 0.986.

The following conclusions were reached: (1) the IL-182 CO-Oximeter is an acceptable instrument for CO analysis; (2) the IL 282 CO-Oximeter is at least equivalent to the IL182 in determining % COHb saturation; (3) both the CO-Oximeter instruments compared favorably with the GC method; and (4) adequate forensic interpretation may be derived from each of these three techniques.

Reference Number: 50106

OLSON, K.N., M.A. HILLYER, J.S. KLOSS, R.J. GEISELHART, AND F.S. APPLE. "Accident or Arson: Is CO-Oximetry Reliable for Carboxyhemoglobin Measurement Postmortem?" *Clinical Chemistry*, 56: 515–520, 2010 (1 table, 1 figure, 10 references)

Abstract: A 46-year-old woman was found dead in the basement of her house after a fire. She showed wounds to her head and soot in her nose, mouth, and trachea. The husband had knocked her unconscious and, believing she was dead, set the house on fire. Her post-mortem COHb, as determined by a six wavelength CO-oximeter was 61.4%. Older UV spectrophotometric CO-oximeter using four or less wavelengths gave inaccurate results due to methemoglobin, sulfhemoglobin, micro-coagulates, lipid-caused turbidity, and putrefaction. But pre-treatment of the sample with sodium dithionite to reduce methemoglobin (MetHb) and oxyhemoglobin and filtration and use of CO-oximetry with wavelengths greater than four allow it to be an accurate and valid method to determine post-mortem COHb as dose GC.

The jury found the husband guilty of second-degree unintentional felony murder. On appeal of the conviction, the defendant challenged the validity of the CO-Oximetry results. During the pretrial to determine the admissibility of the CO results, a clinical chemist provided expert testimony that the CO-Oximetry results obtained with > 4-wavelength technology showed no MetHb interference. On this basis of this testimony, the court ruled that the results were valid and therefore were admissible during the trial.

Reference Number: 50107

KRISTOFFERSEN, L., M. VEVELSTAD, I.A. HANSEN, B. STROMSVAG, AND D.H. STRAND. "Comparative Evaluation of Carboxyhemoglobin Quantification in Postmortem Whole Blood by Co-Oximetry and Headspace Gas Chromatography with Flame Ionization and Atom Absorption Spectrophotometry." *Journal of Analytical Toxicology*, 47: 311–316, 2023 (1 table, 2 figures, 22 references)

Abstract: A comparative evaluation of the CO-oximeter method (using 128 wavelengths) and the traditional headspace GC flame ionization detector and atomic absorption spectrophotometry (HS-GC-FID-AAS) method was conducted on 124 post-mortem whole blood samples over a period of three years. The short-term stability of COHb in post-mortem blood samples stored at four degrees Celsius for up to eight days showed a slight decrease of up to −6.9%. The CO-oximetry method uses 0.1 mL of blood compared to one mL volume of sample required by the HS-GC-FID-AAS method. The coefficient of variation percentage was 1.6% for the CO-oximetry method and 7.7% for the HS-GC-FID-AAS method. Two post-mortem blood samples could not be analyzed by the CO-oximetry method. In one post-mortem blood sample, the CO-oximetry method reported a false negative in a sample containing less than 10% COHb. The Co-oximetry method is fast, requires less resources, and uses one-tenth the blood sample volume as the HS-GC-FID-AAS method. As a result of this evaluation, this forensic laboratory switched to the CO-oximetry method using 128 wavelengths.

> In the first decades of COHb analysis in postmortem blood samples, the accuracy of spectrophotometric methods measuring at three wavelengths was unsatisfactory. The misleading results obtained by spectrophotometry were related to the decomposition of autopsy blood, resulting in the alteration of blood pigments and hence the absorbance spectrum. The development of CO-oximeters measuring at several wavelengths has overcome some of these initial problems with postmortem samples.

5.02 POST-MORTEM CHANGES

> The remarkable carmine-red injection of the liver in carbonic oxide poisoning is indeed a most striking appearance and is quite characteristic. The kidneys too, are usually of an exceedingly florid colour, and the muscles have a beautiful pink coloration, which if once seen, can never be forgotten.
>
> —A. Gamgee, "On Poisoning by Carbonic Acid Gas, and by Charcoal Fumes" (1867) (**40401**)

Post-mortem blood COHb concentrations in drowning deaths may increase by up to 10% (50201). The blood CO concentration decreases greatly when preserved in formalin (50202). Heating blood samples up to 50 degrees Celsius does not affect the blood COHb concentration significantly (50203). Resuscitation attempts in CO-poisoning victims may lower the blood subclavian blood concentration (50204, 50208). Blood is stable for up to two years in sealed containers when refrigerated (50205) but may decrease dramatically if the container is not, or only loosely, sealed (50206, 50209). Pooled blood on the floor may pick up CO, which increases its COHb up to 11% in a fire environment (50207). Blood CO analysis should always be conducted even if the victim does not show a typical cherry red lividity (50212).

Reference Number: 50201

KOJIMA, T., I. OKAMOTO, M. YASHIKI, T. MIYAZAKI, F. CHIKASUE, K. DEGAWA, S. OSHIDA, AND K. SAGISAKA. "Production of Carbon Monoxide in Cadavers." *Forensic Science International*, 32: 67–77, 1986 (6 tables, 3 figures, 16 references)

Abstract: CO, total hemoglobin (Hb), and COHb concentrations were determined in the post-mortem blood or body fluid of 33 deceased who were not exposed to CO or fires. It is not known if the victims were smokers or not. The COHb was determined by headspace GC and measurement of total Hb. In the blood, the highest COHb was 3.6% for death occurring on land, 6.1% in freshwater drownings, and 6.2% in salt-water drownings. In body fluids, the COHb was 10.1%, 44.1%, and 83.7% respectively.

There were many cases in which the HbCO saturations in the body cavity fluids of drowned bodies were more than 10%. The HbCO saturations in the blood samples of all cases, however, were not more than 10%. These results seem to indicate that the interpretation of the HbCO saturation in the blood would not be significantly affected by the postmortem formation of CO, and that body cavity fluid should not be used for CO determination.

Reference Number: 50202

WINEK, W.L., F.M. ESPOSITO, AND D.P. CINICOLA. "The Stability of Several Compounds in Formalin-Fixed Tissues and Formalin-Blood Solutions." *Forensic Science International*, 44: 159–168, 1990 (3 tables, 6 references)

Abstract: The stability of CO and cyanide in 5% and 8% formalin-whole blood samples were determined. The CO concentration, as analyzed by the Conway diffusion cell method, decreased by more than 50% in the 5% solution after one week of storage and after 24 hours of storage for the 8% solution. Cyanide was not detected immediately after spiking the 5% and 8% formalin-whole blood solutions.

The embalming fluids used during the Civil War era often contained large quantities of arsenic, which made it impossible to prove that an individual died from arsenic poisoning, once the body was embalmed. Even though arsenic is no longer used in embalming fluids, a similar problem exists in that methanol, ethanol and isopropanol are commonly found in the embalming fluids of today, along with aldehydes, ketones, esters, perfumes, and colorizing agents. The main constituent of embalming fluid is formalin, which is 37% formaldehyde in water. Formalin complexes with protein found throughout the body producing the fixed effect associated with embalmed tissues. This process postpones decomposition for an extended period of time.

Reference Number: 50203

SETO, Y., M. KATAOKA, AND K. TSUGE. "Stability of Blood Carbon Monoxide and Hemoglobins during Heating." *Forensic Science International*, 121: 144–150, 2001 (4 tables, 2 figures, 44 references)

Abstract: COHb, oxyhemoglobin (HbO_2), and MetHB were measured in control human blood, and 50% and 100% CO-saturated blood using spectrometry (538 and 555 nm wavelength). The COHb results of this method were compared to headspace GC. Under incubation below 54 degrees Celsius, all Hb was stable except for HbO_2, which was partially oxidized to MetHb. The incubation of the blood at 65 degrees Celsius causes coagulation and a brown colour of the blood. The MetHb was denatured, but COHb was relatively resistant to heat denaturation.

Table: COHb Concentrations as Measured by Spectrophotometry and HS-GC in Blood Samples Spiked at 0% and 50% CO and Stored at 4, 37, 54, and 65 Degrees Celsius

Temperature (degrees Celsius)	Initial (spiked) COHb (%)	Appearance of blood	Spectro-photometry COHb (%)	Headspace GS COHb (%)
4°C	0	Red fluid	3.9	0.4
4°C	50	Red fluid	46.8	42.3
37°C (overnight)	0	Dark red fluid	8.0	1.2
37°C (overnight)	50	Red fluid	48.4	42.3
54°C (3 hours)	0	Dark red fluid	4.8	1.5
54°C (3 hours)	50	Dark red fluid	48.7	50.4
65°C (1 hour)	0	Brown coagulated	35.8	18.9
65°C (1 hour)	50	Brown coagulated	52.3	112.3

Source: Adapted from Seto, Kataoka, and Tsuge 2001.

Our in vitro experiment demonstrates that heating of blood sample or Hb solution manifest considerable change of Hb forms and CO levels. Even though under mild heating (50–54C), MetHb was formed from HbO_2, which leads to misjudgment for poisoning from oxidizable gases. COHb was stable against heating and we could detect toxic level of CO gas even in heat-treated blood. Stability of COHb and severe denaturation of Hb-O_2 and MetHb at 65C heating, resulted in artificial increase of % CO-Hb which leads to misjudgments for CO poisoning. The absorbance spectra of heat-treated Hb molecules were deteriorated and this spectral deviation caused wrong spectrophotometric measurement of % COHb which was ascertained in the blood samples of fatal fire casualties. Spontaneous CO production was observed in heated samples.

Reference Number: 50204

RICE, H.M. "Carboxyhemoglobin Dissociation in the Cadaver Following Attempted Resuscitation." *Journal of Clinical Pathology*, 29: 27–29, 1976 (1 table, 2 figures, 3 references)

Abstract: The fatal post-mortem blood COHb concentrations were determined in 184 male and 116 female victims of CO poisoning. There were 17 male and seven female victims with COHb between 40% and 49% who were not elderly (that is, greater than 70 years of age).

Table: Subclavian and Femoral Blood COHb Concentrations in Cases of Resuscitation

Case	Age	COHb (subclavian) (%)	COHb (femoral) (%)
1	14 months	15	31
2	57 years	32	52
3	43 years	42	59
4	5 months	34	48

Source: Adapted from Rice 1976.

I believe that the saturation levels were probably falsely low, due to dissociation of carboxyhemoglobin after death when artificial respiration had been used in attempts at resuscitation. Since blood with a high concentration of carboxyhemoglobin does not coagulate, vigorous attempts at artificial respiration cause movement of blood into and out of the lungs which may amount to actual circulation if sufficient to activate the valvular system. This is more likely to occur when the thorax is highly elastic in young subjects. Gaseous exchange with dissociation of carboxyhemoglobin, occurs and is facilitated by a high oxygen content of the respired gas, as when breathing apparatus is used. The degree of gaseous exchange depends on the vigour and duration of the artificial respiration but will always be greatest in blood from the lungs themselves, the heart, and the great vessels closest to them. It is, therefore, important that carboxy-hemoglobin saturation levels be determined on blood from a more distant site, preferably the femoral vein in the thigh, where diffusion of gases and mixture of highly saturated and partly dissociated blood can have little, if any, effect.

Reference Number: 50205

KUNSMAN, G.W., C.L. PRESSES, AND P. RODRIGUEZ. "Carbon Monoxide Stability in Stored Postmortem Blood Samples." *Journal of Analytical Toxicology*, 24: 572–578, 2000 (1 table, 14 references)

Abstract: The stability of post-mortem blood CO concentrations were determined in samples stored in red (no preservative), grey (sodium fluoride), or purple (ethylenediaminetetraacetic acid) top tubes at three degrees Celsius for two years. The samples were analyzed by either the IL-282 or IL-682 CO-oximeter (a visible spectrophotometer device). A total of 64 medical examiner cases (involving 189 samples) were analyzed over the two-year period. The initial blood COHb concentrations ranged from less than 1% to 82.6%.

Table: Post-Mortem Cases with the Greatest Differences in COHb Percent between Three Types of Blood Collection Tubes

Case number (1996)	Initial CO (%)	Red top (%)	Grey top (%)	Purple top (%)
0022	34.3	39.4	31.7	33.4
0190	2.5	3.5	< 1	4.0
0476	70.1	70.6	70.7	80.2
0574	1.2	< 1	6.6	1.5
1366	1	< 1	< 1	5.1
1458	27.2	7.9	16.0	7.1
1799	7.7	13.1	9.4	9.4

Source: Adapted from Kunsman, Presses, and Rodriguez 2000.

In this study, preservative, volume of headspace, and initial % COHb did not significantly affect the % COHb level in postmortem blood specimens refrigerated for two years following specimen collection at autopsy. It is therefore, acceptable to perform a repeat analysis on a blood specimen, perform a second analysis on a case using a different blood specimen from that used in the initial analysis or perform a first analysis on a stored postmortem blood specimen up to two years after the specimens were collected if they were stored at 3°C over the entire storage period. It is recommended, however, that caution be used in the interpretation of such analyses with samples that have a THb less than 4 g/dL.

Reference Number: 50206

OCAK, A., J.C. VALENTOUR, AND R.V. BLANKE. "The Effects of Storage Conditions on the Stability of Carbon Monoxide in Postmortem Blood." *Journal of Analytical Toxicology*, 9: 202–206, 1985 (5 tables, 3 figures, 13 references)

Abstract: The blood COHb concentrations were measured in six spiked blood and 14 actual post-mortem cases, stored for up to four months in sealed and unsealed tubes stored at 4 degrees Celsius or room temperature for up to four months. The blood COHb was determined by a two-wavelength CO-oximeter (pre-treatment with sodium dithionite) or headspace GC measuring saturated versus unsaturated blood. CO_2 concentrations were also measured and showed that the oxidation of CO to carbon dioxide (CO_2) did not occur in the opened tubes.

Table: Initial Post-Mortem Blood COHb and After Two Months Storage at Room Temperature in Sealed and Uncapped Vials

Mean initial COHb (%)	Mean COHb after room temperature storage in capped vials for two months (%)	Mean COHb after room temperature storage in open vials for two months (%)
94	88	4
77	77	4
64	60	4
49	47	3
35	40	3
19	18	3

Source: Adapted from Ocak, Valentour, and Blanke 1985.

Postmortem carbon monoxide (saturation) may remain stable for at least 4 months in sealed vials. However, if the blood vial is not tightly sealed, the CO may be lost. These losses can be best explained by the passive diffusion of CO gas, although methemoglobin formation may play a minor role. The loss was dramatic when the original % CO saturation was high, possibly due to CO cooperativity and hence facilitated dissociation. When CO saturation was initially low, the loss was not as dramatic.

Reference Number: 50207

THOREN, T.M., K.S. THOMPSON, P.S. CARDONA, A.K. CHATURVEDI, AND D.V. CANFIELD. "In Vitro Absorption of Atmospheric Carbon Monoxide and Hydrogen Cyanide in Undisturbed Pooled Blood." *Journal of Analytical Toxicology*, 37: 203–207, 2013 (1 table, 2 figures, 25 references)

Abstract: In aircraft crashes, the severe trauma can cause body parts of the victim to be scattered and large wounds or pooled blood to be exposed. The pooled blood may then be exposed to a fire atmosphere that contains CO and hydrogen cyanide (HCN). To determine this postmortem effect, human blood samples from non-smokers were exposed to 5,532–33,193 parts per million (ppm) of CO for 30 and 60 minutes and to 100–200 ppm of HCN for 15–60 minutes. The COHb increase ranged from 4.3% to 11.0%, and the blood cyanide from 1.43 micrograms (ug) per mL to 5.01 ug/mL.

> Although the findings of the present study demonstrated that the COHb level in the undisturbed pooled blood exposed to an atmosphere containing CO within the parameters of this experiment would increase only up to 11%, it does not necessarily exclude the possibility that the COHb level in pooled blood from an actual fire victim may be higher than 11% if that blood were exposed to a very high concentration of atmospheric CO from an actual fire. This COHb increase would occur because of the high concentration gradient between blood and CO-rich air. Increases in the COHb level would also be possible if the pooled blood were disturbed, thereby producing a larger dynamic exposure surface area. With HCN however, there is strong evidence that CN [cyanide] levels would easily increase in the undisturbed pooled blood due to the absorption of atmospheric HCN present in the fire environment.

Reference Number: 50208

DINIS-OLIVEIRA, R.J., F. CARVALHO, T. MAGALHAES, AND A. SANTOS. "Postmortem Changes in Carbon Monoxide Poisoning." *Clinical Toxicology*, 48: 762–763, 2010 (2 figures, 3 references)

Abstract: A blood COHb of 50–60% is generally considered fatal, although, in persons with concomitant cardiovascular and lung diseases, it may be lower. Cherry red lividity is not only due to CO poisoning but can also occur in cases of hypothermia and cyanide poisoning if the area of the cadaver was covered with wet clothing or if a recently deceased body is in the morgue's refrigeration. The discolouration may also be lost if the COHb level decreased during resuscitation.

Moreover, establishing the cause of death in a victim who has been in a house fire may be problematic, if samples are not obtained early after the event. Indeed, if the CO-intoxicated victim survives for several hours, postmortem blood samples may fail to show the intoxication through the measurement of COHb concentrations. In these circumstances, blood taken at the time of admission to the hospital is of particular value, in addition to scene investigation and postmortem changes. Particularly, autopsy findings in CO deaths are fairly characteristic and can be important forensic clues in the cause of death.

Reference Number: 50209

CHACE, D.H., L.R. GOLDBAUM, AND N.T. LAPPAS. "Factors Affecting the Loss of Carbon Monoxide from Stored Blood Samples." *Journal of Analytical Toxicology*, 10: 181–189, 1986 (3 tables, 15 figures, 10 references)

Abstract: One-mL blood samples spiked to 80% COHb were stored at 23 degrees Celsius and then exposed to 49 mL of air and stored for 5, 24, and 48 hours. The blood COHb decreased to 60%, 50%, and 50% respectively. For three one mL post-mortem blood samples (mean 83% COHb), stored under the same conditions, there was a lesser decrease to a mean COHb of 64% after 24 hours storage. Samples stored in 49 mL of nitrogen showed no decrease in COHb, and those stored with oxygen showed a greater decrease in COHb. Samples stored at lower temperatures showed a lower decrease in COHb.

Exposure of CO-containing blood to the atmosphere will result in a decrease in the % COHb saturation. The rate of decrease is related to the ratio of the surface area to the volume of blood exposed to the atmosphere, temperature and initial % COHb saturation. Carboxylated blood may lose nearly all of its bound CO if the exposure is sufficiently long.

Reference Number: 50210

CARSON, H.J., AND K. ESSLINGER. "Carbon Monoxide Poisoning without Cherry-Red Livor." *American Journal of Forensic Medicine and Pathology*, 22: 233–235, 2001 (8 references)

Abstract: A 75-year-old white man who lived alone was found dead in his car—a 1969 Ford Limited–with the garage door closed, and the car ignition key on, but the car was no longer running. He left a note regarding the disposition of his property. The environmental temperature of the prior two days was between six to eight degrees Fahrenheit. A cherry red liver was not observed, but the post-mortem blood COHb concentration was 86%. The blood was re-analyzed two weeks later, and the same COHb was obtained.

The case is a good example of the importance of obtaining a carboxyhemoglobin level in any situation where CO poisoning may have played a role, even if there is no obvious external evidence of CO poisoning. Without a CO level, this man's death could have been wrongly attributed to a natural manner of death, coronary artery disease; there was strong family bias to obtain this ruling. The reality of a suicidal manner of death by CO poisoning would have been missed.

5.03 FATAL COHb

[A] poison that troubles the blood by displacing oxygen.

—Claude Bernard (1813–78), cited by G.L. Sternbach and J. Varon, "Claude Bernard: On the Origin of Carbon Monoxide Poisoning" (2003) **(10118)**

The mean post-mortem fatal COHb of fire victims tend to be lower than in pure CO-poisoning victims as there are other toxic agents involved in addition to CO. Victims with pre-existing diseases such as cardiovascular disease or alcohol/drug addictions and the elderly also tend to die at lower COHb concentrations (50301, 50303). One of the highest recorded post-mortem blood COHb concentrations (91%) was found in a man who carefully committed suicide using a charcoal brazier (50304).

Reference Number: 50301

CONWAY, K., O.R. RAYES, T. BROWN, AND W. WEBB. "Atherosclerotic and Hypertensive Cardiovascular Disease Are Associated with Death at Sublethal Carboxyhemoglobin Levels: A Postmortem Study." *Journal of Forensic Sciences*, 65: 855–859, 2020 (4 tables, 2 figures, 22 references)

Abstract: The post-mortem iliac vein blood COHb concentrations were determined in 55 victims of house fires between 2007 and 2019. There were 23 victims without cardiovascular disease (CVD), eight with hypertensive CVD, 11 with atherosclerotic CVD, and 13 with a mixture of CVD. There was a mean 12% decrease in COHb in victims with hypertensive CVD and a 15% decrease in COHb in victims with atherosclerotic CVD. A mean decrease of 3% of COHb was found to occur for very 10 years of age even when adjusted for CVD. This shows the increased susceptibility of the elderly to fire deaths. There was not a statistically significant relationship between COHb and race, sex, body mass index, or cardiomegaly.

Table: Mean Post-Mortem Blood COHb in Fire Victims with Various Cardiovascular Diseases

	Controls	Hypertensive CVD	Atherosclerotic CVD	Mixture of CVD
Mean % COHb (%)	64.4	46.8	37.2	32.8
Mean age (years)	46	43.4	71.6	68.2
Soot in airway (%)	96	100	91	100

Source: Adapted from Conway et al 2020.

The findings here carry important practical significance for the medical examiner as well as important health implications. The effect of cardiovascular disease on fire fatality is significant, and it accounts for significantly lower levels of COHb observed at autopsy and postmortem toxicology. When significant disease is not identified in the setting of sublethal COHb, the contribution of senescence should be considered with care and other factors (such as exposure to hydrogen cyanide, another smoke product) should be investigated).

Reference Number: 50302

TSUNEYA, S., Y. MAKINO, F. CHIBA, M. KOJIMA, M. YOSHIDA, T. KISHIMOTOR, H. MUKAI, S. HATTORI, AND H. IWASE. "Postmortem Magnetic Resonance Imaging Revealed Bilateral Globi Pallidi Lesions in a Death Associated with Prolonged Carbon Monoxide Poisoning: A Case Report." *International Journal of Legal Medicine*, 135: 921–928, 2021 (1 table, 2 figures, 52 references)

Abstract: A man and woman were found dead in a car on a winter's day, with burned coal briquettes near their bodies. The victims had purchased the coal briquettes three days prior, and suicide was suspected. The woman had a post-mortem blood COHb of greater than 70%, and the man had 1%. It is suggested that the man died days later than the woman of pneumonia, dehydration, starvation, and hypothermia.

Table: Post-Mortem Findings of a Man and Woman Found Dead in a Car

Findings	Male victim	Female victim
COHb (%)	1	> 70
Colour of organs, muscles	Normal	Cherry-pink
Rectal temperature (degrees Celsius above ambient)	+20.5°C	+ 9.1°C
Pneumonia	Yes	No
Brain lesions	Yes	No

Source: Adapted from Tsuneya et al. 2021.

> We present a cause of death caused by suspected prolonged CO poisoning with bilateral globi pallidi lesions. This is the first study to report that PMCT [postmortem computed tomography] and PMMR [postmortem magnetic resonance] imaging can be used to detect CO poisoning-related brain lesions. Postmortem imaging, especially PMMR should be considered useful for detecting CO poisoning related brain lesions.

Reference Number: 50303

WEAVER, L., K. DERU, S. CHURCHILL, J. LEGLER, G. SNOW, AND T. GREY. "Carbon Monoxide Poisoning in Utah, 1996–2013." *Undersea and Hyperbaric Medicine*, 43: 747–758, 2016 (2 tables, 5 figures, 51 references)

Abstract: Between 1996 and 2013, there were 7,590 persons in Utah diagnosed with CO poisoning; of whom, 6,469 (85.2%) were treated and released in the emergency room (ER), 596 (7.9%) were admitted to hospital, and 525 (6.9%) died. Less than 10% of the CO victims were treated with hyperbaric oxygen therapy (HBOT). Non-fatal CO poisonings declined following a 2008 law requiring CO alarms in residences. The fatal CO poisonings consisted of 141 (27%) accidental, 361 (70%) suicidal, and two (0.4%) were due to homicides. Victims with CVD had a lower post-mortem mean blood COHb (51.2%) compared to those without CVD (70.8%). The highest mean post-mortem blood COHb occurred in the 20–29-year-old age group (mean 72.5%). Of the 59 victims who died in fires, the mean post-mortem COHb was 59.9%. The mean post-mortem blood COHb was 65.7% in the victims who had a positive alcohol and drug screen, compared to 69.2% in those who did not.

Table: Mean Post-Mortem COHb According to Age

Age group (years)	Mean post-mortem COHb (%)
0–9	65.8
10–19	70.7
20–29	72.5
30–39	70.9
40–49	69.1
50–59	67.2
60–69	64.0
70–79	59.9
80+	63.7

Source: Adapted from Weaver et al. 2016.

The incidence of CO poisoning in Utah is declining, but CO poisoning is still common. Alarm legislation may aid prevention efforts. An educational campaign addressing the many causes and circumstances of CO poisoning is required for prevention.

Reference Number: 50304

FERORELLI, D., G. MANDARELLI, L. SPAGNOLO, F. MISCEO, G. STRISCUILLO, L. POLO, B. SOLARINO, AND A. DELL'ERBA. "An Unusual Suicide by Carbon Monoxide Intoxication: 91% HbCO Saturation in a Sealed Setting." *Emerging Trends in Drugs, Additions, and Health* 1: 100021, 2021 (1 table, 4 figures, 33 references)

Abstract: CO is one of the most common, lethal poisons and has a worldwide death rate of 2.2 per 100,000 CO poisonings. About 60% of the deaths occur at home. In this case, a 60-year-old man was found death inside his house with a strong smell of burning inside. The windows and doors were sealed with scotch tape and a burned charcoal brazier was found inside. A cherry red hypostasis was observed at the rear portion of the body as well as a red colouration of the nails. No drugs or alcohol could be detected, but the post-mortem COHb was 91%.

The value of 91% HbCO saturation we detected in the present case was the highest among the 55 cases in which HbCO assessment was requested at the Institute of Forensic Medicine in Bari. Among these cases, 16 are those in which the cause of death was eventually identified in a carbon monoxide intoxication with an average HbCO saturation of 57.12%. Moreover, to confirm the exceptionality of this data we reviewed two different databases (Pub Med, and Google Scholar) that were questioned using the main keyword-carbon monoxide poisoning that was crossed with HbCO saturation. The value detected in the present case was one of the highest ever reported.

5.04 OTHER TISSUES

Hemoglobin has one strange and dangerous quirk: It vastly prefers carbon monoxide to oxygen. If carbon monoxide is present, hemoglobin will pack it in, like passengers on a rush-hour train and leave the oxygen on the platform. That's why it kills people.

—Bill Bryson, *The Body: A Guide for Occupants* (2019)[2]

Numerous other post-mortem tissues can be analyzed for CO, such as thoracic cavity fluid, spleen, and liver) when post-mortem blood is not available (50401–3).

Reference Number: 50401

CASALI, M.B., L. SIRONI, M. CALIGARA, A. BLANDINO, S. CIRCELLI, D. SCHIAVI, AND C. CATTANEO. "How Reliable Are Parenchymal Tissues for the Evaluation of Carbon Monoxide Poisoning? A Pilot Study." *Journal of Forensic Sciences*, 60: 488–494, 2015 (4 tables, 1 figure, 19 references)

Abstract: The reliability of solid human tissue samples (brain, lung, liver, spleen, heart, psoas muscle, and subcutaneous fat) was evaluated in 11 female and 21 male victims of CO exposure, including nine burned bodies. The blood COHb concentrations were determined by the CO-oxymeter, UV-visible, and GC-TCD methods. The tissue CO concentration was determined using the GC-TCD CO saturation method using 0.500 grams of minced tissue, 0.500 mL of distilled water, and 0.500 *potassium ferricyanide* (K_3 [Fe $(CN)_6$]). The best tissues from burned bodies were lung (r = 0.84), kidney (r = 0.79), spleen (r = 0.92), and liver (r = 0.83). The brain and adipose tissues have the lowest CO tissue/blood CO ratio, which

2 Bill Bryson, *The Body: A Guide for Occupants* (Toronto: Doubleday Canada, 2019).

suggests the existence of a fat barrier that protects against free CO from the blood.

Table: Mean Post-Mortem Tissue/Blood COHb Ratio for 32 Cases

Tissue sample	Mean tissue/blood COHb ratio
Brain	7.1 times
Lung	44.7 times
Liver	11.4 times
Kidney	21.9 times
Spleen	58.4 times
Heart	28.8 times
Adipose tissue	2.0 times

Source: Adapted from Casali et al. 2015.

With specific focus on lung-liver-kidney-spleen tissues, it seems that body charring does not preclude the forensic analysis of a CO poisoning from parenchymal samples.

Reference Number: 50402

MULEYA, J., L. VUKO, J. MADER, AND DAVIES. "Thoracic Cavity Fluid as an Alternative Specimen for Carboxyhemoglobin Measurement in Post-Mortem Toxicology." *Forensic Science International*, 349: 111731, 2023 (1 table, 1 figure, 9 references)

Abstract: Post-mortem peripheral or cardiac blood and thoracic cavity blood was collected from 15 cases of suspected CO poisoning in Cape Town, South Africa. The samples were analyzed for CO and total Hb with an ABL825 flex radiometer CO-oximeter. The blood COHb concentrations ranged from less than 5% to 74%. The instrument could not measure CO in five cases as the total Hb concentration in the thoracic cavity sample was too low (< 1.61 g/dL). A strong positive relationship was observed between the COHb results of the two samples (r = 0.975).

Blood is commonly used for the measurement of COHb in suspected CO poisoning deaths. In many deaths due to charring or exsanguination, blood may not be available. The current study investigated the usefulness of TCF [thoracic cavity fluid] as an alternative to post-mortem blood in instances where the latter specimen is unavailable. The Pearson's correlation coefficient indicated a strong positive linear relationship for COHb between

TCF and blood specimens. This was corroborated by the results of a Bland-Altman analysis, which showed a statistical agreement between the two specimens. These findings showed a statistical agreement between the two specimens. These findings suggest that TCF may be used as an alternative specimen for COHb determination using a CO-oximeter, however further studies with a larger sample size are recommended.

Reference Number: 50403

WU, S.C., B. LEVINE, J.C. GOODIN, Y.H. CAPLAN, AND M.L. SMITH. "Analysis of Spleen Specimens for Carbon Monoxide." *Journal of Analytical Toxicology*, 16: 42–44, 1992 (3 tables, 3 references)

Analysis: Two mLs of post-mortem blood or two grams of spleen were mixed with one mL of CO-oximeter zeroing solution. After vortexing or homogenizing and centrifuging the supernatant, one mL was added to the auto-sampler, and 1.5 mL was saturated with CO and measured by headspace GC. Forty paired blood and spleen samplers were analyzed and compared. The post-mortem blood COHb ranged from less than 1% to 72%. The spleen COHb ranged from 35% to 68%.

For this study, we concluded that spleen CO sat. can reflect blood CO sat. in certain circumstances. Specifically, a spleen CO less than 10% sat. is indicative of normal blood CO. A spleen CO greater than 30% would be highly suggestive of toxic or lethal exposure to CO. A spleen CO between 10 and 30% cannot be interpreted without additional information.

CHAPTER 6

Other Toxic Gases

> While information is the oxygen of the modern age, disinformation is the carbon monoxide that can poison generations.
>
> —Newton Lee, *Counterterrorism and Cybersecurity: Total Information Awareness* (2024)[1]

Although the most ubiquitous toxic gas is carbon monoxide (CO), there are others as well that are mainly produced in fires.

6.01 COMBUSTION TOXICOLOGY

During a fire, there are numerous toxic gases produced depending on the material burning, temperature, and amount of oxygen available. Some of the toxic gases, products, and their sources are:

- CO : carbon-containing compounds
- Nitrogen dioxide (NO_2) : wood
- Hydrogen chloride (HCl) : plastics
- Hydrogen cyanide (HCN) : wool, silk, rubber, polyurethane
- Benzene : petroleum products
- Aldehydes : wood, cotton, paper
- Ammonia : nylons

Chapter 6

1 Newton Lee, *Counterterrorism and Cybersecurity: Total Information Awareness* (London: Springer Nature, 2024).

CO is responsible for about 80% of fire fatalities (60101), but toxic HCN is produced by the burning of plastics, silk, and cotton (60102, 60104, 60106). A post-mortem blood HCN of greater than 0.5 micrograms (ug) per millilitre (mL) indicates exposure to smoke (60106). Other toxic constituents of fires in addition to CO include heat/fire, oxygen depletion, HCN, carbon dioxide (CO_2), and halogenated acids (60105). Heat causes the red blood cells to burst and lowers the COHb concentration (60103).

Reference Number: 60101

COHEN, M.A., AND L.J. GUZZARDI. "Inhalation of Products of Combustion." *Annals of Emergency Medicine*, 12: 628–632, 1983 (2 tables, 42 references)

Abstract: A fire in a closed space consumes oxygen, causing the oxygen level to drop from 21% to 10–15%. The air is also displaced by fire-related gases such as CO, which is present up to 7,500 parts per million (ppm) and is responsible for about 80% of the fire deaths. Other toxic gases found include cyanide, acrolein, HCl, phosgene, and ammonia. Burned plastic generally releases great quantities of dense smoke.

Table: Symptoms and Signs of the Inhalation of Products of Combustion

Central nervous system	Dizziness, headache, confusion, seizures, hallucinations, coma
Respiratory	Rapid breathing, cough, hoarseness, wheezing, sooty sputum, chest pain/tightness
Facial	Singed nasal hair, soot in oropharynx, facial burns
Cardiac	Arrhythmias

Source: Adapted from Cohen and Guzzardi 1983.

The atmosphere of a fire is deadly to breathe. Firefighters or building occupants may be victims of the heat, irritating smoke, depleted oxygen, carbon monoxide, and such other toxic gases as cyanide, hydrogen chloride, and acrolein. Increasing numbers of homes and public buildings are being built and furnished with highly flammable synthetic materials that give off copious smoke and toxic gases when burned. Whether or not there are cutaneous burns, the possibility of inhalation injury must be considered in any fire victim. All victims of a fire environment should be presumed to have CO intoxication and should be treated with 100% oxygen until the COHb level is within normal limits. In an extreme situation, cyanide intoxication should be suspected and administration of sodium thiosulfate may be life saving.

Reference Number: 60102

SON, M.-H., Y. KIM, Y.-H. JO, AND M. KWON. "Assessment of Chemical Asphyxia Caused by Toxic Gases Generated from Rigid Polyurethane Foam (RPUF) Fires." *Forensic Science International*, 328: 111011, 2021 (1 table, 4 figures, 34 references)

Abstract: Rigid polyurethane foam (RPUF) is widely used for thermal and sound insulation due to its low cost, light weight, and low thermal conductivity. When RPUF burns, CO, cyanide, and NO_2 are produced at differing concentrations depending on the oxygen availability. The concentrations of these toxic gases were measured in post-mortem blood from the right heart in 38 male victims of a RPUF fire in a warehouse construction site in South Korea on 29 April 2020. Twenty-five victims had no cutaneous burns (Group 1), and 13 victims had severe burns (Group 2). The mean blood COHb and cyanide concentrations were 1.6 times and 2.2 times greater respectively than in Group 2. The generation of CO in fires is 10 times greater than the cyanide concentration, but CO is one-twenty-fifth as toxic as cyanide. However, there was no correlation between blood CO and cyanide concentrations (regression coefficient = 0.173). NO_2 has 34 times greater toxicity than CO and 1,500 times greater affinity for hemoglobin than oxygen. NO_2 causes intense irritation of the eyes and throat and serious damage to the lungs. Methemoglobin (MetHb) can be increased by oxidizing agents, the inhalation of NO_2, and heat.

Table: Mean Post-Mortem Blood COHb, HCN, and MetHb Concentrations in Victims with and without Burns in RPUF Fires

Toxic gas	Victims with no burns (Group 1)	Victims with severe burns (Group 2)
COHb (%)	29.0	11.1
HCN	2.3 ug/mL	0.6 ug/mL
MetHb (%)	4.7	22.8

Source: Adapted from Son et al 2021.

We investigated the concentrations of COHb, cyanide, and MetHb in the post-mortem blood of 38 fire victims to demonstrate the possibility of inhalation of *toxic gases generated in RPUF fires. In group 1, the concentrations of COHb and cyanide were relatively high, whereas free MetHb was extremely low. This implies a low probability of NO_2* inhalation because of the distance of site from the fire. At the same time, in group 2, 26.8% of victims were found in the basement and 22.5% on the two floors, indicating that there was some possibility of NO_2 inhalation, which meant that victims were found in or near the fire zone. In conclusion, the effects of combined CO, HCN, and NO_2 inhalation should be considered in RPUF fires and this study reiterates that toxic gas poisoning is a major contributor to chemical asphyxia in large fire accidents.

Reference Number: 60103

YOSHIDA, M., J. ADACHI, T. WATABIKI, Y. TATSUNO, AND N. ISHIDA. "A Study on House Fire Victims: Age, Carboxyhemoglobin, Hydrogen Cyanide, and Hemolysis." *Forensic Science International*, 52: 13–20, 1991 (8 figures, 24 references)

The correlation between age, COHb, HCN, oxygen depletion, and hemolysis was studied in 120 fatal house fires. The mean age of the victims was 53.1 years (range 2–90 years), and nearly half (47.8%) of the victims were 60+ years of age. The correlation between age and COHb was r = –0.1746, but only one victim under 30 years of age had a COHb of less than 30%. For both COHb and cyanide, the left blood concentration was higher than the right. The greater the hemolysis of the blood, the lower the COHb.

In comparing, blood carboxyhemoglobin concentration between the right and the left ventricle, the concentration of carboxyhemoglobin from the left ventricle was 1.1 times more than from the right ventricle. This is taken to indicate that blood from the left ventricle was exposed to carbon monoxide gas until death. The rationale for the negative correlation between hemolysis and carboxyhemoglobin is due to the fact that red blood cells (RBCs) burst by heating and carboxyhemoglobin concentrations subsequently decreased by the dissociation of carbon monoxide from carboxyhemoglobin. This paper indicates the rate of hemolysis acted as an indicator of decreased carboxyhemoglobin concentrations by heating. It is necessary therefore to investigate the hemolysis of fire victims in order to determine the dissociation of carbon monoxide.

Reference Number: 60104

ANTONIO, A.C.P., P.S. CASTRO, AND L.O. FREIRE. "Smoke Inhalation Injury during Enclosed-Space Fires: An Update." *Journal of Brasiliero de Pneumologia*, 39: 373–381, 2013 (1 figure, 43 figures)

Abstract: A night club fire in the City of Santa Maria, Brazil, in the early morning hours of 27 January 2013, killed 232 young adults. Of the survivors, 88 were admitted to intensive care unit with severe smoke inhalation injury (SII) and varying percentages of total body surface area burned. Up to one-third of all SII victims can present with acute upper airway inhalation injury. Pneumonia is one of the most common complications of SII (30%), especially in victims over 60 years of age where more than 20% of the total body surface area is burned. About 80% of all SII deaths are due to CO poisoning. Plasma lactate concentrations greater than 90 milligrams (mg) per decilitre (dL) correlate with HCN poisoning. It is recommended that all SII victims with decreased loss of consciousness or signs of hemodynamic instability and respiratory failure be empirically treated with hydroxocobalamin. HCN is formed by the incomplete combustion of carbon and nitrogen-containing compounds such as cotton, silk, and plastics. The lack of ventilation in a closed fire can increase the rate of HCN formation by 10 times.

> Enclosed-space fires should raise the suspicion of SII associated with loss of consciousness, facial burns, and large TBSA [total body surface area] burns. Treatment is basically supportive, consisting of immediate O_2 therapy, rapid administration of an antidote, and protective mechanical ventilation. The use of inhaled heparin in victims of thermal injury is recommended by some experts. In such patients the use of HBOT [hyperbaric oxygen therapy] is highly controversial, as is the use of additional ventilation modes. In addition to immediate management, individuals with SII should be extubated as soon as possible, and attention should be paid to the possibility of early tracheotomy.

Reference Number: 60105

GIEBULTOWICZ, J., M. RUZYCKA, P. WROCZYRISKI, AND D.A. PURSER. "Analysis of Fire Deaths in Poland and Influence of Smoke Toxicity." *Forensic Science International* 277: 77–87, 2017 (4 tables, 8 figures, 35 references)

Abstract: A detailed analysis was conducted of 263 fire deaths in the Warsaw area of Poland between 2003 and 2011. About 80% of the fire deaths occurred in a house/dwelling, and the majority of the victims were found dead in the room of the origin of the fire. In comparison to natural material, many synthetic polymers (plastics) generate more heat more swiftly, spread flames faster, and generate far greater amounts of denser visible smoke and more toxic gases. This has led to the addition of flame retardants that produce even more toxic gases. Seventy percent of the fire victims were having a meal, and 70% had consumed alcohol. The majority of the victims were between 20 and 70 years of age. Approximately 50% of the victims had inhaled lethal doses of toxic gases, while 80% had soot in their airways, which confirms that they were alive for some time during the fire. About 51% of all victims have a COHb of greater than 40%, and about 34 percent of the victims had a low COHb (less than 10–20%).

Table: Common Constituents of Smoke and Their Effect on Humans

Constituent of smoke	Effects
Heat/flame	Body surface burns and burns to the respiratory tract, especially the larynx and trachea
Oxygen depletion	Asphyxia, allows a faster increase in CO
CO	Asphyxia, narcosis, tissue hypoxia, loss of consciousness, organ failure, death
HCN	Asphyxia, tissue hypoxia, loss of consciousness, organ failure, death
CO_2	Asphyxia, respiratory acidosis, distress, narcosis, loss of consciousness, increased respiratory rate
Halogenated acids (HCl, hydrogen bromide, hydrofluoric acid)	Eye and respiratory tract irritation, escape impairment, incapacitation, necrosing airway mucosa, pulmonary edema, and inflammation

Source: Adapted from Giebultowicz et al. 2017.

The two main causes of injury and death in fires are exposure to toxic smoke and burns. From the fire statistics, it is difficult to determine the effects of exposure to optically dense, irritant smoke on escape in fire fatalities, but it is clear from incident investigations and interviews with fire survivors that occupants may become trapped or their escape fatally delayed by exposure to dense smoke. Since furniture fires have been shown to produce dense smoke very quickly following ignition, this is certainly a factor in the failure of smoke fatal victims to escape. In order to determine the effects of burns and asphyxiant gases as causes of incapacitation and death, the % COHb level is an important parameter because it provides a measure of both the extent and duration of exposure during a fire, and the total dose of both CO and other toxic gases to which the victims have been exposed.

Reference Number: 60106

TABIAN, D., G. DROCHIOIU, S.I. DAMIAN, N. GIRLESCU, O.T. GRADINARU, S.I. TOMA, AND D.B. ILIESCU. "Toxic Blood Hydrogen Cyanide Concentration as a Vital Sign of a Decreased Room Fire Victim-Case Report." *Toxics*, 9 (2): 36, 2021 (1 table, 3 figures, 24 references)

HCN is more likely to be encountered in fires in domestic structures, with high temperatures and low oxygen levels, due to its formation from the combustion of nitrogen containing material such as wool, silk, or polyurethane. The cyanide ion binds to the iron of cytochrome a3 oxidase within the mitochondria that inhibit oxidative phosphorylation, resulting in cellular adenosine triphosphate depletion and lactic acidosis. A 51-year-old man and his 77-year-old mother escaped a fire in their house, but the son returned to try to extinguish the fire with a wet coat. The body of the victim was found by firefighters. His mother said that he had consumed alcohol that day. At autopsy, second-degree burns were found over 50% of the body. Soot deposits were noted in the upper and lower respiratory tract and acute brain and pulmonary edema was found. The post-mortem femoral blood COHb was 6.15%. The blood alcohol content (BAC) was 436 mg/100mL and urine alcohol content was 588 mg/100mL. The blood HCN concentration was 1.094 ug/mL. Blood HCN greater than 0.5 ug/mL indicate exposure to smoke and that the victim was alive at the time of the fire.

> First, we considered that the cause of death was the severe second-degree burns on approximately 50% of the body surface, mixed with acute alcohol intoxication. However, relatively high levels of HCN in the blood and in particular the amount of HCN metabolized to HSCN [thiocyanic acid], which provided evidence of a higher initial HCN concentration, suggest a possible cyanide poisoning as a contributive effect.

6.02 CO_2/OXYGEN DEPLETION/CLOSED VENTILATION SYNDROME

> [A] fact worthy of marvel is that from grain which has been long conserved in a closed chamber, for example in underground places as is the custom in Tuscany, arises an exhalation so dangerous as to be sufficient to cause death to anyone who enters such a place to collect the grain, unless the pernicious air is first allowed to escape for a while.
>
> —B. Ramazzini, *De Morbis Artificum Diatriba* (1713)

> Died while venturing into a well at Marton when it was filled with carbonic acid gas [carbon dioxide] or fixed air. From this unhappy accident, let others take warning not to venture into wells without first trying whether a candle will burn. If the candle burns to the bottom, they may be entered safety. If it goes out human life cannot be sustained.
>
> —A epigraph for the death of three miners in Yorkshire, 1812, quoted in J.T. Carter, "Vitiated Air" (1981) **(60201)**

CO_2 is the fourth most common gas in the atmosphere. It is the product of fermentation, combustion, and respiration and is 1.5 times heavier than air. CO_2 acts as an anaesthetic and produces sensory hallucinations in victims and eventually causes death due to hypoxia. It is not useful to determine post-mortem blood CO_2 concentrations as the level increases dramatically upon death (60203–5). It is much easier to detect dangerous ambient CO_2 than CO as CO_2 will cause a candle to go out, but CO will not. In Victorian times, CO was referred to as "carbonic oxide," and CO_2 was "carbonic acid."

In many churches built in that time, ventilators were installed to prevent the formation of vitiated air (60201). CO_2 poisonings and deaths have occurred with dry ice (60206–7, 60213), in breweries (60202), wineries (60210), sewers (60211, 60216), silos (60212), the storage of wood chips and pulp wood (60208), and in the preservation of fruits/vegetables (60215). One house built near an old coal mine resulted in CO_2 (not CO) poisoning for the inhabitants, until corrected (60217). One man used small CO_2 cylinders from an aquarium store and plastic bags to murder his wife and two sons and kill himself (60214). A sudden release of a massive amount of CO_2, which had accumulated in Lake Nyos, Cameroon, due to volcanic activity, caused the deaths of at least 1,700 people on 21 August 1986 (60203).

Reference Number: 60201

CARTER, J.T. "Vitiated Air: A Victorian Villain?" *Journal of the Royal Society of Medicine*, 74: 914–919, 1981 (4 figures, 17 references)

Abstract: A short historic review of the recognition of the differences between CO (carbonic oxide) and CO_2 (carbonic acid) poisoning in Victorian times. The evil repute of CO_2 was enhanced by concerns about the dangers of expired air, which was thought to be due to CO_2 rather than infectious diseases. This resulted in the building of ventilators in most churches and chapels in the second half of the nineteenth century to prevent the formation of vitiated air.

Clear distinctions between the hazards of carbonic acid (carbon dioxide) and carbonic oxide (carbonic monoxide) were rarely made and, as late as 1853, poisoning by carbonic acid [CO_2] from charcoal braziers is noted as a fashionable means of suicide in Paris. The way to a new understanding was led by investigators such as Gustaf Magnus who, in 1837, showed that oxygen was transported to the tissues of the body and carbon dioxide was transported from them to the lungs; hence separate and distinct effects could be expected from oxygen deficiency and carbon monoxide excess. Claude Bernard showed that carbon monoxide was specifically bound to hemoglobin and allowed the effects of the two oxides of carbon to be more clearly identified.

Reference Number: 60202

THOMSON, W.S. "Death from Carbon Dioxide Poisoning in a Brewery." *Lancet*, 31 March 1900, 928.

Abstract: A 26-year-old male employee of Rhymney Brewery entered a beer vat that had to be cleaned of a few inches of sediment or sludge. He removed the cover and ran a water hose for 10 minutes and entered the vat with a lighted candle. He was found dead at the bottom of the vat.

And surely the fact that a candle will burn in an atmosphere containing 10 per cent or 12 per cent of carbon dioxide, whereas death will occur in one containing only 4 per cent or 5 per cent, makes the candle test no sufficient guarantee of the fitness of an atmosphere for respiration. the jury brought in a verdict of Death by misadventure and added a rider that more stringent precautions ought to be taken to test the purity of the air and that a watch should always be placed outside the vat.

Reference Number: 60203

KLING, G.W., M.A. CLARK, H.R. COMPTON, J.D. DEVINE, W.C. EVNAS, A.M. HUMPHREY, E.J. KOENIGSBERG, J.P. LOCKWOOD, M.L. TUTTLE, AND G.N. WAGNER. "The 1986 Lake Nyos Gas Disaster in Cameroon, West Africa." *Science*, 236: 169–175, 1987 (6 figures, 1 table, 37 references)

Abstract: A sudden catastrophic release of CO_2 from Lake Nyos, Cameroon, in West Africa on 21 August 1986, caused the deaths of at least 1,700 people in the surrounding area. Survivors reported a rumbling sound lasting 15–20 seconds, seeing a white cloud rising from the lake. Many people smelled the odour of rotten eggs or gunpowder and felt a warm sensation before losing consciousness. These sensations occurred at high ambient CO_2 concentrations. Post-mortem toxicology on several victims showed no elevated COHb concentration, cyanide, or sulfhemoglobin.

If we assume that Lake Nyos was saturated before the event, it could have released about 1.2 km^3 of CO_2, an amount slightly greater than the minimum volume estimate of a 100% CO_2 cloud that would account for all fatalities. The trigger mechanism that released gas from the lake is unknown; if waters were nearly saturated before the event, any small disturbance of the water column would have initiated degassing. Interviews with survivors and the pathologic studies indicated that victims rapidly lost consciousness and that death was caused by CO_2 asphyxiation. At nonlethal levels, CO_2 acts like an anaesthetic and can produce sensory hallucinations, such that many people exposed to CO_2 will report the odor of sulphur compounds when none are present. Skin lesions found on survivors represent pressure sores and in a few cases exposure to a heat source, but there is no evidence of chemical burns or of flash burns from exposure to hot gases. All other findings on the deceased victims can be attributed to postmortem decomposition.

Reference Number: 60204

OK, Y.B., Y. KIM, K.J. BAEK, K.R. LEE, D.Y. HONG, S.O. PARK, J.W. KIM, AND S.Y. KIM. "Case of Cardiac Arrest Due to Carbon Dioxide Poisoning Following an Explosion of a Carbon Dioxide Tank." *Clinical and Experimental Emergency Medicine*, 7: 234–237, 2020 (1 table, 3 figures, 9 references)

Abstract: CO_2 is the fourth most common gas in the atmosphere, and, as such, the public generally regard it as safe. Two supervisors were called to a CO_2 tank explosion. They entered the affected room without any precautions and immediately became unconscious. Emergency responders provided cardiopulmonary resuscitation (CPR) to the victims, and one 66-year-old man with hypertension suffered a cardiac arrest and was admitted to hospital. His initial arterial blood analysis showed a potential of hydrogen (pH) of 6.63, CO_2 of 201.6 millilitres of mercury (mmHg), oxygen of 8.8 mmHg, and a capillary oxygen saturation of 3%. Seven days later, he was transferred to another hospital.

Our case is unique in view of the rarity of cardiac arrest caused by highly concentrated carbon dioxide exposures in a confined space over a short period. Had he been knowledgeable about the risk of carbon dioxide accumulation and the means to prevent carbon monoxide poisoning, our patient could have taken evasive action when the tank exploded. Even though being in charge of the tanks posed a hazard, he seemed ignorant of safety precautions. This ignorance is probably not limited in this case. In Korea, the potential hazards of carbon dioxide exposure are generally underestimated due to the paucity of reports of serious exposure-related incidents.

Reference Number: 60205

PERMENTIER, K., S. VERCAMMEN, S. SOETAERT, AND C. SCHELLEMANS. "Carbon Dioxide Poisoning: A Literature Review of an Often Forgotten Cause of Intoxication in the Emergency Department." *International Journal of Emergency Medicine,* 10 (1): 14, 2017 (19 references)

Abstract: CO_2 is a product of combustion, fermentation, and respiration and in normal room air is at a concentration at around 0.04%. CO_2 is 1.5 times heavier than air. There are approximately 90 deaths per year in the United States due to CO_2 (that is, confined space hypoxic syndrome). The post-mortem diagnosis of CO_2 poisoning is difficult as the external examination is often unremarkable, except for bruises and hematomas from CPR or other rescue attempts. Post-mortem blood CO_2 determination is often of limited value as the blood CO_2 rapidly increases after death.

Carbon dioxide poisoning, having a role both as an asphyxiant and as a toxicant, is a rare but not to miss diagnosis. Special attention is needed for pre-hospital responders, who should stay alert for the possibility of a CO_2 intoxication for their own safety, especially in cases involving dry ice or confined spaces. When confronted with a victim, he/she should be removed from the dangerous area as fast as possible and oxygen should be administered. Without adequate treatment, victims may show acute reduced cognitive performance, respiratory failure, and circulatory arrest.

Reference Number: 60206

SRISONT, S., T. CHIRACHARIYAVEJ, AND A.V.M.V.A. PEONIM. "A Carbon Dioxide Fatality from Dry Ice." *Journal of Forensic Sciences*, 54: 961–962, 2009 (3 figures, 15 references)

Abstract: A 20-year-old male worker was fleeing a confrontation in a dry ice manufacturing plant and hid in a 1.5 x 1 x 1 metre plastic container, which had dry ice inside. There was a mattress over the dry ice and no warning sign that it contained dry ice. Within five minutes, a factory security guard opened the containers and observed the young man convulsing inside for a minute and then stopped. Emergency medical services was unable to resuscitate him. At the autopsy, the victims showed only moderate congestion of the lungs and mild edema and congestion of the brain.

Autopsy findings were unremarkable. The probable cause of the convulsions was carbon dioxide intoxication due to both the dry ice sublimination and the small confined space in which he was hiding. This report emphasizes that significance of scene investigation in establishing a cause of death.

Reference Number: 60207

TUCKER, M., B. EICHOLD II, K. MICHER, J.P. LOFGREN, J. SCHIER, M. BELSON, M. MATEL, AND C. RUBIN. "Acute Illness from Dry Ice Exposure during Hurricane Ivan- Alabama 2004." *Morbidity and Mortality Weekly Report*, 53: 1182–1183, 2004 (3 references)

Abstract: In September 2004, a 34-year-old man bought a 100-pound block of dry ice in Mobile, Alabama, in anticipation of a power failure due to the approaching Hurricane Ivan. He cut the block into four pieces and placed them in paper bags in the front of his truck. The windows were closed, and the air conditioner was set on recirculation. As he was driving, he developed shortness of breath and called his wife to call 9-1-1. He parked his truck and lost consciousness. His wife found the truck and opened its door, and, immediately, her husband began to awake. He recovered completely, although he complained of a headache for the next 24 hours.

In the case described in this report, the man did not receive any warnings from the ice house regarding the potential danger of CO_2 exposure from dry ice. If the air conditioner had not been set to recirculate air inside the cab of the truck, the CO_2 poisoning symptoms might not have occurred. In addition, placing the ice in the bed of the man's truck would have reduced exposure.

Reference Number: 60208

SVEDBERG, U., C. PETRINI, AND G. JOHANSON. "Oxygen Depletion and Formation of Toxic Gases Following Sea Transportation of Logs and Wood Chips." *Annals of Occupational Hygiene*, 53: 779–787, 2009 (1 table, 6 figures, 16 references)

Abstract: The CO_2 and oxygen concentrations were measured in 76 stairwells before the hatches were removed in 41 shipments of wood chips and pulpwood transported by 10 bulk carriers between 2007 and 2008. The mean oxygen concentration was 10% (range 0–20.9%). Nearly one-fifth of the samples had an oxygen concentration of zero and was less than 6% in one-third. The mean CO_2 levels was 7.5%, ranging from 0.5% to 15% (normal ambient air CO_2 is about 0.035%). The mean CO concentration was 46 ppm (range 2–17.4 ppm). The cargoes had only been onboard for 37–66 hours.

Table: Mean Bacteria and Fungi Concentrations in Wood Pellets, Wood Chips, and Bark in Colony-Forming Units per Gram

Sample	Bacteria (colony forming units per grams)	Fungi (colony forming units per grams)
Dry wood pellets	< 9	< 9
Wet wood pellets	< 13	< 13
Fresh wood chips	940,000	710,000
Bark	5,700,000	4,900,000

Source: Adapted from Svedberg, Petrini, and Johanson 2009.

In conclusion, the measurements show that transport of logs and wood chips in confined spaces may result in rapid and severe oxygen depletion and CO_2 formation. Thus, apparently harmless cargoes may create potentially life-threatening conditions. The oxygen depletion and CO_2 formation are seemingly primarily caused by microbiological activity, in contrast to the oxidative processes with higher CO formation that predominate in cargoes of wood pellets. Improved technical and organizational measures are considered necessary to prevent future accidents. Recommendations given regarding safe entry procedures and technical preventive methods may also apply to other oxygen depleting products.

Reference Number: 60209

HOFMAN, W.I. "Suicidal Carbon Dioxide Asphyxiation: Case Report." *Military Medicine*, 141 (7): 477, 1976 (6 references)

Abstract: A depressed 24-year-old man was found dead face down in a bathtub at his home. The water in the bathtub was still warm, and the tap was partially running. Bath towels were stuffed against the cracks in the bathroom to make it airtight. Two fire extinguishers were found beside the tub. A suicide note was found. The victim induced unconsciousness by turning the fire extinguishers (which contain CO_2) on, and he slid under the water and drowned.

Carbon dioxide, a natural component of air (three parts per 10,000), is a colorless gas possessing a faint pungent smell and a slightly acid taste. Its physical properties are similar to those of nitrous oxide. This readily available gas has wide industrial applications and it is surprising that few deaths have been reported. It is available in the solid form at low temperatures as dry ice as well as in the gaseous state. Symptoms are believed to begin when oxygen deprivation has fallen to 12 per cent, with headaches and deep respirations at eight per cent. Its most important effect is to stimulate the medullary respiratory centre and it has clinical significance in the treatment of newborn infants.

Reference Number: 60210

GUILLEMIN, M.P., AND B. HORISBERGER. "Fatal Intoxication Due to an Unexpected Presence of Carbon Dioxide." *Annals of Occupational Hygiene*, 38: 951–957, 1994 (2 figures, 8 references)

Abstract: A 26-year-old worker entered a 100-cubic-metre storage tank that had residues of rose wine and activated charcoal (used to change the colour of wine) at the bottom to clean it. The bottom access door was opened for two hours prior to entry, but the inspection hole at the top of the tank was not also opened for ventilation as it was frozen shut. The worker entered the tank using the access door and was found dead a few minutes later in the sludge. The cause of death was due to CO_2 poisoning. It was estimated that the CO_2 concentration inside the tank was between 30% and 60%. A CO_2 concentration of 5% is dangerous to life and health, and concentrations above 10% rapidly produce a loss of consciousness.

Opening the door of the entering hole 2 h before the worker entered the tank could not guarantee safe air quality inside the vessel. This was mainly due to the fact that the inspection hole at the top of the tank was not also open. Carbon dioxide is about 1.5 times heavier than air, so that the concentration of CO_2 will theoretically be higher near the bottom of the tank, although it probably came to some sort of equilibrium during the 10 days it was closed.

Reference Number: 60211

MANNING T.J., G. FFIGUEROA, K. ZIMINSKI, L. LUKASH, AND A. HYMAN. "Methane Deaths? Was It the Cause?" *American Journal of Forensic Medicine and Pathology*, 2: 333–336, 1981 (4 figures, 3 references)

Abstract: In a baseball field on school grounds, three maintenance workers discovered an uncovered drainage pit that was 12 feet deep in which the lid could be seen lying at the bottom. Each in turn descended to recover the fallen grate or rescue their fellow worker, but, upon descent, each immediately became unconscious and died within minutes. It was originally thought that the deaths were due to methane. But air samples taken in a similar pit (there were three pits on the school grounds) showed an increasing CO_2 concentration up to 40% and oxygen decreasing to only 0.5%.

Table: Oxygen, Methane, and CO_2 Concentrations in a Drainage Pit at Various Depths

Depth below ground level	Oxygen concentration (%)	Methane concentration (%)	CO_2 concentration (%)
1 feet	20	2	48
3 feet	14	2.8	12
5 feet	4	1.8	24
7 feet	2.5	0.2	40
9 feet	3	0.1	40

Source: Adapted from Manning et al. 1981.

Obviously, these three victims died from CO_2 asphyxiation. The high levels of CO_2 and the low levels of O_2 were probably caused by the decomposition of leaves and vegetation which were present in the pit. The pit on the first base side of the infield did not show abnormal levels of CO_2 or O_2 because it contained water which prevented leaf and vegetation decomposition. From the data presented, it becomes apparent that in certain death situations, more than toxicological tissue analysis is needed to provide the exact cause of death. In this particular instance, toxicological analysis definitely ruled out the initially assumed (but erroneous) cause of death. This led to a more detailed investigation and further atmospheric analysis of the death site ambience, which ultimately provided the correct circumstances and cause of death.

Reference Number: 60212

TROISI, F.M. "Delayed Death Caused by Gassing in a Silo Containing Green Forage." *British Journal of Industrial Medicine*, 14: 56–58, 1957 (9 references)

Abstract: On 20 May 1956, two boys ages 18 and 12 years descended into a silo on a small farm. The silo had stored green forage mixed with beans, barley, and oats but had been nearly empty and sealed for a month. Shortly after reaching the bottom, they collapsed. A 61-year-old man attempted to rescue the boys but also collapsed. They were recovered with grappling hooks, but the boys were already dead, and the man near death. He died in hospital five days later. Fermentation of the green forage produces high CO_2 concentrations and small amounts of hydrogen and methane.

Table: CO_2 Produced within 24 Hours in Grain of Different Moisture Contents Stored at 18 Degrees Celsius

Moisture content of grain (%)	CO_2 produced within 24 hours (mg)
14–15	1.4
16.9	123
33	2000

Source: Adapted from Troisi 1957.

In the incident now described the cause of the fatal toxic effect in the three employees was evidentially the carbon dioxide produced by fermentation of forage remaining in the silo activated by the higher spring temperatures and accumulating on account of its higher density in the lower part of the closed silo. Chemical examination of a sample of the forage taken from the silo revealed active fermentation; it was thought that weed killers and insecticides used in the fields might have left a residue which could have given rise to arsine or phosphine, but arsenic and phosphorous were absent from the sample.

Reference Number: 60213

GILL, J.R., S. ELY, AND Z. HUA. "Environmental Gas Displacement: Three Accidental Deaths in the Workplace." *American Journal of Forensic Medicine and Pathology*, 23: 26–30, 2002 (4 figures, 30 references)

Abstract: Asphyxia is from the Greek words meaning no or absence of throbbing or to be without a pulse. Three fatal asphyxia cases are discussed in which the oxygen was displaced by CO_2 and nitrogen. The CO_2 source was due in one case to a large amount of dry ice in a closed unventilated laboratory space and, in the other due, to the activation of a CO_2 fire extinguisher in an accidentally closed bank vault. The source of the nitrogen in the third case causing asphyxia was from liquid nitrogen cannisters used in magnetic resonance imaging.

All three deaths described here demonstrate avoidable occupational hazards, the recognition of which could help prevent future fatalities. Deaths resulting from work-related environmental gas displacement are not limited to industrial settings. Such knowledge must increase workers' vigilance when they handle potentially lethal gases in otherwise safe office based and academic laboratory. Like carbon monoxide, the gases involved in these deaths are odorless and colorless, underscoring the need for workplace precautions. Oxygen detectors must be used by workers using large quantities of liquid nitrogen. Dry ice in a research setting should be kept in small quantities in insulated cold box/freezer that should be maintained at -70 degrees Celsius in an open, ventilated laboratory. Employee education and prominent signs posted near fire alarms describing the risks associated with carbon dioxide antifire systems also are needed. Alternative antifire systems that do not pose a threat of asphyxiation in closed unventilated spaces also should be considered.

Reference Number: 60214

SAUTTER, J., R. GUPERT, M. TSOKOS, AND L. OESTERHELWEG. "Murder-Suicide by Carbon Dioxide (CO_2) Poisoning: A Family Case from Berlin, Germany." *Forensic Science Medicine and Pathology*, 10: 97–102, 2014 (4 figures, 19 references)

Abstract: The body of a 69-year-old husband and his 26-year-old wife and two sons (ages three and six years respectively) were found in an apartment in Berlin in 2012. The wife and two sons were found dead in their beds, and the husband dead in a supine position with a plastic bag over his head. Two 500 gram (g) single use, nearly empty CO_2 cylinders (sold as a CO_2 fertilizer for aquarium plants) was found as well as two inhalation face masks and tubing. It is hypothesized that the husband placed the inhalation mask over his sleeping wife and sons and opened the CO_2 cylinder. The rapid increase in CO_2 would render the victim helpless and consequently cause death.

The cerebral circulation has a limited self-protection mechanism, the autoregulation mechanism. It acts by increasing cerebral blood circulation during hypoxia and hypercapnia through direct myogenic effects without nervous stimulation, while it decreases the circulation during hypercapnia. Circulation is increased to ensure that enough oxygen reaches the brain in cases of low oxygen and high CO_2 blood concentrations; however, the mechanism fails when pCO_2 levels are too high. The inhalation of a gas mixture high in CO_2 concentration leads to a lowering of the intracellular O_2 concentration within a few seconds. This is called hyperacute anoxia, which is also commonly found in hanging. Severe anoxia leads to a breakdown of cognitive functions, unconsciousness, and after a few minutes irreversible brain damage.

Reference Number: 60215

DEDOUIT, F., G. TOURNEL, A.B. ROBERT, P. DUTRIEUX, V. HEDOUIN, AND D. GOSSERT. "An Apple a Day Does not Always Keep the Doctor Away." *Journal of Forensic Sciences*, 53: 1444–1436, 2008 (2 figures, 10 references)

Abstract: One method of preserving fruits and vegetables is to have a high CO_2, low oxygen atmosphere while packaging the product. A 45-year-old male employee of such a packaging company was found dead in a confined room used for packaging. A suicide note was found. A digital CO tester found 600 ppm of CO and 3% oxygen. But analysis using an airtight syringe showed no CO and high levels of CO_2. His post-mortem blood COHb was 2%, and the body lacked any signs of cherry red lividity.

> Deaths by intentional CO_2 intoxication are rare. Generally, deaths are suicides by intentional inhalation of automobile exhaust gases. To the best of our knowledge, this case is the first describing this manner of suicide in the modified atmosphere of an apple packaging room attributed to asphyxiation caused by CO_2 intoxication and notably O_2 depletion.

Reference Number: 60216

WIGMORE, J.G., AND F. MCAULEY. "Case Report: Two Deaths Due to Asphyxiation in a Manhole and Analysis of the Manhole Gases." *Canadian Society of Forensic Science Journal*, 24: 1–4, 1991 (1 table, 2 figures, 6 references)

Abstract: A group of boys gained access to a manhole in a vacant lot that was connected to the city's sewers. Two boys (ages 11 years) entered the manhole and collapsed and fell to the bottom. Another boy entered and started to climb down but felt dizzy and climbed back up. Firefighters arrived shortly with self-contained breathing apparatus recovered the two boys but both eventually died. Analysis of the gases in the manhole showed a decrease of oxygen from approximately 21% volume per volume (v/v) at the top of the sewer to 4.8% v/v at the bottom. The CO_2 concentration increased from 0.3% v/v near the top of the manhole to 4.3% v/v at the bottom. The pathologist determined the cause of death to be asphyxiation due to a lack of oxygen.

> Further investigation revealed that a man, a City employee died in the same manhole eleven years earlier. A Coroner's inquest was held into the deaths of the two boys. Among the jury's recommendations was that:

Privately and publicly owned manholes and sewers that are no longer in use or situated in a secluded area should be secured by a latching device; welded shut, bolting them shut or backfilling. The manhole was backfilled shortly after the inquest.

Reference Number: 60217

KREISS, K., J.M. HARRISON, S.C. KAYDOS-DANIELS, AND L.G. BENAISE. "Investigation of a Home with Extremely Elevated Carbon Dioxide Levels, West Virginia, December 2003." *Morbidity and Mortality Weekly Report*, 53: 1181–1182, 2004 (3 references)

Abstract: In June 2001, and two male and female smokers aged 42 years moved into a newly built two-storey house in West Virginia. They developed repeated episodic, mild confusion, poor concentration, and headache while in the basement. No CO or methane gases were detected in the basement by the fire department. The man entered the crawlspace and became breathless in 2003. He attended hospital and, four hours later, had a COHb of 6%, which is normal for smokers. Again, the fire department reported no CO, methane, or other explosive gases. A specialist was called to investigate further and detected only 14% oxygen in the crawlspace. Further investigations showed a CO_2 concentration of 9.5% in the crawlspace, 11% in the crawlspace gravel, and 12% in the basement floor drain (normal air was 0.035% CO_2). Carbon isotopic analysis of the air samples indicated a carbonate source of the CO_2. The house was built on an old coalmine site. Changes to the crawlspace redirected the CO_2, which had decreased to 0.2% in the basement.

The results of this investigation underscore the need for heightened public awareness and special training for emergency response and utility workers, careful environmental measurements to assess potential risks and precautions to avoid incapacitation and prepare for rescue during immediate dangerous conditions. Building codes that mandate preventive construction including sealing cracks, maintaining positive pressure with the structure and subsurface ventilation for new buildings over landfills, caves, and abandoned mines might also be appropriate public health actions.

6.03 CYANIDES

If poisons were ponies, I'd put my money on cyanide.

—Alan Bradley, *The Sweetness at the Bottom of the Pie* (2009)[2]

In murder mysteries, the detective usually diagnoses cyanide poisoning by the scent of almonds wafting from the corpse.

—Diana Lutz, "Beware the Smell of Bitter Almonds" (2010)[3]

When carbon forms a triple bond with oxygen, it forms CO; when it forms a triple bond with nitrogen, it forms cyanide. Both are highly toxic poisons that prevent the body from using oxygen. Cyanide can exist as a gas (HCN), a liquid or solid—that is, sodium or potassium cyanide (60304). Cyanide is 20 times more toxic than CO and has a bitter almond smell, which 40% of the population cannot smell. Cyanide is also present in some plant seeds such as apricot pits, apple seeds, and cassava (60308). [Note that when I first joined the Centre of Forensic Sciences in 1976, one of the first things I was tested on was whether I could smell cyanide (I cannot), which was a useful skill in the old forensic labs, where the smell of bitter almonds in a post-mortem stomach content could indicate the type of poison used].

HCN occurs in mainly enclosed space fires due to the burning of polyurethane, rubber, wool, silk, and nylon (60302, 60308). Fatal COHb concentrations (greater than 50%) occurred in 32% of fire victims compared to 4% of the victims with a fatal blood cyanide concentration (greater than two ug/g) (60305). But, typically, blood COHb and cyanide concentrations are significantly correlated in fire victims (60306). HBOT is not an effective treatment for cyanide poisoning; it is recommended that cobalt-containing drugs such as hydroxocobalamin be employed (60301, 60303–4). Blood cyanide concentrations decrease rapidly after death, and it is recommended that blood and tissues samples be collected as soon as possible and analyzed as soon as possible to obtain an accurate result (60205–7).

---

2 Alan Bradley, *The Sweetness at the Bottom of the Pie* (Toronto: Doubleday Canada, 2009).

3 Diana Lutz, "Beware the Smell of Bitter Almonds: Why Do So Many Plants Contain Cyanide?" *The Source* (Washington University), 20 July 2010.

Reference Number: 60301

LAWSON-SMITH, P., E.C. JANSEN, L. HILSTED, AND O. HYLDEGAARD. "Effect of Hyperbaric Oxygen Therapy on Whole Blood Cyanide Concentrations in Carbon Monoxide Intoxicated Patients from Fire Accidents." *Journal of Trauma Resuscitation and Emergency Medicine*, 18: 32–38, 2010 (2 tables, 28 references)

Abstract: Fire victims can be poisoned by both CO and cyanide gases. When fire temperatures reach 315 degrees Celsius (600 degrees Fahrenheit), cyanide develops in the form of hydrogen cyanide (HCN), which binds to cytochrome oxidase enzymes, stopping the mitochondrial respiratory chain. Nineteen male and 18 female patients (mean age 52 years) of whom 25 were victims of fire and 12 of CO poisoning alone, had their blood cyanide measured (Conway diffusion method) before and after HBOT. Twelve of the fire victims were administered a dose of hydroxocobalamin (OHCob). The toxic blood cyanide concentrations have been reported to be 39 micromole (umol) per litre (L) and higher and 100 umol/L can be fatal. Once cyanide is absorbed, about 50% will be in the blood, 25% in the muscle tissues, and 25% in other organs (mainly the brain and the liver).

Table: Mean Whole Blood Cyanide and CO Concentrations before and after HBOT with and without Receiving OHCob.

	Mean whole blood-cyanide at arrival (umol/L)	Mean whole blood-cyanide difference before and after HBOT (umol/L)	Mean COHb before HBOT (%)	Mean COHb after HBOT
12 fire patients receiving OHCob	15.4	2.1	27.9	2
13 fire patients not receiving OHCob	14.33	1.9	23	1.9
12 non-fire patients	12.5	1.1	25.9	2.9

CN [cyanide] concentrations in blood from patients admitted to hospital with CO intoxication and smoke inhalation exposure did not differ significantly from controls. Accordingly, we were not able to detect any changes in CN concentrations in blood after treatment with HBO.

Reference Number: 60302

TABIAN, D., D.B. ILIESCU, T. LOV, B. BARNA, S.I. TOMA, AND G. DROUCHIOIU. "Hydrogen Cyanide and Carboxyhemoglobin Assessment in an Open Space Fire Fatality." *Journal of Forensic Science*, 66: 1171–1175, 2021 (1 table, 1 figure, 22 references)

Abstract: An 80-year-old man was burning garbage in his yard when his clothes caught fire. He managed to move several metres away from the fire where he collapsed and was later found dead in that position. The autopsy of the man was conducted 24 hours later. Grade 3 and 4 burns were found in 85–90% of the body. The post-mortem femoral blood showed a COHb concentration of 73.7% and a HCN of 1.3 ug/L (measured spectrophotometrically at a wavelength of 493 nanometres). The urine alcohol concentration was negative.

Usually, HCN intoxication occurs in enclosed space fires. It is known that fire smoke produced in combustion of wood does not contain HCN. The case we talk about is atypical and raises a lot of questions because combustion of vegetation debris, victim's clothes and of household waste in open space liberated HCN at high concentration, which was found in the blood of the deceased. Frequently in rural areas from Romania, people burn plastic bottles, and other house waste made from plastic, along with vegetation debris, resulting in high amounts of HCN liberated in the fire smoke. Materials such as wool, silk, polyurethane, polyamide and polyacrylonitrile are known to produce HCN when combusted. The victim's clothing and body were extensively burned and the extremely high COHb saturation indicates that much of the combustion of the clothing likely occurred while the man was alive. Taking these into account, we can suggest that the man's clothing could have played an important role in the production of HCN in the reported open space fire.

Reference Number: 60303

MARCAL, A.L., D. COEHLO, AND R. ARAUJO. "Carbon Monoxide and Cyanide Intoxication: An Association to Remember." *Acta Médica Portuguese*, 33: 147–148, 2020 (11 references)

Abstract: A 78-year-old male victim of a house fire was intubated at the scene and transported to hospital. An airway burn was confirmed via video laryngoscopy, and his blood COHb was 18.7%. He was administered two sessions of HBOT. On the third day, he developed agitation, hypertension, and tachycardia, which was suspected to be due to concomitant cyanide inhalation. He was administered five g of hydroxocobalamin and 12.5 g of thiosulphate IV. His symptoms then completely resolved.

> In the presented case, the context of domestic fire and airway burn, combined with the initial mental status changes and soot in the airway should have raised the suspicion of CN [cyanide] intoxication, which was however, not raised. We believe that the later and less severe presentation could be explained by the use of HBO to treat the CO poisoning since HBO seems to mobilize CN from the tissue to the circulation, facilitating its metabolism and excretion. HBO is not currently recommended as the therapeutic option for CN poisoning.

Reference Number: 60304

ANSEEUW, K., N. DELVAU, G. BURILLO-PUTZE, F. DE IACO, G. GELDNER, P. HOLMSTROM, Y. LAMBERT, AND M. SABBE. "Cyanide Poisoning by Smoke Inhalation: A European Expert Consensus." *Emergency Medicine*, 20: 2–9, 2013 (2 figures, 56 references)

Abstract: A fire causes a complex toxic environment involving smoke, heat, oxygen deficiency, CO, and other toxic gases such as CO and HCN. Although cyanide exists in several forms such as potassium, sodium, copper, and mercury slats, HCN gas from fire smoke is the most common cause of cyanide poisoning in developed countries. Cyanide presents a major hazard to firefighters entering a fire. The early signs of cyanide poisoning are neurological and respiratory stimulation in an attempt to compensate for hypoxia and results in giddiness, confusion, vertigo, nausea, and vomiting. Later symptoms are due to depressive effects as there is an inability to prevent hypoxia and include hypotension, bradycardia, seizures, coma, and death. Cyanide poisoning can also elevate plasma lactate concentrations. The major antidote for cyanide poisoning are cobalt compounds such as dicobalt edeate and hydroxocobalamin. Cobalt has a strong affinity for cyanide and hydroxocobalamin chelates with cyanide to form vitamin B12, which is excreted in the urine.

> There is growing evidence of cyanide toxicity in fire smoke victims presenting to the emergency department. Reducing mortality and morbidity attributed to cyanide inhalation primarily depends on early recognition and the timely administration of an antidote. Several specific cyanide

antidotes are available but given the practical and safety advantages, hydroxocobalamin is the preferred antidote.

Reference Number: 60305

STAMYR, K., G. THELANDER, L. ERNSTGARD, J. AHLNER, AND G. JOHANSON. "Swedish Forensic Data 1992–2009 Suggest Hydrogen Cyanide as an Important Cause of Death in Fire Victims." *Inhalation Toxicology*, 20: 194–199, 2012 (3 tables, 4 figures, 30 references)

Abstract: About 65–138 people (7–15 per million population) die annually in Sweden due to fires. Between 60% and 80% of the fire deaths are related to the release of toxic fumes, mainly CO. CO is produced by the incomplete combustion of carbon-containing compounds, whereas HCN is formed by the incomplete combustion of nitrogen-containing compounds such as silk, wool, polyurethane foam, or synthetic rubber. Post-mortem blood CO and HCN concentrations were determined in 816 fire victims between 1992 and 2009. The median COHb was 22% (range 0.5–87%). A blood COHb greater than 50% was found in 32% of the samples. The median blood cyanide concentration was 0.15 ug/g (range 0–7.6 ug/g) A blood cyanide concentration greater than two ug/g can be fatal and was found in 4% of the samples. The blood cyanide concentration was positively correlated with the blood COHb concentration. The post-mortem blood cyanide concentration decreased on average by 48% after four months of storage at −20 degrees Celsius.

It is difficult to judge the extent of HCN and CO formation in fires. Both gases are likely to vary widely depending e.g. on the composition of the burning material, the temperature, and the oxygen supply. A typical feature in fires with nitrogen containing materials, e.g., furniture with polyurethane foam is that HCN in the air rises abruptly some minutes earlier than CO. Thus, one may speculate that the initial rise in HCN impairs the ability to escape the smoke, where after the victim dies from asphyxiation by CO or by CO and HCN combined. It should be noted that heavy smoke will also affect the escape behaviour and possibly impair the escape efficiency.

Reference Number: 60306

BAUD, F.J., P. BARROT, V. TOFFIS, B. RIOU, F. VICAUT, Y. LECARPENTIER, R. BOURDON, A. ASTIER, AND C. BISMUTH. "Elevated Blood Cyanide

Concentrations in Victims of Smoke Inhalation." *New England Journal of Medicine*, 325: 1761–1766, 1991 (1 table, 3 figures, 37 references)

Abstract: Blood cyanide concentrations were determined in 109 fire victims in blood samples obtained at the scene of the residential fire (before treatment) and 114 controls (40 persons who died of drug intoxication, 40 who died of trauma, and 29 who died of pure CO poisoning from a malfunctioning gas water heater or coal stove) using a colorimetric micro-diffusion method. COHb was determined in the fire and CO poisoning of the victims by an infrared method. Forty-three of the 109 fire victims died, and the mean blood cyanide concentration was 116.4 umol/L in those who died and 21.6 umol/L for those who survived. The mean half-life for cyanide before HBOT or hydroxocobalamin treatment. There was a significant correlation between blood COHb and cyanide concentrations.

Table: Mean Post-Mortem Blood Cyanide Concentration in Various Post-Mortem Cases

Group	Mean post-mortem blood cyanide concentration (umol/L0
Controls	5.0
CO poisoning only	4.3
Fire victim who survived	21.6
Fire victim who died	116.4

Source: Adapted from Baud et al. 1991.

Several potential pitfalls have been noted in the measurement of blood cyanide concentrations in patients with either fatal or nonfatal poisoning. Cyanide may disappear rapidly from body tissue after death from acute cyanide poisoning and its half-life in blood is short in cases of nonfatal cyanide poisoning in dogs and humans. Our findings that the half-life of cyanide is approximately one hour agrees with the results of previous studies. Such rapid disappearance precludes accurate evaluation of cyanide poisoning by delayed blood sampling in surviving fire victims. Furthermore, the production of cyanide has been described in various tissues including blood. The potential for postmortem cyanogenesis is uncertain, however, since some have failed to demonstrate it.

Reference Number: 60307

MCALLISTER, J.L., R.J. ROBY, B. LEVINE, AND D. PURSER. "Stability of Cyanide in Cadavers and in Postmortem Stored Tissue Specimens: A Review."

Journal of Analytical Toxicology, 32: 612–620, 2008 (3 tables, 5 figures, 29 references)

Abstract: Exposure of HCN concentrations greater than 100 ppm can cause rapid incapacitation and inability to escape a fire, although CO appears to be the major cause of death. Poorly ventilated or smoldering fires produce high yields of CO and HCN. In general, the whole blood cyanide concentration decreases over time even when the sample is refrigerated or frozen. The average decreases of post-mortem blood cyanide concentrations within 24 hours between autopsy and toxicology analysis showed a maximum of 60%. There can also be a slight increase in blood cyanide concentration due to microbial contamination that is inhibited by sodium fluoride.

Table: Post-Mortem Blood Cyanide Decrease in Samples Stored for Eight Weeks at 4 and −20 Degrees Celsius

Sample	Temperature stored (°C)	Initial cyanide concentration (mg/L)	Percent decrease in cyanide concentration (%)
1	4	5	30
2	−20	5	31
3	4	2.5	17
4	−20	2.5	17

Source: Adapted from McAllister et al. 2008.

The main recommendations from this review are that samples of blood and other tissues should be taken as soon as possible after death and analyzed for cyanide as soon as possible after collection. Storage of samples at 4°C provide relatively stable results; samples are best preserved in deep freezer (−20°C). Additionally, the time intervals between death, sampling and analysis should be recorded to allow for better interpretation of toxicological results. In general, repeatable (non-case study) tests that assessed increased cyanide in samples showed that cyanide increases were an order of magnitude less that those levels typically found in lethal cyanide exposures (3 mg/L). Therefore, when levels of cyanide in blood are within lethal ranges and samples are collected and analyzed within 24 h, artificial increases in cyanide can be considered negligible. If, however, the sample cyanide concentrations are an order of magnitude less than the lethal range, these levels should be compared against the COHb blood level and the known facts of the case regarding fuel loads and their relative burning characteristics prior to data interpretation.

Reference Number: 60308

KOSCHEL, M.J. "Where There's Smoke, There May Be Cyanide." *American Journal of Nursing*, 102: 39–42, 2002 (10 references)

Abstract: In the United States, smoke inhalation may account for 10,000 deaths per year. HCN is a common product of fires involving the burning of nylon, wool, silk, polyurethane, and rubber. HCN has a bitter almond smell, but 40% of the population may not be able to detect it. Cyanide is 20 times more toxic than CO and can cause an immediate cardiac arrest. Cassava, a starchy root used to make flour in Africa, Indonesia, and Brazil, also contains cyanide, and some varieties contain so much cyanide that they can cause death. Nigerians call one variety "chop and die." Long-term consumption of sub-lethal cyanide-containing cassava can cause a disease called konzo (tied legs), which is the irreversible paralysis of the legs.

Patients who have suffered smoke inhalation in fires in enclosed spaces and who have soot in the mouth or nose, altered mental status or hypotension may have significant blood cyanide concentrations of 1 mg/L. It may be only seconds or minutes before symptoms of cyanide poisoning through inhalation emerge. Cyanide inactivates cytochrome oxidase and halts cellular respiration even in the presence of adequate oxygen stores. Hence, as in carbon monoxide poisoning, body tissues that have high oxygen requirements such as the brain, heart, and liver are most profoundly affect. Hypoxia without cyanosis is a hallmark of cyanide toxicity.

Reference Number: 60309

TURRINA, S., C. NERI, AND D. DE LEO. "Effect of Combined Exposure to Carbon Monoxide and Cyanides in Selected Forensic Cases." *Journal of Clinical Forensic Medicine*, 11: 264–267, 2004 (20 references)

Abstract: Two case reports involving post-mortem CO and cyanide levels as a result of fires are described. In the first case, a 35-year-old woman was found dead in a seated position in a train compartment as a result of inhaling toxic fumes from a fire in an adjacent compartment. There was a presence of soot on the face and nostrils of the victim, and the organs were a cherry red colour. Her post-mortem blood COHb was 80%, and her cyanide concentration was 9 mg/L. In the second case, a 42-year-old homeless man was found unconscious in an underground tunnel of a bus station as a result of a deliberately set fire consisting of plastic material, bags, and mattresses. He survived for more than one hour after the fire, including 20 minutes of 100% oxygen therapy, before he died. His post-mortem blood COHb and cyanide concentration were 33% and two mg/L respectively. His BAC was 131 mg/dL. The half-life of cyanide is about one hour.

> In our first case the elevated concentration of carboxyhemoglobin (80%) is accompanied by an elevated concentration of cyanide (9 mg/L). Given the immediate exposure to cyanide fumes in a highly restricted and non aerated environment, it could be possible the victim initially suffered a knock-down effect, typical of cyanides, being unable to escape the action of CO. The cause of death is therefore attributable to CO, while only a contextual toxic action can be traced back to cyanides though obviously not without relevance as far as the lethal outcome is concerned.

6.04 HYDROGEN SULFIDE

> Hydrogen sulfide is as lethal, molecule for molecule as cyanide. This may explain why humans evolved such an exquisite sensitivity to its smell.
>
> —Mary Roach, "Passing Gas" (2013)[4]

Hydrogen sulfide (H_2S) is a very toxic (about 4 times that of CO), rapidly acting, highly flammable, poisonous gas that is widely used in the chemical industry and found in sewers, manure pits, geothermal/volcanic outcrops, and hot springs (60401, 60406). The effect of H_2S is similar to cyanide as it also binds to cytochrome-c oxidase as a cellular asphyxiant (60405). H_2S is also known as sewer gas and has the smell of rotten eggs, which makes it much easier to detect than CO, although, at high concentrations, the olfactory senses are paralyzed by H_2S (60401). Common accidental deaths due to H_2S occur in manure pits and sewers as it is produced by microorganisms growing there, but CO is not produced (60401–3). After CO, H_2S is the second most common gas that is associated with the deaths of construction workers (60403). "Detergent" suicides have become more common as the combination of an acid such as toilet bowl cleaners and a sulphur source such as bath salts produce large amounts of H_2S (60405–7).

4 Mary Roach, "Passing Gas: A Modern Scientific History," *Salon*, 7 April 2013.

A common sign of H_2S poisoning is a greenish colour of the brain and other tissues (60402, 60403, 60407). Another important indicator is an elevated post-mortem blood thiosulfate concentration due to the metabolism in the cells from H_2S. H_2S cannot be used as an indicator as it is not stable and leaves the body rapidly after death (60403, 60407, 60409). However, thiosulfate may not form if death occurs very rapidly (60403), and it may increase in post-mortem blood due to putrefaction (60409). Another indicator of H_2S exposure is the black tarnishing of silver jewellery worn by the victim (60407). Two men died of H_2S poisoning in the geothermal baths in Rotorua, New Zealand (60408).

Reference Number: 60401

CHATURVEDI, A.K., D.R. SMITH, AND D.V. CANFIELD. "A Fatality Caused by Accidental Production of Hydrogen Sulfide." *Forensic Science International*, 123: 211–214, 2001 (1 table, 1 figure, 18 references)

Abstract: H_2S is a very toxic, rapidly acting poisonous gas that is used in the chemical industry. It is usually detected by the odour of rotten eggs, but high concentrations may paralyze the olfactory nerves and not be detected. A 55-year-old male truck driver was found dead at the scene when he accidentally transferred sodium hydrogen sulfide from his tanker truck into a tank containing 4% sulfuric acid and iron sulphate, which produced H_2S gas. Another employee evacuated the area when he detected the odour of gas. A post-mortem blood sulfide ion concentration of 1.68 ug/mL was detected compared to a normal concentration of 0.03 ug/mL.

> In the view of the findings, it is concluded that the cause of death was hydrogen sulfide poisoning associated with a hazardous material accident in an industrial situation.

Reference Number: 60402

DONHAM, K.J., L.W. KNAPP, R. MONSON, AND K. GUSTAFSON. "Acute Toxic Exposure to Gases from Liquid Manure." *Journal of Occupational Medicine*, 24: 142–145, 1982 (20 references)

Abstract: Liquid manure storage is common in swine, beef, dairy, and veal operations in the United States, involving 500,000 people. Anaerobic and other microbes produce about 150 gases in stored liquid manure. The main toxic gases produced are methane, ammonia, CO_2, and H_2S. CO is not produced by liquid manure but is due to the heating systems in the buildings. Since 1975, 14 deaths and nine illnesses have been reported due to toxic gases formed by liquid manure. Nearly all the cases were due to the rapid release of H_2S due to the agitation of the liquid manure.

> Despite the limited descriptive data, our interpretation of the available evidence leads us to conclude that H_2S is the primary agent involved in these acute intoxications. The symptoms are compatible with acute H_2S toxic reaction in that the gases cause rapid unconsciousness with relatively little forewarning. The autopsy information was compatible in that the major finding was extensive pulmonary edema. There have been other reports in the literature describing a greenish tint to the viscera in people poisoned with H_2S, however this was not noted in any of the cases reported herein. Although no data was available on atmospheric H_2S levels at the time the short-term exposures occurred, measurements made in one instance during reenactment of the accident showed H_2S levels to be well in excess of 400 ppm. Subsequently measurements of dissolved sulfides in several manure pits suggested that, acutely, toxic levels of H_2S could be released into the immediate area.

Reference Number: 60403

SMITH, S.S., D.L. CANNON, K. FAGAN, AND C.P. WEIS. "Occupational Hydrogen Sulfide Fatalities and Thiosulfate Levels." *American Journal of Forensic Medicine and Pathology*, 38: 47–48, 2017 (11 references)

Abstract: A 31-year-old man was standing atop a truck near the hatch, monitoring the pumping of manure into the truck. He was not wearing a respirator or a H_2S monitor. As the truck reached capacity, he lost consciousness and slumped over the top of the truck. In spite of resuscitation attempts, he was declared dead. No thiosulfate, an indicator of H_2S poisoning, was detected in the post-mortem blood, but there was a green discolouration of his brain.

> In this case, our opinion is that the knockdown occurred because of a concentrated plume of H_2S from the agitation of the manure as it reached the filling point. The immediate death prevented metabolism of H_2S to thiosulfate; thus, no thiosulfate was detected. Pathological and toxicological findings in the published reports reviewing intentional exposure to H_2S provide insight as to why blood thiosulfate levels are so variable.

Reference Number: 60404

DOREVITCH, S., L. FORST, L. CONROY, AND P. LEVY. "Toxic Inhalation Fatalities of US Construction Workers 1990 to 1999." *Journal of Occupation and Environmental Medicine*, 44: 657–662, 2002 (3 tables, 23 references)

Abstract: An analysis was conducted of 87 construction workers deaths in the United States related to toxic gas inhalation between 1990 and 1999. The majority of deaths occurred in confined spaces. Water, sewer, and utility works have three times the risk of toxic gas exposure.

Table: Main Toxic Gases Associated with the Deaths of Construction Workers in the United States between 1990 and 1999.

Toxic gas	Percent of deaths (%)
Carbon monoxide	21.8
Hydrogen sulfide	18.4
Nitrogen	10.3
Methane	4.6
Argon	4.6
Carbon dioxide	2.3

Source: Adapted from Dorevitch et al. 2002.

Like occupational fatalities in general, these poisoning deaths should be completely preventable. The use of direct reading instruments with alarms for hydrogen sulfide, carbon monoxide, and oxygen could have prevented the majority of poisoning fatalities of construction workers identified in this study. Training, the use of air-supplied respirators and carbon monoxide alarms could prevent the deaths resulting from indoor use of fuel-powered tools.

Reference Number: 60405

REEDY, S.J.D., M.D. SCHWARTZ, AND B.W. MORGAN. "Suicide Fads and Characteristics of Hydrogen Sulfide Suicides in the United States." *Western Journal of Emergency Medicine*, 12: 300–304, 2011 (2 tables, 13 references)

Abstract: The effect of H_2S is similar to cyanide as it also binds to cytochrome-c oxidase as a cellular asphyxiant. Typical occupational exposures involve petrochemical plants, sewers, volcanoes, and manure pits.

Suicide using homemade H_2S (detergent suicides) have become a fad on Japanese websites since 2007. The H_2S is produced by mixing an acidic toilet bowl cleaner with an insecticide or baths salts as a sulphur source. Overall, 30 H_2S suicides in the United States between 2008 and 2010 were evaluated. The majority of victims were white males less than 30 years of age who were found dead in a car.

Table: Examples of Household Chemicals That, When Mixed Together, Produce H_2S Gas

Acid source	Sulphur source
Lysol toilet bowl cleaner (9.5% HCl)	Dandruff shampoos (0–15% zinc sulfide)
Sno Bol toilet cleaner (15% HCl)	Pesticides (5–30% calcium polysulfides)
The Works toilet bowl cleaner (15–25% HCl)	Spackling paste (1-2% zinc sulfide)
Kaboom shower and tile cleanser (20.5% phosphoric acid)	Garden fungicides (5–90% sulphur)
Tile stone cleaners (1–30% HCl)	Selected bath salts (25–25% sulphur)

Source: Adapted from Reedy, Schwartz, and Morgan 2011.

To date, the incidence by intentional inhalation of homemade H_2S gas is probably under-estimated by physicians and public health officials. It seems likely that we can expect these rates to increase as H_2S suicides are popularized in the media coverage of these deaths and more websites appear offering guidance on this issue.

Reference Number: 60406

KOCAK, U., A.H. ALPASLAN, M. YAGON, AND E. OZER. "Suicide by Homemade Hydrogen Sulfide in Turkey: A Case Report." *Bulletin of Legal Medicine*, 21: 189–192, 2016 (28 references)

Abstract: H_2S is a colourless and flammable gas found in volcanoes, hot springs, and geothermal sources. It is also produced by the putrefaction of sulphur containing organic material and causes the rotten egg smell in sewage systems, tanneries, and swamps. It has been reported that a single inhalation of H_2S at about 1,000 ppm will cause a knock-down, loss of consciousness, rapid apnea, and death. Suicide by the inhalation of homemade H_2S was reported for the first time in Japan in 2007. In this case report, a 24-year-old male university student was found dead in his bathroom due to H_2S produced by the addition of bleach to orpiment (yellow-orange arsenic sulfide mineral) in the bucket beside him. He left a suicide note and several notes warning of H_2S gas. Toxicological analysis reported 8.8% methemoglobin but no CO, alcohol, or drugs.

In conclusion, most of the reported deaths due to H_2S gas have involved sewer systems, or industrial accidents in the world. By the late 2000s, suicide by inhalation of homemade H_2S was introduced and it has rapidly become widespread.

Reference Number: 60407

GILBERT, J.D., AND R.W. BYARD. "Tarnished Jewellery and Skin: A Subtle External Marker Indicating Exposure to Hydrogen Sulfide." *Forensic Science Medicine and Pathology*, 17 July 2024, 1–4 (3 figures, 16 references)

Abstract: A 39-year-old woman with a history of anxiety and depression was found dead in the rear seat of her car with a sign warning of HCN gas. The attending police officers noticed the small of rotten eggs and bottles of brick/tile cleaner and lime sulphur, which when combined produce large amounts of H_2S. H_2S reacts with metallic silver causing it to produce a black tarnishing, which occurred on the ring on the victim's finger. The victim's brain and other tissues had the typical greenish colour of H_2S poisoning.

Although silver is relatively nonreactive, in the presence of hydrogen sulfide gas it will develop a black tarnish due to the formation of silver sulfide (Ag_2S). We are not aware of any other reports of skin discoloration in the forensic literature from silver jewellery exposed to hydrogen sulfide and suggest that this may be a subtle external marker to add to other findings in cases of lethal or non-lethal hydrogen sulfide exposure. This finding may be of particular use in focusing investigations if hydrogen sulfide exposure has not been suspected.

Reference Number: 60408

BASSINDALE, T. AND M. HOSKING. "Deaths in Rotorua's Geothermal Hot Pools: Hydrogen Sulfide Poisoning." *Forensic Science International*, 207: e28–e29, 2011 (8 references)

Abstract: In late 2007 and early 2008, two healthy men, one 77 years old and the other 88 years old, were found dead near several enclosed hot pools fed with Rotorua's geothermal waters. Thiosulfate is used as a marker of H_2S poisoning as H_2S is unstable in post-mortem blood and is rapidly eliminated from the body at death. The post-mortem blood thiosulfate concentration, as determined by the gas chromatograph/mass spectrometer (MS) method, were highly elevated at five and four mg/L respectively. The coroner concluded that the deaths were related to H_2S poisoning.

Rotorua is a small city in central North Island of New Zealand of approximately 50,000 inhabitants. The area is one of the few settlements in the world to be situated on an actively degassing volcanic field. The main gases emitted by the field are carbon dioxide and hydrogen sulfide, which are both denser than air. Hydrogen sulphide (H_2S) results in a typical rotten egg smell hanging over the city. Some of the main tourist attractions of the local area are hot pools and thermal areas including bubbling mud and geysers. Many of the city's hotels and motels contain hot pools and there are municipal pools in the city, some open air and some enclosed.

Reference Number: 60409

SUZUKA, M., S. JIN, A. TAKEUCHI, M. MURAKAMI, K. TAKAHASHI, AND K. MATOBA. "Influence of Blood Thiosulfate Produced by Postmortem Changes for the Diagnosis of Hydrogen Sulfide Poisoning in Forensic Autopsy." *Asian Biomedicine: Research, Reviews, and News*, 18: 281–286, 2024 (1 table, 3 figures, 19 references)

Abstract: The thiosulfate concentration in post-mortem blood is an important indicator of H_2S poisoning. However, thiosulfate also forms during decomposition. The post-mortem blood thiosulfate concentration was determined by liquid chromatography and MS in 19 severely decomposed, 19 partially decomposed, and 19 non-decomposed deceased who did not die from H_2S poisoning. The mean blood thiosulfate concentrations were 70, 16.3, and 1.1 umol/L respectively.

Thiosulfate was found to be produced in the blood by postmortem decomposition. The elevation of blood thiosulfate concentration caused by postmortem decomposition is often as high as observed for death by hydrogen sulfide poisoning, which may have a significant impact on its diagnosis. Therefore, when measuring thiosulfate as an indicator of hydrogen sulfide poisoning, the effect of decomposition should be carefully considered.

APPENDIX 1

Historical Highlights of Carbon Monoxide since the Eighteenth Century

Year	Study/Finding
1772	Joseph Priestly synthesized CO via CO_2 reduction.
1778	Troja coined the phrase "cherry red" to describe the colour CO gives to the blood.
1793	James Watt (inventor of the steam engine) suggested that coal fumes may displace oxygen in the blood.
1815	Coal (and water) gas lighting containing high CO concentrations was first used in the Chestnut Street Theatre in Philadelphia. Its use quickly spread and, for the next 150 years, caused untold thousands of deaths and chronic illness until replaced by electricity and natural gas.
1842	LeBlanc confirmed the presence of CO in coal fumes.
1857	Laude Bernard, a French physiologist, describes how CO displaces oxygen from the blood.
1910	Respiratory physiologist, J.B.S. Haldane recommended miners use canaries or mice to detect CO in mines.
1913	The worst mining disaster in Britain occurred at the Senghenydd mine, which killed 440 men, mainly due to CO poisoning.
1914	The First World War produced many new methods of CO poisoning due to shells, machine guns in poorly ventilated pillboxes or dugouts, tanks, airplanes, and other motorized vehicles, including ambulances.
1949	Torgny Sjostrand demonstrated endogenous CO production from hemoglobin catabolism.
1970	CO is first detected by astronomers in interstellar clouds
1975	Catalytic converters in passenger motor vehicles decreased the CO in exhaust from 7% to 12% to as low as 0.1%. The number of non-fatal CO poisonings decreased by 63% over 33 years.

Year	Study/Finding
1993	First battery-powered home CO detectors become commercially available.
1993	CO is characterized as a neurotransmitter (gasotransmitter).
2005	First clinical trial that evaluated CO for the treatment of inflammation.

Source: J.G. Wigmore, "A History of Carbon Monoxide," *Forensic Science Review,* 36: 90–91, 2024.

Tables of Blood COHb with Ambient CO Concentrations and Time

Time in minutes to reach 10% COHb

CO concentration inhaled ppm (%)	Resting (minutes)	Light activity (minutes)	Heavy work (minutes)
50 ppm (0.005)	667	400	182
500 ppm (0.050)	67	40	18
5,000 ppm (0.5)	6.7	4.0	1.8
10,000 ppm (1)	3.3	2.0	0.9
20,000 ppm (2)	1.7	1	0.5
40,000 ppm (4)	0.8	0.5	0.2
100,000 ppm (10)	0.3	0.2	0.1

Time in minutes to reach 20% COHb

CO concentration inhaled ppm (%)	Resting (minutes)	Light activity (minutes)	Heavy work (minutes)
50 ppm (0.005)	1,334	800	362
500 ppm (0.050)	133	80	36
5,000 ppm (0.5)	6.7	8.0	3.6
10,000 ppm (1)	13.3	4.0	1.8
20,000 ppm (2)	6.6	2.0	0.9
40,000 ppm (4)	1.7	1.0	0.4
100,000 ppm (10)	0.7	0.4	0.2

Time in minutes to reach 40% COHb

CO concentration inhaled ppm (%)	Resting (minutes)	Light activity (minutes)	Heavy work (minutes)
50 ppm (0.005)	2,668	1,600	728
500 ppm (0.050)	268	160	73
5,000 ppm (0.5)	26.8	16	7
10,000 ppm (1)	13.2	8.0	3.5
20,000 ppm (2)	6.8	4.0	1.8
40,000 ppm (4)	3.3	2.0	0.9
100,000 ppm (10)	1.3	0.8	0.4

Source: Adapted from W.H. Forbes, F. Sargent, and F.J.W. Roughton, "The Rate of Carbon Monoxide Uptake by Normal Men," *American Journal of Physiology* 143 (4): 594–608, 1945.

Table of the Effect of Ambient CO Concentrations

CO concentration in air ppm (%)	Effects of exposure
0.01 ppm (0.00001)	Normal atmospheric concentration
1–2 ppm (0.0001)	Might be normal, from cooking, outdoor traffic
2 ppm (0.0002)	Suggested maximum long-term allowable exposure
> 2 ppm	Raises questions as why CO is elevated
9 ppm (0.0009)	Maximum allowable concentration for short-term exposure in a living area
15–20 ppm (0.002–0.003)	Impaired performance in time discrimination (COHb, 2%), decrease in absolute exercise time (COHb 2.5%), shortened time to angina (COHb 1.9%), vigilance decrement (COHb 3%).
27 ppm (0.0027)	21% increase in cardio-respiratory complaints
30 ppm (0.0030)	Earlier onset of exercise-induced angina (COHb 4.96%)
35 ppm (0.0035)	Maximum allowable limit for eight-hour continuous exposure
75 ppm (0.0075)	Significant decrease in oxygen reserve to the heart (COHb 10)
200 ppm (0.02)	Slight headaches, tiredness, dizziness, nausea after 2–3 hours
400 ppm (0.04)	Frontal headaches within 1–2 hours, life threatening after 3 hours
800 ppm (0.08)	Dizziness, nausea, and convulsions within 45 minutes, unconsciousness within 2 hours, and death within 2–3 hours
1,600 ppm (0.16)	Headache, dizziness, and nausea within 20 minutes, death within 30 minutes

CO concentration in air ppm (%)	Effects of exposure
3,200 ppm (0.32)	Headache, dizziness, and nausea within 5–20 minutes, quickly impaired thinking and response, death within 30 minutes
6,400 ppm (0.64)	Headache, dizziness, and nausea within 1–2 minutes, impaired thinking and response, death within 10–15 minutes
12,800 ppm (1.28)	Death within 1–3 minutes

Source: Adapted from J. Modic, "Carbon Monoxide and COHb Concentration in Blood in Various Circumstances," *Energy and Buildings,* 35: 903–907, 2003.

Table of the Effect of Ambient H_2S Concentrations

H_2S concentration (ppm)	Symptom/Effect
0.001–1.5	Odour threshold of rotten egg smell; odour becomes more offensive at 3–5 ppm and, above 30 ppm, becomes sweet or sickeningly sweet
2–5	Prolonged exposure may cause nausea, tearing of the eyes, headache, or loss of sleep
20	Possible fatigue, loss of appetite, headache, irritability, poor memory, dizziness
50–100	Slight conjunctivitis ("gas eye") and respiratory tract irritation after 1 hour
100	Coughing, eye irritation, loss of smell after 2–15 minutes. Altered breathing, throat irritation, drowsiness after 15–30 minutes
100–150	Loss of smell (olfactory paralysis)
200–300	Marked conjunctivitis, pulmonary edema from prolonged exposure
500–700	Staggering, collapse in 5 minutes; serious damage to the eyes in 30 minutes; death after 30–60 minutes
700–100	Rapid unconsciousness, "knockdown" or collapse within 1–2 breaths
1,000–2,000	Nearly instant death

Source: Adapted from *Occupational Safety and Health Administration*, https://www.osha.gov/hydrogen-sulfide/hazards.

APPENDIX 5

Famous People Who Allegedly Died from CO Poisoning

Rosie Nis Adams	Singer, songwriter
Ahn Jae-hwan	South Korean actor
Barbara Bates	Actor
Rene Le Begue	Race car driver
Taylor Brooke	Actor
Kevin Carter	Photojournalist
Micky Cave	English footballer
Cleopatra ?	
Brad Delp	US musician
Jack Dunlop	Soviet spy
Merrit A. Edson	US Marine Corps general
Bernard Floud	British member of parliament
Bob Grant	English actor
Morris K. Jessup	Ufologist
Emperor Jovian	
Empress Fausta	
John Porter East	US senator
Benny Frey	Major League Baseball pitcher
Vitas Gerulaitis	Tennis star
Libby Holman	Actor
Dow Hower	Last New York State executioner
William Inge	Playwright
Yasunari Kawabata	Author

Capt. George Kerr	Canadian, Victoria Cross recipient from First World War
Ross Lockridge Jr.	Author
Ron Luciana	Major League Baseball umpire
Bobby McKean	Scottish footballer
Rosamond Pinchot	Actor
Sylvia Plath	Author
Edgar Allan Poe?	Author
Bill Robinzine	National Basketball Association athlete
Anne Sexton	Poet
Seneca	Roman scholar
Matthias Sindelar	Austrian footballer
Viacheslav Solodukhin	Soviet professional hockey player who played against Canada in the 1972 Summit series
Sir Bernard Spilsbury	Forensic pathologist
Mabel Stark	Tiger trainer
Steve Stonebreaker	National Football League linebacker
Thelma Todd	Actress
John Kennedy Toole	Writer
Dan James White	Assassinated Harvey Milk
Assia Wevill	Second wife of Ted Hughes
Don Wilson	Major League Baseball pitcher
Colin Winkelmann	BMX rider
Randall L. Wreghitt	Broadway producer
Zurab Zhvania	Prime minister of Georgia
Emile Zola	Author

Abbreviations

ADL	activity of daily living
ADR	age-adjusted death rate
AKI	acute kidney injury
ARI	acute respiratory infection
ARS	air raid shelter
BAC	blood alcohol content
bpm	beats per minute
CAD	coronary artery disease
CH_2Cl_2	methylene chloride
CHF	congestive heart failure
cm	centimetre
CO	carbon monoxide
CO_2	carbon dioxide
COHb	carboxyhemoglobin
COPD	chronic obstructive lung disease
CPAP	continuous positive airway pressure
CPK	creatine phosphokinase
CPR	cardiopulmonary resuscitation
CT	computed tomography
CVD	cardiovascular disease
DEACMP	delayed encephalopathy after acute carbon monoxide poisoning
dL	decilitres
DLCO	CO diffusing test

DNS	delayed neurological sequelae
DVT	deep vein thrombosis
EC	electronic cigarette
ECG	electrocardiogram
ED	emergency department
EMS	emergency medical services
ER	emergency room
ft	feet
FiO_2	fraction of inspired oxygen
g	grams
GC	gas chromatograph
GCS	Glascow Comma Scale
H_2O	water
H_2S	hydrogen sulfide
HACO	hookah associated carbon monoxide
Hb	hemoglobin
HbCO	carboxyhemoglobin
HbO_2	oxyhemoglobin
HBO_2	hyperbaric oxygen
HBOT	hyperbaric oxygen therapy
HCl	hydrogen chloride
HCN	hydrogen cyanide
Hg	mercury
hp	horsepower
HR	hazard ratio
ICU	intensive care unit
IH	isocapnic hyperventilation
kg	kilograms
km	kilometres
kW	kilowatt
L	litre
LPG	liquid petroleum gas

m^3	cubic metres
MACE	major adverse cardiovascular event
MetHb	methemoglobin
mg	milligrams
mL	millilitres
mm	millimetres
mmol	millimoles
mph	miles per hour
MRI	magnetic resonance imaging
MS	mass spectrometer
MVC	motor vehicle collision
MVEG	motor vehicle exhaust gas
NIH	non-isocapnia hypoventilation
ng	nanograms
NO	nitric oxide
NO_2	nitrogen dioxide
nm	nanometre
NBO_2	normobaric oxygen
NBOT	normobaric oxygen therapy
nm	nanometre
NO	nitric oxide
NO_2	nitrogen dioxide
O_2	oxygen
OHCob	hydroxocobalamin
pH	potential of hydrogen -a measure of acidity
PD	Parkinson Disease
PE	pulmonary embolism
PG	propylene glycol
$PM_{2.5}$	particulate matter
PNS	permanent neurological sequelae
pO_2	partial pressure of oxygen
ppm	parts per million
PVC	polyvinyl chloride
r^2	regression coefficient
RPUF	rigid polyurethane foam

SAC	serum alcohol concentration
SHS	second-hand smoke
SNHL	sensorineural hearing loss
spCO	carboxyhemoglobin concentration
spO_2	oxygen concentration
SUV	sport utility vehicle
TCD	thermal conductivity detector
ug	micrograms
umol	micromole
UV	ultraviolet
VAS	visual analogue scale
VG	vegetable glycerin
WBGT	dry and wet bulb temperature

Subject Index

About the Author

James Wigmore has worked as a forensic toxicologist for over 29 years at one of the foremost forensic laboratories in North America — the Centre of Forensic Sciences in Toronto, Ontario. He has testified in over 700 criminal cases throughout Canada and in numerous personal injury civil cases and coroner's inquests. He has published over seventy scientific articles in forensic toxicology that have been cited by the Supreme Court of Canada and the High Court of South Africa. James has been interviewed numerous times on the CBC and other media on the issues arising out of the legalization of cannabis in Canada and on vaping. He was on Health Canada's expert panel on public information regarding cannabis. He received the prestigious Prix Derome Award from the Canadian Society of Forensic Sciences for his outstanding contributions to the field of forensic science. His book series on the medicolegal aspects of alcohol, cannabis, and nicotine has received international acclaim.

FSC
www.fsc.org
MIX
Papier | Fördert gute Waldnutzung
FSC® C083411

Kolibri 360 GmbH
Ferdinand-Jühlke-Straße 7
99095 Erfurt, Deutschland
produktsicherheit@kolibri360.de